Innovative Wastewater Treatment Technologies

The INNOQUA Project

Costel Bumbac, Eoghan Clifford,
Jean-baptiste Dussaussois, Alexandre Schaal
and David Tompkins
(Editors)

now
the essence of knowledge

Published, sold and distributed by:
now Publishers Inc.
PO Box 1024
Hanover, MA 02339
United States
Tel. +1-781-985-4510
www.nowpublishers.com
sales@nowpublishers.com

Outside North America:
now Publishers Inc.
PO Box 179
2600 AD Delft
The Netherlands
Tel. +31-6-51115274

ISBN: 978-1-68083-780-3
E-ISBN: 978-1-68083-781-0
DOI: 10.1561/9781680837810

Front cover photos: Centre image courtesy of Julia Knop. Other images (clockwise from top) courtesy of Tatjana Schellenberg, Nicolas Salmon, Victoria Salvadó, Victoria Salvadó, Pietro De Cinque

Suggested citation: Costel Bumbac, Eoghan Clifford, Jean-Baptiste Dussaussois, Alexandre Schaal and David Tompkins (eds.). (2021). *Innovative Wastewater Treatment Technologies – The INNOQUA Project.* Boston–Delft: Now Publishers

Table of Contents

Dedication

This book is dedicated to the memory of Professor Chris Buckley, who devoted his life to providing water and sanitation services to the unserved.

Foreword

This decade will set the course for the future of present and upcoming generations. The impact of today's decisions reaches further into the future than those of earlier decades.

The concept of Planetary Boundaries and the collective efforts to avoid tipping points where human activities could give rise to planetary-scale ecological regime shifts were important markers on the road to the Paris Agreement under the United Nations Framework Convention on Climate Change, negotiated in 2015 and signed in 2016.

Five years on, we talk a lot about climate change mitigation. We are also making progress with the instruments, policies (e.g. "Fit for 55" package) and its perception among the population – including the business community – that urgent action is required now, and that we need to speed up.

We talk far too little about biodiversity. A very inconvenient fact indeed from the perspective of climate change mitigation itself, because part of the solution to the problem lies in the protection of biodiversity. Here are two examples:

- About half of the greenhouse gas emissions are absorbed by ecosystems – the oceans, the soil and forests. At the same time, however, we are weakening the sink capacity of ecosystems – that is, their ability to bind greenhouse gases – by, for example, cutting down forests or draining peatlands and then turning them into agricultural land. More greenhouse gases and less biodiversity: we are burning the candle from both ends!
- The second point is discussed even less: All the scenarios that we have directing and enabling us to limit global warming to between 1.5 and 2 degrees show us: we need negative emissions from 2030, 2035 onwards. To get there, we depend on the ecosystems we have now to a considerable extent – but

> ecosystem stabilisation is not enough, we will need expansion: of global forests and wetlands.

Sustainable water management is a crucial prerequisite for the stabilisation of our ecosystems and vice versa – its loss impacts our water security adversely. In this context, how we safely manage wastewater and faecal sludge in our habitat becomes key – the climate dimension (and its financial instruments) is adding to the wealth of knowledge around public and environmental health.

This is why the INNOQUA project and this book are so timely. Demonstrated at numerous locations worldwide, INNOQUA convincingly innovates the structure and process engineering of wastewater treatment in terms of "Nature Based Solutions" – an eye opener!

Congratulations and thanks to all members of the project team – one of who sadly passed away just before this publication. It is with gratitude that we remember our academic advisor and friend Prof. Chris Buckley from the WASH R&D Centre at the University of KwaZulu Natal. His commitment to the INNOQUA project was invaluable.

The time for wastewater innovation is now – INNOQUA shows us a way.

Stefan Reuter
Member of the Supervisory Board of the Faecal Sludge Management Alliance and Independent Advisor, Germany

DOI: 10.1561/9781680837810.ch1

Chapter 1

Introduction to the INNOQUA Project

By Jean-Baptiste Dussaussois

1.1 Introduction

Access to safe sanitation, wastewater treatment and drinking water are key pillars of societal development and individual quality of life. However, despite significant efforts over recent decades, there remain many shortcomings in the provision of sanitation and wastewater treatment and the water cycle is still managed ineffectively. Natural water resources, potable water and wastewater are all components of this cycle which interconnect with a larger water system that, in turn, supports life on earth.

Much work has been done to solve the worldwide sanitation issue. However, despite these efforts, United Nations SDG (Sustainable Development Goal) 6 ("Ensure availability and sustainable management of water and sanitation for all") has yet to be achieved (United Nations, 2021). Two-thirds of urban wastewaters are discharged without treatment into lakes, rivers and coastal waters, while poor hygiene and sanitation contribute to more than 1,000 daily deaths of children under the age of five from diarrhoeal diseases (UNICEF, 2020). Estimates suggest that the lack of sanitation costs upwards of 200 billion USD every year – although for every 1 USD invested in water and sanitation, there is more than a 4 USD return in the form of reduced health care costs (WHO, 2014). With less than 10 years left until 2030, the rate at which sanitation coverage is increasing will need to quadruple if the world is to achieve the SDG sanitation targets (UNICEF and WHO, 2020).

This is not an issue that is confined to developing countries, by which we mean those in the 'Global South', as defined by the United Nations Development Programme (UNDP, 2004). Although wastewater treatment is broadly considered to be well established in the urban areas of many developed countries, the situation is not actually so deep-rooted. Many countries of the Global North encounter difficulties in providing sustainable sanitation for their whole populations. For example, rural areas, small or remote communities, and low-income areas are challenged in using conventional technologies to treat wastewater due to their high operational and capital costs, and associated societal issues. Even when sanitation facilities are in place, their management and maintenance can present ongoing issues. In many cases this means that wastewater ends up being discharged directly into water bodies (or discharged following minimal treatment), resulting in significant environmental damage and leading to health, social and economic impacts.

Many wastewater treatment solutions are already available to address these sanitation challenges, ranging from highly technical centralised systems fed by extensive sewer networks to simple low-tech decentralised units suitable for single households that offer on-site treatment and disposal. Decentralised technologies are often considered as a temporary solution until sewer-based systems can be built, although more than 2 billion people rely on on-site systems such as pit latrines, septic tanks or soakaways. It is expected that by 2030 this number will increase to around 5 billion people (Cairns-Smith *et al.*, 2014). In this context, decentralised wastewater treatment can be viewed as flexible, rapidly-deployable and durable – not a stop-gap, but rather a sustainable long-term option.

Sustainability is an increasingly relevant factor, not only in terms of delivering treated wastewater of a quality appropriate to the receiving environment, but also in financial terms and achieving zero net carbon emissions. Centralised wastewater treatment systems not only require skilled staff, continuous observation and management – but they are significant users of electricity. Estimates range from 0.3 to 2.1 kWh per m^3 of wastewater treated in the EU, and between 0.41 to 0.87 kWh per m^3 of wastewater treated in the USA (Gandiglio *et al.*, 2017). Outside the utilities the percentage of energy devoted to wastewater treatment can be underappreciated (Capodaglio and Olsson, 2020).

Decentralised wastewater treatment solutions present their own challenges. Poor construction and maintenance – combined with inadequate user training – can all lead to substandard treatment. Nonetheless, the potential to intensify natural depuration processes and deliver sustainable decentralised solutions remains significant. The challenges and opportunities for such innovative nature-based wastewater treatment solutions ('NBS') are explored in this book through a variety of lenses that include: regulatory barriers and drivers, experience with existing solutions in the Global South, the importance of resource recovery – and the INNOQUA project, which set out to develop and field test a suite of new NBS. These solutions are defined as "inspired and supported by nature, which are cost-effective, simultaneously provide environmental, social and economic benefits and help build resilience. Such solutions bring more, and more diverse, nature and natural features and processes into cities, landscapes and seascapes, through locally adapted, resource-efficient and systemic interventions" (European Commission, 2021). INNOQUA drew on the diverse experiences of partners, highlighting the value of international collaboration in delivering innovative research and technologies in 12 countries:

- Ecuador
- France
- Germany
- India
- Ireland
- Italy
- Peru
- Romania
- Spain
- Tanzania
- Turkey
- United Kingdom

1.2 Status of Sanitation Provision

The World Health Organization defines safely managed sanitation as:

> "... *improved sanitation facilities that are not shared with other households and where excreta are safely disposed of in situ or transported and treated offsite. Improved sanitation facilities include flush/pour flush to piped sewer systems, septic tanks or pit latrines: ventilated improved pit latrines, composting toilets or pit latrines with slabs*" (WHO, 2021).

In this section we briefly examine access to improved sanitation within INNOQUA partner countries in the Global North and Global South – as indicators of the need for increased sanitation provision and therefore potential market sizes for novel solutions such as those developed by INNOQUA.

1.2.1 The Global North

Although access to safely managed sanitation is generally extremely high in the Global North, it is by no means universal, and access for urban populations is frequently higher than access for rural populations (Table 1.1). Overall, data for Europe and North America indicate that more than 80% of the total population has access to a sewer system (at the end of which some form of wastewater treatment takes place), while roughly 10% of the population rely on decentralised or on-site solutions such as septic tanks (UNICEF and WHO, 2020). As explored in Chapter 3, septic tanks are by no means fool proof and can act as point sources of environmental pollution unless appropriately located, constructed, maintained and operated.

Table 1.1. Populations of INNOQUA partner countries in the Global North, the percentage of the total populations using safely managed sanitation and the percentage of the rural populations using safely managed sanitation. Data from The World Bank (2021) for 2017. NR = Not Reported.

2017 World Bank Data	Population (Millions)	% Population Using Safely Managed Sanitation	% Rural Population Using Safely Managed Sanitation
France	66.9	88.4	NR
Germany	82.7	97.2	91.8
Ireland	4.8	82.4	72.2
Italy	60.5	96.2	96.1
Romania	19.6	76.5	NR
Spain	46.6	96.6	96.7
UK	66.1	97.8	91.8

Data summarised in Table 1.1 also highlight the differences between total and rural sanitation provision in Europe. Installation of collective sewer networks and associated centralised wastewater treatment can be inappropriate for dispersed rural populations. In their place, septic tanks, package plants and other decentralised solutions are required. This can lead to diffuse pollution over wide areas unless there is adequate central government or local authority oversight – and sufficient user training or other communication that supports robust operation and maintenance of such solutions. For example, the estimated contribution of water pollution to eutrophication of France's coastal waters, the associated loss of tourism revenue and the direct cost to remove algae may be in the order of 100 to 200 million USD per year (OECD, 2017). Tourism is also relevant as a potential contributor to environmental contamination. As more and more "eco-friendly" lodges and other tourism facilities are built in natural and sensitive environments, delivery of sustainable wastewater treatment facilities becomes essential. These remote areas are excellent locations to install on-site/decentralised sanitation systems to preserve the local environment.

This does not mean that urban developments provide no opportunities for decentralised wastewater treatment solutions. For example, although regulatory instruments such as the European Urban Waste Water Treatment Directive suggest that collective sewerage systems be provided where population centres exceed specified thresholds, alternative approaches are permitted (OJEU, 1991). As urban populations continue to grow in the Global North, and climate change goals force re-examination of the long-term viability of centralised wastewater collection and treatment, there are likely to be many opportunities for future integration of novel decentralised solutions in both urban and rural environments.

1.2.2 The Global South

Provision of sanitation is far less extensive in the Global South. For example, when considered at a regional level, only 31.3% of the whole population of Latin America and the Caribbean have access to safely managed sanitation – while this number drops to just 18.7% of the population of Sub-Saharan Africa (2017 data from The World Bank (2021)). When INNOQUA partner countries are considered, then provision can be seen as highly variable (Table 1.2).

These headline figures obscure a number of important aspects. For example, in Quito (Ecuador), city authorities have overseen the installation of an extensive sewerage network. However, 95% of the collected wastewater from a city of 2.5 million inhabitants is discharged to the environment without treatment. The local municipal agency – EPMAPS – has initiated the "River Depollution Program" (EPMAPS, 2020), launched in November 2016 with the commissioning of a new treatment plant located in the south of the city (Quitumbe) that was designed to treat around 5% of the city's wastewater flow at that time (Davis *et al.*, 2016).

In Peru, according to the last census of 2017 the population was ~29 million, with 79% in urban areas and 21% in rural areas (INEI, 2021). In terms of access to sanitation systems, 72.7% of the country has a sewer service (INEI, 2015) but only one third of the population has access to safe sanitation (The World Bank, 2021). Another important factor to consider is that while production per capita of potable water is approximately 216 L/inhabitant/day, transmission losses of 20% occur across the water distribution network (SUNASS, 2015). Moreover, in Peru, out of 253 population centres, 89 do not have a system for wastewater treatment.

Table 1.2. Populations of INNOQUA partner countries in the Global South, the percentage of the total populations using safely managed sanitation and the percentage of the rural populations using safely managed sanitation. Data from The World Bank (2021) for 2017. NR = Not Reported.

2017 World Bank Data	Population (Millions)	% Population Using Safely Managed Sanitation	% Rural Population Using Safely Managed Sanitation
Ecuador	16.8	42.0	57.2
India	1,339	NR	39.1
Peru	31.4	42.8	NR
Tanzania	54.7	25.4	22.4
Turkey	81.1	65.2	NR

The Water and Sanitation Program (part of World Bank Group's Water Global Practice) released a study in 2011 that estimated the total economic impacts of inadequate sanitation in India to be 53.8 billion USD a year – the equivalent of 6.4% of India's GDP in 2006, or 48 USD per person (Tyagi, 2011). The same study presents data on access to sanitation and reports that 70.6% of the Indian population did not have access to improved facilities – and among these, 55.3% were defecating in the open (compare with the more recent data on safe sanitation access amongst India's rural population, Table 1.2).

In Africa, sanitation access is also a major issue with more than 70% of the population in Eastern and Southern Africa (340 million people) having no access to basic sanitation services (UNICEF, 2021). Among these, 98 million people (19%) practise open defecation, 179 million use unimproved facilities and 63 million shared sanitation facilities. Ethiopia, Uganda, Kenya and Tanzania have by far the largest number of people in the region with no access to basic sanitation services. Access to basic sanitation services in some communities has only increased by 6% since 2000 and projections show that at current rates only 36% of the population of these countries will have access to basic sanitation services by 2030 (UNICEF and LIXIL, 2020). To achieve universal access to at least basic sanitation by 2030, global rates of progress would need to double (UNICEF and WHO, 2020).

It has been estimated that achieving full access to safely managed sanitation would cost in the order of 105 billion USD (Hutton and Varughese, 2016). Meanwhile poor sanitation and water supply result in annual economic losses estimated at 260 billion USD in developing countries – equivalent to about 1.5% of their GDP. This gap provides a huge stimulus for the development and promotion of new approaches for sustainable, cost-effective and low maintenance wastewater treatment solutions (Hutton, 2013).

1.3 Barriers to the Implementation of Sanitation

Globally, ~12% of the population resorts to open defecation and a further ~20% only have access to basic sanitation. This equates to about 2.5 billion people around the world without access to proper sanitation (WHO and UNICEF, 2017). Solutions to bridge this gap already exist (Rahmasary *et al.*, 2019), but there are a number of barriers:

- Resource Availability: Wastewater treatment systems require the use and availability of materials and energy for construction; skilled personnel at design, construction and operation phases and resource consumption during operation. Electricity is usually a core resource requirement to ensure stable and robust operation, and the electrical requirements of wastewater treatment are significant at a global scale. Access to steady electricity supplies is not always possible;
- Technical: Commonly used wastewater treatment plants can be adapted to various situations. Nevertheless, all collective wastewater treatment plants require an associated sewerage network, with implied maintenance and operation costs. Despite the benefits of removing wastewater from the source of production to the treatment plant, this doesn't allow subsequent reuse of treated wastewater on site;
- Economic: Any treatment system requires sufficient financing, mostly by government or local authorities in the case of a collective system. Where potable and wastewater costs are bundled into single utility charges, this can disincentivise the implementation of on-site systems, where treatment is not the responsibility of the central utility. Where provision of sewage treatment systems is borne by the individual this can also provide financial challenges;
- Social and educational: Social acceptance is a key parameter for the implementation of any wastewater treatment system. Communicating the goals of the system to key stakeholders is essential in order to obtain a "social licence" to install and maintain a system – and facilitates any requirement for the system to raise revenue to pay back the investment. Furthermore, in the case of water reuse, demonstration of previous successful examples has been shown to improve public perception when compared to standardised educational workshops/programmes (Rupiper and Loge, 2019);

- Space requirement: Availability of space is one of the limitations for any treatment system – whether centralised or decentralised. However, architectural integration of decentralised systems can be achieved in many projects, delivering positive aesthetic and environmental benefits at a community scale that are not possible with centralised approaches and
- Governance: Issues such as competing political agendas and priorities, inconsistent policy and management and institutional inertia can all impede the development and roll out of successful sanitation programmes (Rahmasary *et al.*, 2019).

These limitations can be illustrated through the example of India. By 2024 India is expected be the most populous country in the world with a population greater than 1.3 billion people (United Nations, 2017). Poor sanitation has significant health costs, and untreated faecal sludge and septage from cities is the single biggest source of water resource pollution in India. India's biggest cities have large, centralised sewerage systems with vast underground pipelines, pumping stations and huge treatment plants. These systems are expensive to build and even more expensive to operate, as they require continuous electrical power, access to skilled operators and extensive electro-mechanical maintenance. It is for these reasons that India's 7000+ small towns (those with populations of less than 100,000) do not have such systems and are unlikely to be covered by centralised sewerage systems in the near future (Government of India Ministry of Urban Development, 2017). When looking more closely at approaching challenges, India's fresh water resources amount to 4% of the global total, while the population amounts to 14% of the global total (Dhawan, 2017). The need to conserve water resources by minimising consumption – and maximising reuse of treated wastewater is extremely pressing. Decentralised, nature-based solutions can contribute to meeting these challenges and India is the perfect country to broadly deploy such solutions. Furthermore, with a huge variation in climate zones (including tropical, arid, temperate and even polar [mountainous] regions), very large urban and rural populations and a broad range of social classes, India provides a significant challenge but also a wide range of scenarios within which novel approaches to sanitation and wastewater treatment can be developed and successfully translated for use in other countries.

1.4 The INNOQUA Project

It was within the context of these challenges that the INNOQUA project consortium was established in 2016, to develop modular, sustainable, affordable and innovative wastewater treatment solutions for both rural and urban contexts. The consortium comprised 20 organisations with diverse backgrounds and skills, and included Non-Governmental Organisations (NGOs), Universities, Commercial Research Organisations, Technology Providers, Business Development and other entities from 12 different countries. The project was supported by the European Commission through the Horizon 2020 research and innovation programme under grant agreement No. 689817, with a budget of €6.9 million.

The key aim of the INNOQUA project was to integrate individual modular, low-cost, sustainable and biologically-based water treatment and reclamation technologies in configurations matched to local contexts and markets. The INNOQUA solution comprised four treatment technologies, which were combined in different configurations to suit local conditions. Remote system monitoring and operation was used where required to both enable more comprehensive system management and also determine its likely robustness in long-term monitoring of systems in various scenarios. The four technologies deployed were:

- The Lumbrifilter which acts as a primary and secondary treatment system. It comprises organic and inorganic media which provide a development substrate for physical and biological filtration. Earthworms consume and digest organic solids, while dissolved pollutants are consumed and transformed within a microbial biofilm;
- The Daphniafilter which provides tertiary treatment through a combination of microbial biofilm, floating macrophytes and Daphnia;
- Biosolar Purification system (BSP) which acts as a tertiary treatment solution through the activities of a microalgal biofilm. Designed for use in climates with high insolation, the thin layer cascade design of this unit also delivers partial disinfection through exposure to sunlight; and
- UV disinfection, which acts as a quaternary system using pulsed ultraviolet radiation to disinfect pre-treated wastewater before its reuse or discharge into the environment.

Figure 1.1 provides an overall schematic for the INNOQUA solution, showing how the different technologies can be integrated.

The project passed through different phases to clearly understand, optimise and evaluate the system:

1. Laboratory scale tests allowed the team to understand the limits of the nature-based solutions in terms of wastewater treatment efficiency, and also provided opportunities for process optimisation through adjustments to filter media and bed depth/loading rates etc;

2. On-site prototype tests enabled the team to test and monitor the different technology integrations at two pilot sites under real (but controlled) conditions, receiving diverted flows from municipal wastewater treatment facilities. This phase allowed the performance of technologies to be tested with real wastewater under different hydraulic load and climatic conditions; and
3. During the final phase, various technology combinations were installed at demonstration facilities in 11 countries, as shown in Table 1.3. The systems were operated and monitored under real conditions to analyse their treatment efficiency and any required maintenance. This monitoring also informed Life Cycle and Life Cycle Cost Assessments to ensure the relevance of the solutions in their intended markets.

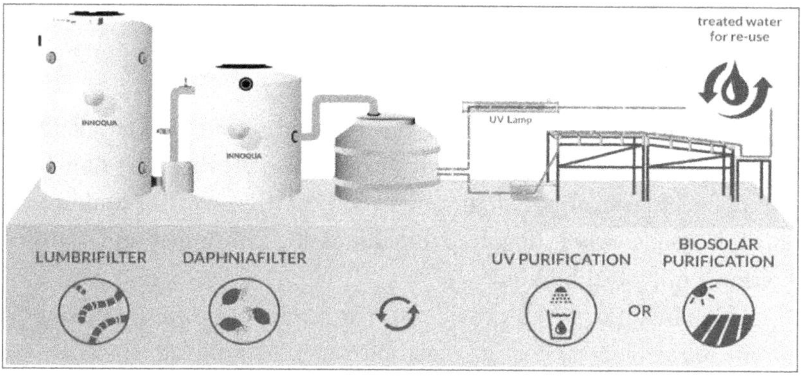

Figure 1.1. The INNOQUA technologies.

Table 1.3. Locations of the INNOQUA demonstration sites and the different modular integrations used at each.

Countries	Lumbrifilter	Daphniafilter	UV	BSP
Ireland	✓	×	×	×
Italy	✓	×	✓	×
France	✓	✓	×	×
UK	✓	✓	✓	×
Turkey	✓	✓	✓	×
Romania	✓	✓	×	×
Ecuador	✓	✓	×	×
Peru	✓	✓	✓	✓
Tanzania	✓	✓	✓	×
India	✓	✓	✓	✓

The demonstration sites were selected as representative of a broad range of environmental conditions, influent characteristics, regulatory requirements and discharge or reuse needs from the end users. Filter media (both biological and mineral) as well as the various required biological organisms (earthworms, Daphnia, microalgae, etc.) were all sourced locally to each demonstration facility. It should be noted that in addition to the Lumbrifilter and Daphniafilter demonstration site in France, a second demonstration facility was installed at an inland aquaculture facility to test the potential for lumbricomposting of the industrial sludges. The results from this second French installation are not described in this book.

In their various combinations, the INNOQUA modules delivered effective wastewater treatment at low cost and with minimal sludge production. This showed how the impact of wastewater on the environment could be reduced sustainably – protecting rivers and wider biodiversity. Beyond these important quality aspects, the INNOQUA system decreases stress on water resources by facilitating water reuse. Although the project concluded in November 2020, monitoring remains on-going at the majority of the demonstration sites and will contribute to future commercial development of the INNOQUA technologies, as described in Chapter 9.

The INNOQUA project worked for 4.5 years from early concept to full deployment of the technologies at the demonstration sites. Given the broad variety of wastewater characteristics, climatic conditions, regulation and social considerations across the different sites, the project was not without challenges. This book considers these challenges, the solutions implemented, and lessons learned. Twenty project partners all brought different skills and experience, with varying knowledge of wastewater treatment as required by the different project roles. Delivering 2 pilot, 10 demonstration sites and 1 showcase facility in 11 different countries required extensive collaboration from local partners – both in terms of identifying sites where equipment could be installed and monitored over a period of years, and also in practical terms of delivering and monitoring the systems. A key aspect of the project was ensuring a two-way information flow. In this, lessons learned from sites in one country could be adapted and used to improve system operation in another country. These exchanges also highlighted the creative possibilities when a team integrates experiences from the Global North and Global South, developing solutions with truly global potential but from a context of very different local market conditions.

Reactors for the 10 main demonstration sites were all shipped from a single facility in Europe (as were the UV treatment units), while availability of suitable pumps and other ancillaries varied greatly between different partner countries, together with access to commercial laboratory services with adequate quality controls to allow partners to check the accuracy of field-collected data. In some cases, installations were dependent on construction activities that were unrelated to the project and completely outside our control, while in others the wastewater characteristics or flows varied considerably between initial testing and design – and system implementation. However, the simplicity of the selected nature-based solutions meant that their implementation could readily be flexed on the ground – whether through changing filter media and dosing rates or even the dosing systems. The COVID-19 pandemic hit the project during later stages of field testing and for some months it was not possible to even visit the installations. Nonetheless, treatment efficiencies quickly rebounded. In some cases more sophisticated technical elements (in-line flow meters, remote monitoring units, dosing pumps) failed under the stressful operational conditions, while the biological treatment elements continued to deliver.

The following chapters provide both a comprehensive overview of the need and opportunities for nature-based sanitation solutions, as well as detailed explorations of the theory and practice of each modular INNOQUA solution.

References

Cairns-Smith, S., Hill, H. and Nazarenko, E. (2014). *Working Paper. Urban Sanitation: Why a portfolio of solutions is needed.* Available at: https://image-src.bcg.com/Images/December_2014_Sanitation_WORKING_PAPER_FINAL_tcm9-79574.pdf (Accessed: 16 February 2021).

Capodaglio, A. G. and Olsson, G. (2020). Energy Issues in Sustainable Urban Wastewater Management: Use, Demand Reduction and Recovery in the Urban Water Cycle. *Sustainability*, 12(1), 266. https://doi.org/10.3390/SU12010266

Davis, M. J. M., Gutiérrez, M. L. and Serrano, J. (2016). 'Know your people: social research and water recycling system design with communities in the Amazon', *Procedia Engineering*, 145, pp. 158–1266.

Dhawan, V. (2017). *Water and Agriculture in India. Background paper for the South Asia expert panel during the Global Forum for Food and Agriculture (GFFA) 2017.* Available at: https://www.oav.de/fileadmin/user_upload/5_Publikationen/5_Studien/170118_Study_Water_Agriculture_India.pdf (Accessed: 16 February 2021).

EPMAPS (2020). *Programa para la descontaminación de los ríos de Quito.* Available at: https://www.aguaquito.gob.ec/programa-para-la-descontaminacion-de-los-rios-de-quito/ (Accessed: 19 March 2021).

European Commission (2021). *Nature-based solutions.* Available at: https://ec.europa.eu/info/research-and-innovation/research-area/environment/nature-based-solutions_en (Accessed: 16 February 2021).

Gandiglio, M. *et al.* (2017). 'Enhancing the Energy Efficiency of Wastewater Treatment Plants through Co-digestion and Fuel Cell Systems', *Frontiers in Environmental Science*. Frontiers Media S.A., 5, p. 70. doi: 10.3389/fenvs.2017.00070.

Government of India Ministry of Urban Development (2017). *NATIONAL POLICY ON FAECAL SLUDGE AND SEPTAGE MANAGEMENT (FSSM)*. Available at: http://amrut.gov.in/upload/newsrelease/5a5dc55188eb0FSSM_Policy_Report_23Feb.pdf (Accessed: 16 February 2021).

Hutton, G. (2013). *What costs the world $260 billion each year?* Available at: https://blogs.worldbank.org/water/what-costs-the-world-260-billion-each-year (Accessed: 29 March 2021).

Hutton, G. and Varughese, M. (2016). *The Costs of Meeting the 2030 Sustainable Development Goal Targets on Drinking Water, Sanitation, and Hygiene*. Available at: https://openknowledge.worldbank.org/bitstream/handle/10986/23681/K8543.pdf?sequence=1&isAllowed=y (Accessed: 16 February 2021).

INEI (2015). *Perú: Síntesis Estadística 2015*. Available at: https://www.inei.gob.pe/media/MenuRecursivo/publicaciones_digitales/Est/Lib1292/libro.pdf (Accessed: 16 February 2021).

INEI (2021). *PERÚ Instituto Nacional de Estadística e Informática*. Available at: https://www.inei.gob.pe/ (Accessed: 16 February 2021).

OECD (2017). *Diffuse Pollution, Degraded Waters. Emerging Policy Solutions*. OECD (OECD Studies on Water). doi: 10.1787/9789264269064-en.

OJEU (1991). *Council Directive of 21 May 1991 concerning urban waste water treatment*. Available at: https://eur-lex.europa.eu/legal-content/EN/TXT/PDF/?uri=CELEX:31991L0271&from=EN (Accessed: 17 February 2021).

Rahmasary, A. N. *et al.* (2019). 'Overcoming the Challenges of Water, Waste and Climate Change in Asian Cities', *Environmental Management*. Springer New York LLC, 63(4), pp. 520–535. doi: 10.1007/s00267-019-01137-y.

Rupiper, A. M. and Loge, F. J. (2019). 'Identifying and overcoming barriers to onsite non-potable water reuse in California from local stakeholder perspectives', *Resources, Conservation and Recycling: X*. Elsevier B.V., 4, p. 100018. doi: 10.1016/j.rcrx.2019.100018.

SUNASS (2015). *DIAGNÓSTICO DE LAS PLANTAS DE TRATAMIENTO DE AGUAS RESIDUALES EN EL ÁMBITO DE OPERACIÓN DE LAS ENTIDADES PRESTADORAS DE SERVICIOS DE SANEAMIENTO*. Available at: https://www.sunass.gob.pe/doc/Publicaciones/ptar.pdf (Accessed: 16 February 2021).

The World Bank (2021). *World Development Indicators | DataBank*. Available at: https://databank.worldbank.org/source/world-development-indicators (Accessed: 16 February 2021).

Tyagi, A. (2011). *Economic Impacts of Inadequate Sanitation in India.* Available at: https://www.wsp.org/sites/wsp.org/files/publications/WSP-esi-india.pdf.

UNDP (2004). *Forging a Global South.* Available at: http://www.undp.org/content/dam/china/docs/Publications/UNDP-CH-PR-Publications-UNDay-for-South-South-Cooperation.pdf.

UNICEF (2020). *Diarrhoea.* Available at: https://data.unicef.org/topic/child-health/diarrhoeal-disease/ (Accessed: 19 March 2021).

UNICEF (2021). *Sanitation and hygiene | UNICEF Eastern and Southern Africa.* Available at: https://www.unicef.org/esa/sanitation-and-hygiene (Accessed: 19 March 2021).

UNICEF and LIXIL (2020). *'Make A Splash!' Partnership Progress Report.* Available at: https://www.unicef.org/esa/media/7421/file/UNICEF-LIXIL-2020-Make-A-Splash-Progress-Report.pdf (Accessed: 19 March 2021).

UNICEF and WHO (2020). *State of the World's Sanitation: An urgent call to transform sanitation for better health, environments, economies and societies.* Available at: https://apps.who.int/iris/bitstream/handle/10665/336688/9789240014473-eng.pdf (Accessed: 16 February 2021).

United Nations (2017). *World Population Prospects: The 2017 Revision, Volume I: Comprehensive Tables (ST/ESA/SER.A/399).* Available at: https://population.un.org/wpp/Publications/Files/WPP2017_Volume-I_Comprehensive-Tables.pdf.

United Nations (2021). *Goal 6 | Department of Economic and Social Affairs.* Available at: https://sdgs.un.org/goals/goal6 (Accessed: 16 February 2021).

WHO (2014). *UN-Water Global Analysis and Assessment of Sanitation and Drinking-Water GLAAS 2014 Report.* Available at: https://apps.who.int/iris/bitstream/handle/10665/139735/9789241508087_eng.pdf;jsessionid=0C91336A7ACE170E376A8A6A782E39D4?sequence=1 (Accessed: 16 February 2021).

WHO (2021). *Population using safely managed sanitation services (%).* Available at: https://www.who.int/data/gho/indicator-metadata-registry/imr-details/4820 (Accessed: 17 February 2021).

WHO and UNICEF (2017). *WHO/UNICEF Joint Monitoring Program for Water Supply, Sanitation and Hygiene (JMP) – 2017 Update and SDG Baselines.* Available at: https://www.unwater.org/app/uploads/2020/04/WHOUNICEF-Joint-Monitoring-Program-for-Water-Supply-Sanitation-and-Hygiene-JMP—-2017_ENG.pdf (Accessed: 16 February 2021).

DOI: 10.1561/9781680837810.ch2

Chapter 2

Drivers and Barriers Towards a Sustainability Transition in the Wastewater Sector – A Nexus Perspective with Nature-based Solutions

By Tatjana Schellenberg

Clean water and sanitation are key elements of society and represent fundamental pillars to poverty alleviation. Despite their significance, service delivery challenges remain after many decades. With the absence of progress predominantly rooted in lack of funding, debates and discussions continue and usually concentrate around the question of finding the right business case for a sector, which is understood as an essential inclusive public necessity and often evolves as a non-profit space with limited revenue streams.

While challenges in the overall development of the wastewater sector continue, pressures are rising in water scarcity, food insecurity and ongoing climate change coupled with increasing energy demand. Although the potential for the wastewater sector to contribute to reduced water scarcity has long been recognized, it also holds huge potential in terms of nutrient and energy resources. This potential could not only help to address the various acknowledged pressures but also fundamentally change the wastewater sector into a profitable resource centre, driving forward a sustainable transition and sectoral development.

This study examines the drivers and barriers towards a sustainability transition in the wastewater sector under a nexus perspective in 12 different countries. Nexus approaches integrate environmental management and governance across multiple sectors and scales. Although pressures can be observed in all sectors at a landscape level, water insecurity forms the major driver towards a circular approach and adaptive policies at regime levels, while direct regulations towards the recycling or recovery of nutrients and energy or GHG mitigation mostly lag behind. Furthermore, high deviations can be observed between the Global North and Global South. Despite typically higher pressures in the Global South, the implementation of adaptive and innovative policies is more commonly observed in the Global North.

2.1 Introduction

The sanitary revolution in the 19th century in Europe is considered to be one of the most important medical milestones since 1840 (BMJ, 2007). Despite subsequent advances in technology and access to wastewater services, about 55% of the global population still do not have access to safely managed sanitation facilities (WHO, 2019). This causes a significant burden on public health and contamination of the environment. These issues are now exacerbated by new challenges. More than 2 billion people are living in countries with high water stress (UN, 2018a). Meanwhile global water demand is reported to grow at twice the rate of population (UN Water, n.d). During the last two decades land area use for agriculture has increased by 16% and the area of irrigated agricultural land has doubled (UNCCD, 2017). The agricultural sector represents globally the major water user accounting for over 70% of all water withdrawals (UNCCD, 2017; WRI, 2018). However, despite the observed agricultural expansion and achievements in tripling agricultural production, two billion people are still affected by food insecurity (FAO, 2017). Unsustainable land management practices are reported to cause an annual loss of soil in the range of 24 billion tonnes each year (UNCCD, 2017). In order to achieve production targets, immense amounts of fertiliser are applied. Production of these is energy intensive and their use accounts for one of the major causes of water pollution (Gellings and Parmenter, 2004; FAO and IMWI, 2017). While available phosphorus resources are facing depletion, most conventional wastewater treatment plants are generally designed to remove (rather than recover) potentially valuable nutrients, and often at high costs (Daneshgar *et al.*, 2018; Larsen *et al.*, 2013). Climate change and an increasing demand for bioenergy are adding to the complexity and sectoral conflicts over resources.

In response to these accelerating pressures, there is a growing consensus that society cannot afford anymore to waste valuable water resources, nutrients, and the energetic benefits that wastewater can provide. Given the deeply interrelated dynamics of water, land and energy management systems – and the recognized benefits of wastewater – integrated resource management combining all sectors is needed in order to achieve sustainable and circular management (Brundtland, 1987; Capodaglio and Olsson, 2019). The nexus approach forms the newest integrated management paradigm for such a concept (WEF, 2011; Simpson and Jewitt, 2019; Benson *et al.*, 2015; Al-Saidi and Elagib, 2016).

Historically the main function for sanitation systems was the protection of human health by pollution control, whereas nowadays they are perceived and presented as resource centres for water, nutrients and energy. This new perspective not only provides the prospect of a circular and more sustainable handling of resources but furthermore can immensely support the broader development of the wastewater sector through additional revenue streams. Decentralised and integrated approaches with nature-based technologies are particularly recognized as possible solutions for achieving this reality with high or very high potential to contribute to Sustainable Development Goals (SDGs) 1, 2, 8, 9, 11, 12, 13, 14 and 15, in addition to the main SDG6 (Larsen *et al.*, 2013; Lüthi *et al.*, 2011; Capodaglio, 2017; UN WWAP, 2017; UN, 2018a; UN WWAP, 2018).

Other than wastewater management, only few sectors offer equivalent promise for delivering a sustainable transition. However, in order to achieve this transition, continuously evolving research on innovative and sustainable technologies for targeted treatment at the niche level, innovative risk management approaches with responsive policies in socio-technical regimes and a wider societal behaviour change at the broader landscape level are required.

The context of this study evolved during the INNOQUA project. As a response to current challenges the core objectives of this research project were the development and demonstration of innovative decentralised nature-based technologies that can provide a more sustainable solution for the development of the wastewater sector and enable resource recycling.

Given the nature of innovation research, such project investigations can typically be classified as niche level activities in sustainability transitions. This study uses a nexus approach and multi-level perspective framework to assess drivers and barriers for more sustainable development of the wastewater sector in terms of resource recycling. The assessment is structured in two parts. Within the first, pressures and drivers are analysed at landscape level. The second part assesses possible barriers at the regime level and includes a policy review highlighting the need for adaptive regulations that support reuse, recycling and recovery of water, nutrients and energy in the wastewater sector. The scope includes a comparative analysis of the 12 countries involved in the INNOQUA project and discusses the state of possible alignment with a further outlook towards a sustainability transition.

2.2 Methodology

The multi-level perspective (MLP) serves as a framework for analysis along the different levels identified in transition research, and their alignment. The assessment is built upon three segments, where (a) a generic review of global trends serves to provide an overview with information on key observed benefits, drivers and barriers. This review builds the foundation for subsequent analysis of (b) environmental pressures and related drivers from the wastewater sector, at landscape level and helps to (c) identify possible barriers in the form of inflexible policies at the regime level. The results of the assessment inform discussion on the current status and potential for alignment leading to more sustainable development in the wastewater sector.

a. Review of global trends and drivers

A review based on published literature, databases and reports of international organizations such as the United Nations, World Bank, FAO (The Food and Agriculture Organization of the United Nations) and IPCC (The Intergovernmental Panel on Climate Change) was carried out to assess global progress towards the paradigm shift in the wastewater sector that was conceived by the United Nations in their World Water Development Report in 2017. Particular attention was given to the identification of (a) the contextual benefits of wastewater as a resource, (b) the global agenda and recognized drivers and (c) the major reported barriers preventing these drivers being taken forward. This study focuses on pressures at the landscape scale and reviews regulations at the regime level relating to water recycling, nutrient recycling or recovery and energy efficiency or recovery. The results of this review serve as the basis for further assessment in this study. Figure 2.1 describes the established format for this assessment, which was based on the principles of the pressure-state-response concept (FAO, 1996).

b. Analysis of environmental pressures at landscape level

With related benefits identified in the fields of water, energy and the food/land sector, the nexus approach serves as an assessment framework to provide an integrated perspective for quantitative analysis of environmental pressures at landscape level in the different fields. Qualitative and quantitative data on these fields were identified through reviews of published literature. A comparative analysis was

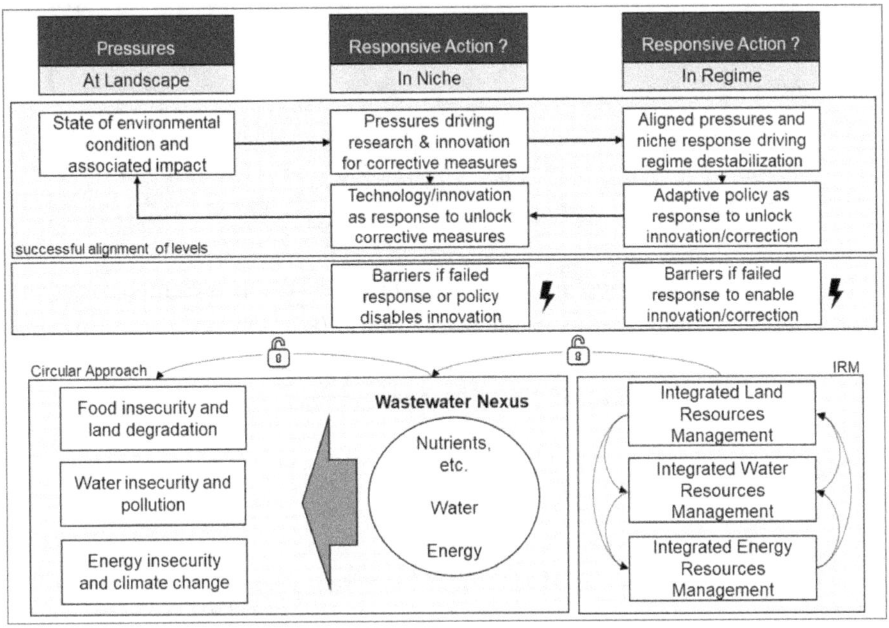

Figure 2.1. Scope of analysis under pressure-response concept.

carried out on the state of identified pressures and their related drivers in various countries. Indicators for quantitative analysis were selected based on their respective representation and data availability, and cover (a) diarrheal impact in the field of public health, (b) access to basic and safely managed sanitation, (c) a water stress index combined with sectoral water withdrawal and resource potential in wastewater, (d) land degradation status and nutrient resources in wastewater and (e) a climate risk index combined with energy resource potential in wastewater.

c. Analysis of possible barriers at regime level

The successful performance of wastewater treatment technologies was evaluated with reference to regulatory standards. These typically are based on risk management approaches focused on discharges to the aquatic environment. One of the major current barriers towards circular approaches and innovation in the wastewater sector can be the lack of adaptive policy and regulations. The assessment of possible barriers at the regime level follows a review of regulations in regard to (a) progress in integrated (water) resources management, (b) existing regulations for water recycling with respective ranges of pollution control measures for diverse application areas, (c) existing regulations for nutrient recycling or recovery and (d) existing regulations for energy recovery or greenhouse gas mitigation. Qualitative and quantitative data on these aspects were identified through literature and policy reviews.

Subsequent results on the state of pressures, drivers and barriers in the different countries are discussed and conclusions drawn with respect to their alignment and overall potential to change.

2.3 Assessment

Within the following section a short overview of the MLP is presented along with an assessment of global trends in the wastewater sector. Pressures and drivers are analysed at a landscape level, and the state of adaptive policy at the regime level is reviewed within the confines of the selected geographies.

2.3.1 The MLP on Sustainability Transitions

Ever since the publication of "limits to growth" by the Club of Rome in 1972, research around sustainability has gained increased attention (Meadows *et al.*, 1972). However, while innovation has led to a wide range of technological advancements, sustainable development in many sectors is still lagging and overexploitation of resources continues. The lack and failure in understanding the parameters and pathways relevant to initiation and shifting of a system towards a more sustainable state has led to the rapidly developing research field of sustainability transition studies.

Every sector can be conceptualized as a socio-technical system which (a) consists of different actors, institutions, materials, artefacts and knowledge, and in which (b) all elements interact and provide a system function to society (Geels, 2004; Markard and Truffer, 2008). These different elements are closely interrelated and with their co-dependencies incrementally influence the dynamic function of a system (Finger *et al.*, 2005; Markard, 2011). Transition of socio-technical systems encompasses a set of processes that lead to changes along different dimensions, and thus a more sustainable state in production and consumption patterns within the system (Markard *et al.*, 2012). The MLP provides a contextualized view and framework for analysing the potential and paths for such systemic change (Geels, 2002; Rip and Kemp, 1998; Smith and Stirling, 2010).

The MLP conflates the concepts of evolutionary economics, sociology of innovation and institutional theory. The framework structures a system into three levels, namely the niche, the socio-technical regime and the socio-technical landscape, and builds upon the core concept that transitions can only occur if interactions between these levels are taking place and are aligned (Schot and Geels, 2008; Geels, 2004).

- Niches represent the micro-level, in which radical innovations develop in a protected space. They can include social movements, research and development of novel technologies or demonstration projects, and their predominant character lies in their deviation from the current regime norms. Niches build the fundament for transitions by providing elements that can support systemic change.
- The socio-technical regime forms the broader system structure in which system functions and activities are coordinated through cognitive routines, beliefs or regulatory structures – which may typically be manifested through the policies of governmental institutions. They shape elements such as system infrastructure, which can lead to locked-in states that slow or prevent transition.
- The socio-technical landscape represents the exogenous macro level, which encompasses system functions in the broader environment with deep cultural patterns, macro-economics, macro-political development or environmental conditions. The illustration by Geels, 2011 shows the interactions between these different levels (Fig. 2.2).

Within the MLP, the core notion is that transitions can only be enabled through interactions and alignment of networks and processes at all three levels, where (a) innovations at the niche level build up an internal momentum along with (b) pressures on the regime formed through changes at the landscape level and (c) leading to a destabilization of the socio-technical regime and finally the creation of windows of opportunity for niche innovations to diffuse into the regime and form new and more sustainable norms (Geels, 2011). The difficulties in initiating structural systemic change are rooted in a range of lock-in mechanisms, be they shared beliefs or discourses, institutional commitments and (global) power relationships, consumer lifestyles or (commonly) heavy investment in infrastructure. The latter is often the case with centralised wastewater treatment.

2.3.2 The Wastewater Sector and Global Trends Towards the Paradigm Shift

The UN water development report on "Wastewater – The Untapped Resource" in 2017 calls for a paradigm shift in the wastewater sector and summarises benefits that can be realized in the fields of water resources, energy and nutrients, combining the core disciplines of the nexus approach. Apart from enabling more sustainable practices through resource recovery, this transformation would provide new possibilities in revenue streams, which could help to address current financial constraints and stimulate overall development of the wastewater sector (UN WWAP, 2017). Given

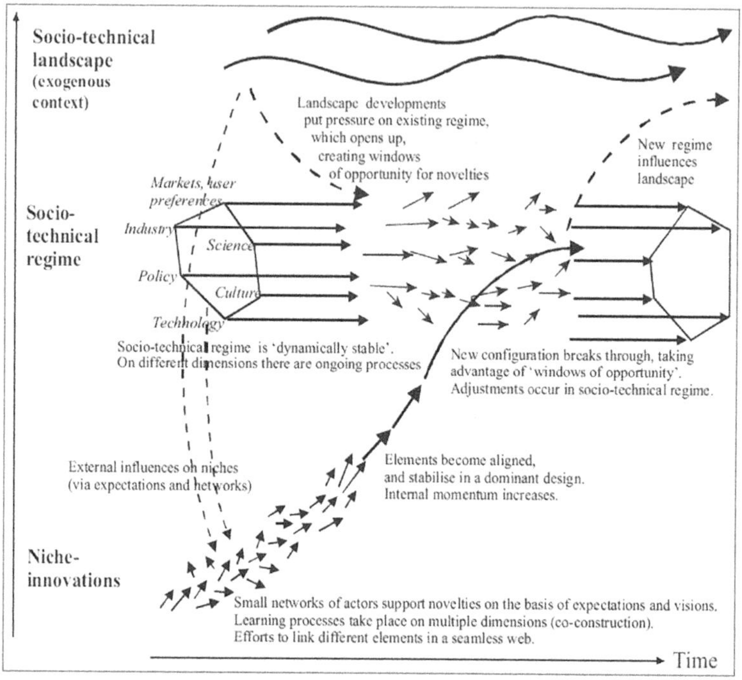

Figure 2.2. Structuration in the MLP (Geels, 2011).

the ongoing public health concerns coupled with growing pressures on water, land and energy resources, a global consensus can now be observed amongst international organizations, action plans and guidelines for the shift towards more sustainable resource management. While wastewater plays a major role in all conceptual models of the dynamic metabolism of closed (urban) cycles, the further formulation of the SDGs by the UN forms perhaps the most common and followed vision uniting global agendas and development frameworks under a broad variety of actors and governments.

SDG 6 (Clean Water and Sanitation) encompasses seven closely interlinked targets. Targets 2 and 3 focus directly on access to sanitation and safe reuse of wastewater, while integrated water resource management is described as a prerequisite for the acceleration towards water security (UN, 2018a). Furthermore, UN water development reports indicate that large-scale centralised treatment systems may no longer represent the most viable option. Decentralised systems have relevant advantages with (in many cases) significantly decreased costs, while facilitating resource recycling (UN WWAP, 2017). In addition, nature-based solutions are expected to play a central role in the achievement of the SDGs by 2030 (UN WWAP, 2018) and together with a wider polycentralised treatment strategy are described as a requirement to effectively solve water problems and ensure water safety (UNESCO WSSM, 2019). Various other SDGs encompass different aspects of resource efficiency, while the main driver behind SDG 6 is water scarcity, rather than resource recovery. Similar trends can be observed for national agendas, where limited water resources are described as a major driver for evolving wastewater recycling strategies, while the potential of further beneficial resources plays either no or only a minor role in such considerations (Schellenberg *et al.*, 2020; Alcalde-Sanz and Gawlik, 2017; US EPA, 2012; Hanseok *et al.*, 2016; FAO, 2010; UNESCO WSSM, 2019). Although wastewater reuse is not a new phenomenon and has been practiced for over 5,000 years in some parts of the world, it still represents a comparatively new concept for institutionalization, regulations or in practice (Angelakis and Gikas, 2014; Alcalde-Sanz and Gawlik, 2017; Hanseok *et al.*, 2016).

Wastewater is associated with a wide range of hazards and thus requires a profound understanding of risk and risk management, especially when considered for reuse. Specific guidelines and regulations are of utmost importance to provide guidance on safety measures and can act as drivers to enable windows of opportunity for reuse. Conversely, the lack of specific regulations or overall prohibition can form an indirect or direct barrier to reuse. The most common global guidelines have been formulated by the WHO, US EPA, FAO or ISO – all of which bring together a wide range of expertise in public health, environment, agriculture and technology (US EPA, 2004, 2012; WHO, 2006; FAO, 1992; ISO 16075, 2015; ISO 23056, n.d.). However, the recommended norms can differ widely – particularly when the WHO and US EPA guidelines are compared. This is due to the different risk management approaches adopted: a best available technology (BAT) and one barrier approach or a multi-barrier approach that allows for alternative best practices where there are economic limitations (US EPA, 2012; WHO, 2006; Krantzberg *et al.*, 2010). The range of reuse practices varies from potable to non-potable applications such as irrigation, industrial cooling, toilet flushing or car washing, reflecting the whole variety of common water uses (Tortajada and Ong, 2016; IMPEL, 2018). Despite the growing variety in reuse applications, it is observed that guidelines and established limits mostly focus on reuse scenarios in agriculture. It is also the case that standards often comprise single sets of limits that do not distinguish between discharge and the receiving environment or different reuse purposes (Jimenez and Asano, 2008; Schellenberg *et al.*, 2020).

With growing environmental pressures, research and social movements have responded with innovative technologies and approaches – and also supported the formulation of a global vision for sustainability agendas through development of diverse sets of guidelines. Although there is an increasing trend in policy adaptation towards more circular approaches in the wastewater sector, and alignment of the landscape and niche level can now be observed, the number of countries with successful wastewater recycling strategies (especially beyond water resources) remains low (Alcalde-Sanz and Gawlik, 2017; Hanseok *et al.*, 2016).

Barriers are reported to arise largely from three fields: the lack of appropriate risk management approaches and adaptation of national regulatory measures; lack of technological options and lack of societal acceptance (Salgot *et al.*, 2018; Angelakis and Gikas, 2014; Perraton *et al.*, 2014; Bichai *et al.*, 2018; Voulvoulis, 2018; Schellenberg *et al.*, 2020). However, as the range of innovative technologies expands, the primary barrier may no longer lie with the availability of technologies but rather with the willingness of politicians or planners to create an environment that enables comprehensive reuse, recycling or recovery (Guest *et al.*, 2009; Salgot *et al.*, 2018; Baumann, 1983; Wester *et al.*, 2015).

Since there are major identified benefits of wastewater as a resource in the water, food and energy sectors, our nexus perspective continues in the following assessment of pressures and drivers at the landscape level.

2.3.3 Pressures and Drivers at the Landscape Level

Within this section, pressures and drivers have been analysed and are presented in the context of the 12 INNOQUA partner countries. A range of indicators has been examined, including data on access to sanitation, related public health impacts, water resources, land resources, climate change and energy.

Access to Sanitation

Access to sanitation remains a big challenge worldwide. Many people still lack basic sanitation services, while overall wastewater treatment coverage also remains low. Data evaluated by the World Bank for the overall global state and the different geographies are illustrated in Table 2.1.

In 2017, 73.4% of the population used at least basic sanitation. However, the percentage of the population using safely managed sanitation services differed widely between countries, and at a global level represented only 45%. The lowest access amongst the analysed countries is observed in Tanzania and India with 30% and 60%, respectively of the population using basic sanitation. While some countries in the Global South achieve a relatively higher basic coverage, safely managed services lag behind, such as in the case of Ecuador with 88% against 42% respectively. The coverage of safe services for the Global North is overall very high.

Financial limitations represent a major challenge for the establishment of broader wastewater treatment coverage. In the World Water Development report in 2017 the UN describes that decentralised and nature-based treatment systems can

Table 2.1. Access to sanitation services in 2017, *Source*: World Bank.

Countries	People Using at Least Basic Sanitation Services (% of Population)	People Using Safely Managed Sanitation Services (% of Population)
GLOBAL	73.4	45.0
Ecuador	88.0	42.0
France	98.7	88.4
Germany	99.2	97.2
India	59.5	no data
Ireland	91.2	82.4
Italy	98.8	96.2
Peru	74.3	42.8
Romania	84.3	76.5
Spain	99.9	96.6
Tanzania	29.9	25.4
Turkey	97.3	65.2
UK	99.1	97.8

represent more viable solutions under these circumstances, with multiple benefits reported, including:

- investment costs estimated at only 20–50% of those compared to conventional systems
- operation and maintenance costs estimated in the range of 5–25% compared to conventional systems
- recycling enables resource provision and related potential revenue streams
- sewerage infrastructure can be reduced, which forms one of the major investments in centralised systems and further requires vast water resources to act as a transport medium

Public Health and Environment

Given the lack of access to basic sanitation services and overall treatment of wastewater, the impact on public health remains a significant burden. Diarrheal disease incidence represents one of the most common indicators. Table 2.2 illustrates disability adjusted life years (DALYs) attributable to diarrheal disease as well as deaths for children below five years and all ages for analysed countries.

Table 2.2. Deaths, episodes and DALYs attributable to diarrheal disease in 2015 by country. *Source*: The Lancet, 2017, numbers >10 were rounded in order to ease reading.

	Children <5 Years			All Ages				
	Death			Deaths			DALYs	
Countries	Numbers	Mortality Rate*	% Change*	Numbers	Mortality Rate*	% Change*	Numbers	% Change*
GLOBAL	**498,889**	**74.3**	**−34.3**	**1,312,128**	**17.8**	**−20.8**	**71,589,511**	**−27.2**
Ecuador	228	14.1	-64.5	485	3	-50.5	39,815	-49.4
France	37	0.9	-36.7	2,392	3.7	46.4	26,396	13.3
Germany	17	0.5	5.1	3,829	4.6	167.6	40,695	110.9
India	104,643	84.2	-43.2	489,000	37.3	-21.7	20,666,210	-32.1
Ireland	2·6	0.7	-4.4	25	0.5	43	599	15.5
Italy	11	0.4	-15.7	559	0.9	110.8	7,065	48.7
Peru	230	7.6	-60.3	627	2	-39.9	56,184	-38.8
Romania	29	3.1	-71.7	58	0.3	-57.2	4303	-62.6
Spain	9	0.4	-45.8	769	1.6	41.2	8576	9.5
Tanzania	5,852	62.9	-28.9	23,428	43.9	-1.9	1,168,719	-12.3
Turkey	244	3.8	-71.4	549	0.7	-59.0	69,082	-47.7
UK	30	0.8	-28.1	1484	2.3	-18.5	22,069	-18.6

*Mortality rate per 100,000; % change for period from 2005 to 2015.

Results show that a significant decrease of −34.4% and −20.8% in deaths caused by diarrhoea among children and all ages can be observed on a global scale. However, 1.3 million people (of which 500,000 are children) still die every year because of diarrhoea, and DALYs amount to 72 million globally. Diarrhoea only represents one of the many additional health burdens related to WASH (water, sanitation and hygiene) and thus the overall impact caused by the lack of sanitation will remain at significantly higher levels. It can be seen that more than a third of total deaths attributable to diarrheal disease worldwide are observed in India, and there are big disparities in mortality rates ranging from 0.4 to 84.2 (per 100,000) for children in Spain and India and 0.3 up to 43.9 (per 100,000) in Romania and Tanzania for persons of all age classes.

In 2019 a World Bank study developed a large database on water quality in terms of BOD, nitrogen and electrical conductivity as major proxies defined by SDG 6.3.2. ("Proportion of bodies of water with good ambient water quality"). The map in Figure 2.3 presents the results of this study on a water quality risk basis.

With wastewater being predominantly discharged to water bodies, most studies focus on water contamination, and while an overall impact on both public health and the environment is widely recognized, the magnitude of public health consequences and impacts on the wider environment are difficult to assess. Barriers to detailed understanding include the long-term financial commitments required to develop appropriate datasets, together with the ever-increasing range of hazards to

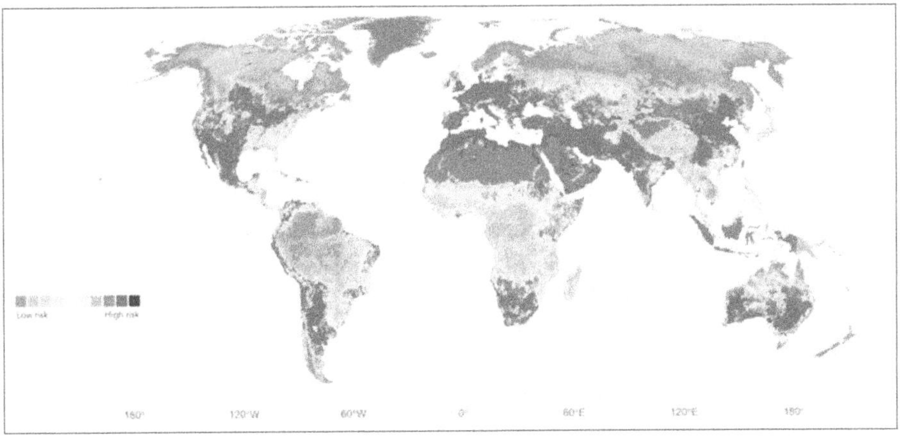

Figure 2.3. Water quality risk in regard to biological oxygen demand, nitrogen and electrical conductivity (Damania *et al.*, 2019).

be considered and the challenges associated with assessing these in complex environments and over the long term (Alcalde-Sanz and Gawlik, 2017; Freeman *et al.*, 2020; Adeel *et al.*, 2017; Sajid *et al.*, 2015; Petrie *et al.*, 2015).

Water Insecurity

While there is no global water shortage in overall terms, spatial and temporal deviations combined with changing environmental conditions confront individual countries with accelerating water stress. According to recent studies by the World Resources Institute (WRI), this growing scarcity is described as one of the greatest challenges and crises in current times, which is not adequately addressed. New hydrological models reveal that global water withdrawals have doubled since 1960 and are growing at more than twice the rate of population, putting even locations which are usually characterised as water abundant at risk (UN Water, n.d; WRI, 2013). While physical scarcity historically existed in different regions due to climatic conditions, the rising imbalance in supply and demand is exacerbated by climate change as well as growing and dynamic population with increased water consumption patterns. These pressures are resulting from and often coupled with inefficient water consumption, rising distribution and resource-allocation difficulties in large agglomeration centres, and failures to ensure equitable and affordable access throughout the year. Burek *et al.* (2014) analysed the impact across different sectors and verified the need for an integrated approach and management. Given the increase in socio-economic growth and changing water use patterns, as well as the consequent increase in wastewater production and use in fertiliser for improved agricultural productivity, the continuous deterioration of valuable water resources and dependent ecosystems is expected to drastically intensify.

WRI data further indicate that 51% of the population, accounting for 3.6 billion persons, is potentially experiencing severe water scarcity for at least one month annually. This will increase up to 5.7 billion in 2050 and 73% of the affected population will be located in Asia. As illustrated in Fig. 2.4, WRI projections indicate that 33 countries will face high water stress in 2040. Water scarcity levels at a global scale and for specific water stress categories in the scope of given geographies are shown in Table 2.3.

The analysis shows that by 2050 Peru, Spain and the UK may be severely affected by water scarcity, with half (or more) of their populations living in areas with limited water resources. While for most of the assessed countries an increase in the percentage of the affected population can be observed, data for Turkey and Tanzania indicate a decrease or no impact as in the case of Ireland. The reasons for

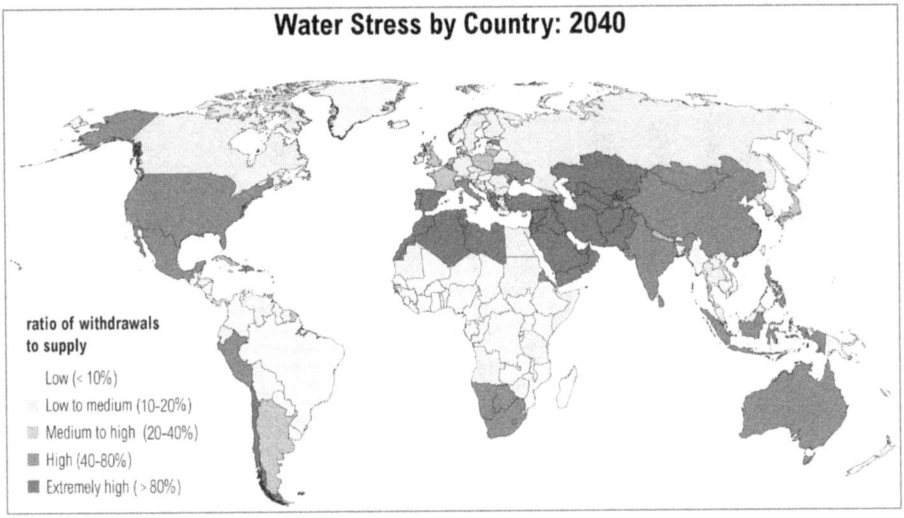

Figure 2.4. Global map on water stress by country (WRI, 2015).

projected downward pressures are likely to indicate positive interactions between population, distribution infrastructure and local climate change impacts. Despite some positive indications, overall results for the assessed geographies indicate that the population living in water stress and scarcity will increase from 578 million to 935 million by 2030.

When analysing overall water withdrawals, agriculture accounts for the highest water consumption worldwide with 70% of all water withdrawals, followed by industrial demands of 19% and municipal and domestic demands of 11%, respectively (WRI, 2018). Water withdrawals and irrigation requirements for agriculture in the different countries are shown in Table 2.4.

The overall municipal water withdrawal for all analysed countries amounts to 269 million m^3 daily. Sectoral withdrawal shows high deviations amongst the countries such as for municipal withdrawals accounting for 7% and 74%, respectively for India and the UK while agricultural withdrawals account for 1.2% in Germany and up to 90.4% in India. Similar variations can be observed for industrial withdrawals, which account for 0.5% in Tanzania and 80.8% in Germany (where ~90% are reported to be used for cooling processes (Statista, 2020)). The big differences in withdrawal are a result of various developmental factors.

In the Global North the sectoral shift to industrial activities in the economy is prevalent, while economies in the Global South are more commonly driven by an extensive agricultural sector. Furthermore, the agricultural sector in the Global North is generally rain fed and seasonal (World Bank, 2020). While warmer

Table 2.3. Population living in water scarce areas in INNOQUA countries. *Source*: World Data Lab, n.d..

	% of People Living in Water Scarce Areas			Water Stress (1,000–1,700 m^3/capita)		Water Scarcity (500–1,000 m^3/capita)		Absolute Scarcity (<500 m^3/capita)	
Countries	2000	2020	2030	2000	2030	2000	2030	2000	2030
Ecuador	8	22	22	354,458	1,401,324	1,065,569	348,289	3,355	3,728,887
France	6	8	8	1,437,137	13,300,498	2,672,117	3,810,240	968,804	2,112,535
Germany	14	16	17	10,263,223	11,463,608	9,461,478	11,413,477	2,586,486	2,381,660
India	30	33	34	98,689,116	168,914,346	165,857,058	232,241,955	158,211,556	298,095,742
Ireland	0	0	0	1,711,672	1,927,276	0	0	0	0
Italy	20	19	19	3,678,069	10,495,783	4,439,609	1,519,073	6,847,141	9,770,284
Peru	45	57	58	291,831	73,306	1,508,225	3,518,399	10,392,169	15,809,057
Romania	8	8	9	910,595	248,678	1,790,251	1,212,935	0	472,759
Spain	41	48	49	3,898,498	6,612,490	8,531,463	5,091,190	8,109,389	20,007,862
Tanzania	30	13	16	1,638,067	12,184,624	1,175,262	6,935,262	9,123,180	4,441,391
Turkey	35	27	30	5,019,318	14,127,326	10,349,455	7,361,016	12,062,413	19,681,146
UK	39	46	50	12,270,194	9,367,752	12,155,749	18,247,047	10,878,477	16,693,523

Table 2.4. Water withdrawal per sector 2013-2017, data by FAO Aquastat and WRI, 2018. Data for Tanzania cover the period 2000-2005.

	Agriculture		Industry		Municipal/Domestic		Total	Irrigation
Countries	**(10^9 m³/yr)**	**%**	**(10^9 m³/yr)**	**%**	**(10^9 m³/yr)**	**%**	**(10^9 m³/yr)**	**(10^9 m³/yr)**
WORLD		70.0		19.0		11.0		
Ecuador	8.08	81.4	0.55	5.50	1.29	13.0	9.92	8.08
France	3.11	11.8	18.2	68.6	5.18	19.6	26.4	2.59
Germany	0.30	1.20	19.8	80.8	4.39	18.0	24.4	0.18
India	688	90.4	17.0	2.20	56.0	7.40	761	–
Ireland	0.18	20.8	0.05	5.92	0.63	73.3	0.86	0.003
Italy	17.0	49.7	7.70	22.5	9.49	27.8	34.2	16.0
Peru	13.1	81.4	0.21	1.30	2.80	17.4	16.1	–
Romania	1.49	22.0	4.23	62.5	1.05	15.5	6.77	0.44
Spain	20.4	65.2	5.97	19.1	4.89	15.7	31.2	18.6
Turkey	50.1	84.9	2.90	4.92	6.01	10.2	59.0	42.2
Tanzania	4.63	89.4	0.03	0.50	0.53	10.2	5.18	–
UK	1.18	14.1	1.01	12.0	6.20	74.0	8.42	0.08
Total	**808**		**77.5**		**98.5**		**983**	

conditions support agricultural activities throughout the year, the generally more complex precipitation patterns and dependence on irrigation paints an emerging crisis for these regions.

Mekonnen and Hoekstra (2011) report that in total 2,320 Gm^3 of virtual water are exported annually, with the largest share of 76% accounted for by international trade in crops and crop products. In Fig. 2.5 countries with negative water balances are represented in green, while positive water balances through import in red can be found in Europe, Mexico, Japan and North Africa. India represents the third biggest virtual water exporter with 125 Gm^3/year, despite being the country with one of the highest water scarcities. The increasing flows of resources and resulting conflicts

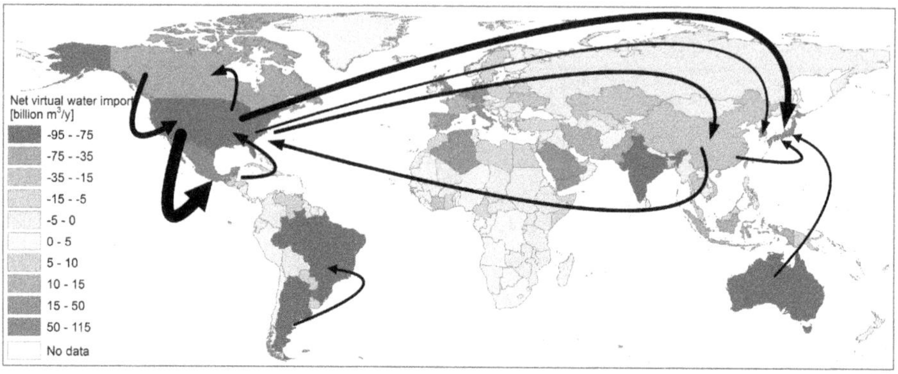

Figure 2.5. Global virtual water flows (Mekonnen and Hoekstra, 2011; Hoekstra and Mekonnen, 2012).

amongst (water) users, especially in the context of widespread poverty and access to water or food, are now driving debates for integrated management strategies at a global scale.

Land Degradation and Food Security

The growing globalization and ever-increasing trend in emigration to agglomeration centres are accelerating the spatial disconnection of production and consumption systems and thus a disruption of sustainable natural processes and closed resource cycles. The constant export of nutrients from land, if not replenished, represents a serious imbalance in soil stocks with far-reaching impacts on productivity or nutritional quality of food crops. In addition, valuable resources accumulated in wastewater are disposed to water bodies, where they pose a threat to the aquatic environment.

Tan *et al.* (2005) reported soil nutrient deficits on a global scale at an average rate of 18.7 N, 5.1 P and 38.8 K kg/ha/year with an annual total deficit of nutrients of 5.5 Tg N, 2.3 Tg P and 12.2 Tg K. The study further indicated a total potential production loss of 1,136 Tg annually (one teragram [Tg] = one megatonne [Mt]). In order to sustain intensive production and yields, fertilization in agricultural systems is usually focused on macronutrients such as N, P and K. Given the excessive use of chemical fertilisers and associated nutrient leaching, nitrate nowadays forms the most common chemical contaminant in groundwater resources. Despite initial peaks in agricultural production following the Green Revolution, yield responses in India have since declined from 13.4 kg grain/kg nutrient to 3.7 kg grain/kg nutrient in 2005 (Sarma and Sharma, 2010). Jones *et al.* (2013) further report that the extent of these impacts has been seriously underestimated and that long-term nutrient-stripping, particularly of micronutrients such as zinc, copper or selenium from locations with highly matured soils is associated with decreased food quality – leading to increased concerns in terms of public health (FAO, 2017; UN WWAP, 2013).

The UN Convention to Combat Desertification (UNCCD) reviewed land degradation profiles in 21 countries in 2018. Their report states that the global economy is predicted to lose 23 trillion USD by the year 2050 as a result of land degradation. It is further concluded that the costs for taking immediate action to prevent and reverse degradation amount to 4.6 trillion USD and thus only a fraction of the predicted losses (UNCCD, 2018). It is highlighted that Asia and Africa will be impacted by the highest costs of 65 billion USD annually. While the overall global loss amounts to 9% of GDP, in Africa total losses may be as high as 40% of GDP. Globally, small family farming enterprises represent 80% of all farms. With 1.5 billion people dependent on land that is subject to continuous degradation pressures, extreme poverty continues to accelerate (UNESCO WWAP, 2019). The status of land degradation in analysed countries along with its causes is illustrated in Table 2.5.

These data show that apart from Ireland, where none to slight degradation is reported, all other countries are affected by land degradation to some degree. In total 80% of the land area in these countries is degraded to some extent, with 46% severely or very severely degraded. Deforestation and agriculture represent the main causes, with water erosion and chemical deterioration as major degradation types.

Most parts of the world experience land degradation, and various studies report nutrient stripping of soil. The combined pressures again point to wastewater as a potentially viable resource. Estimated ranges for macro and micronutrients present in urine and faeces are indicated in Tables 2.6 and 2.7.

These estimates show that on a daily basis globally between 74 kt and 119 kt of nitrogen, 5 kt and 34 kt of phosphorus, 8.9 kt and 39 kt of potassium, 1–29 kt of calcium, 1.7–3.5 kt of magnesium and 1–1.5 kt of sulphur are produced in urine and faeces.

These global estimates show that daily production of micronutrients in faeces amounts to 683 t for Cl, between 223 t and 7,594 t for Fe, 182 t and 683 t for Mn, 15 t and 30 t for Mo, 7.7 t and 15.9 t for Cu, 38 t and 101 t for Zn and from 0.6 t to 2.2 t for Ni.

Table 2.5. Land degradation status in INNOQUA countries. *Source*: Nachtergaele *et al.* FAO (2000).

Countries	Total Area	None		Light		Moderate		Severe		Very Severe		Land Area Degraded: Total	Land Area Degraded: Severe	Cause and Type
	1,000 km^2	1,000 km^2	%	1,000 km^2	%	1,000 km^2	%	1,000 km^2	%	1,000 km^2	%	%	%	
World	134,907	46,066	34	24,292	18	27,389	20	27,036	20	7,971	6	64	26	
Ecuador	283	14	5	188	66	64	23	7	2	11	4	95	6	D W,C
France	552	0	0	320	59	173	32	50	9	0	0	100	9	A-W
Germany	357	10	3	111	31	123	35	112	31	0	0	97	31	A,I-C,W, P
India	3,157	1,184	37	16	1	111	4	1,352	43	495	16	64	59	D,A-W,C
Ireland	70	47	68	22	32	0	0	0	0	0	0	32	0	O-W
Italy	301	0	0	0	0	216	72	84	28	0	0	100	28	D,A-W,P
Peru	1,281	254	20	335	26	270	21	409	32	13	1	80	33	D,O-W,C
Romania	238	0	0	0	0	0	0	212	89	25	11	100	100	D,A-W
Spain	505	11	2	103	20	198	39	175	35	16	3	97	38	I,A-C,(P)
Tanzania	937*	114	12	289	31	295	31	228	24	11	1	87	25	A,O-W
Turkey	779	5	1	0	0	2	0	535	69	235	30	99	99	O,D,A-W,N,P
UK	245	59	24	40	17	98	40	47	19	0	0	76	19	D-W
INNOQUA	**8,705**	**1,698**	**20**	**1,424**	**16**	**1,550**	**18**	**3,211**	**37**	**806**	**9**	**80**	**46**	

A	agriculture	W	water erosion
O	overgrazing	N	wind erosion
D	deforestation	C	chemical deterioration
V	overexploitation of vegetation	P	physical deterioration

*An error in the total area for Tanzania was identified in the source report (which stated the area as 17,000 km^2). The figure for total area listed in this table is calculated by summation of figures for area in all land degradation categories.

Table 2.6. Primary and secondary macronutrients from faeces (F) and urine (U), presented at global and country level[1] as derived from ranges reported by Rose *et al.* (2015).

	C	N	Total P	Total K	Ca	Mg	S
	7 (F)	**1.8 (F) 8–13.9 (U)**	**0.35–2.7 (F) 0.35–1.8 (U)**	**0.2–2.52(F) 0.97–2.6(U)**	**0.1–3.6(F) 0.03–0.23(U)**	**0.15–0.34(F) 0.07–0.12(U)**	**0.13–0.2(F)**
Countries	(ranges indicated above in g/capita/day (F) and g/L (U); multiplied up below to country level as t/day)						
World	53,158	74,421–119,226	5,316–34,173	8,885–38,957	987–29,085	1,671–3,493	987–1,519
Ecuador	120	167–268	12–77	20–88	2–65	4–8	2.2–3.4
France	469	656–1,052	47–301	78–344	9–257	15–31	8.7–13.4
Germany	580	812–1,302	58–373	97–425	11–318	18–38	10.8–16.6
India	9,468	13,256–21,236	947–6,087	1,583–6,939	176–5,181	298–622	176–271
Ireland	34	48–76	3–22	6–25	1–19	1–2	0.6–1
Italy	423	592–949	42–272	71–310	8–231	13–28	7.9–12.1
Peru	224	313–502	22–144	37–164	4–123	7–15	4.2–6.4
Romania	136	191–306	14–88	23–100	3–75	4–9	2.5–3.9
Spain	328	459–735	33–211	55–240	6–179	10–22	6.1–9.4
Tanzania	394	552–884	39–253	66–289	7–216	12–26	7.3–11.3
Turkey	576	807–1,292	58–370	96–422	11–315	18–38	10.7–16.5
UK	465	651–1,043	47–299	78–341	9–255	15–31	8.6–13.3
INNOQUA	**13,218**	**18,505–29,645**	**1,322–8,497**	**2,209–9,687**	**246–7,232**	**415–869**	**245–378**

[1] Iteration based on World Bank data for population in 2018.

C values as indicated by Snyder *et al.*, 1975; N as to Rose *et al.*, 2015; TP as to Vinneras *et al.*, 2006 and Wignarajah *et al.*, 2003 for F and Jönsson *et al.*, 1997 and Ban and Dave, 2004 for U; TK as to Calloway and Margen, 1971 and Vinneras *et al.*, 2006 for F and Beler-Baykal *et al.*, 2004 and Putnam, 1971 for U; Ca as to Wignarajah *et al.*, 2003 and Calloway and Margen, 1971 for F and Diem and Lentner, 1970 and Jana *et al.*, 2012 for U; Mg as to Eastwood *et al.*, 1984 and Goldblith and Wick, 1961 for F and Diem and Lentner, 1970 and Tilley *et al.*, 2008 for U; S as to Goldblith and Wick, 1961 and Meinzinger and Oldenburg, 2009 for F.

Table 2.7. Micronutrients in faeces, presented at global and country level,[1] as derived from ranges reported by Rose *et al.* (2015).

Countries	Cl	Fe	Cu	Mn	Mo	Zn	Ni
	90	**30–1000**	**1.02–2.1**	**24–90**	**2–4**	**5–13.3**	**0.08–0.3**
	(ranges indicated above in mg/capita/day; multiplied up below to country level for Cl, Fe and Mn as t/day, while Cu, Mo, Zn and Ni are presented as kg/day)						
World	**683**	**228–7,594**	**7,746–15,947**	**182–683**	**15,188–30,376**	**37,970–101,076**	**608–2,278**
Ecuador	1.54	0.51–17.1	17.4–36	0.41–1.54	34.2–68	85–227	1.4–5.1
France	6.03	2.01–67	68–141	1.6–6	134–268	335–891	5.4–20.1
Germany	7.46	2.49–83	85–174	1.99–7.46	166–332	415–1,103	6.6–24.9
India	122	40.6–1,353	1,380–2,840	32.4–121.7	2,705–5,410	6,763–18,003	108–406
Ireland	0.44	0.15–4.9	5–10.2	0.12–0.44	9.7–19.5	24.3–65	0.4–1.5
Italy	5.44	1.81–60	61.6–127	1.45–5.44	121–242	302–804	4.8–18.1
Peru	2.88	0.96–32	32.6–67	0.77–2.88	64–128	160–426	2.6–9.6
Romania	1.75	0.58–19.5	19.9–41	0.47–1.75	38.9–78	97–259	1.6–5.8
Spain	4.21	1.4–47	47.7–98	1.12–4.2	94–187	234–623	3.7–14
Tanzania	5.07	1.69–56	57.4–118	1.35–5.07	113–225	282–750	4.5–16.9
Turkey	7.41	2.47–82	84–173	1.98–7.41	165–329	412–1,096	6.6–24.7
UK	5.98	1.99–66	68–140	1.6–5.98	133–266	332–885	5.3–19.9
INNOQUA	**170**	**57–1,888**	**1,926–3,965**	**45.3–169.9**	**3,776–7,553**	**9,441–25,132**	**151–566**

[1] Iteration based on World Bank data for population in 2018.

Cl as to range indicated by Goldblith and Wick, 1961; Fe as to Goldblith and Wick, 1961 and Wignarajah *et al.*, 2003; Cu as to Goldblith and Wick, 1961 and Wignarajah *et al.*, 2003; Mn and Mo as to Wignarajah *et al.*, 2003; Zn as to Eastwood *et al.*, 1984 and Vinneras *et al.*, 2006; Ni as to Hansen and Tjell, 1979 and Schouw *et al.*, 2002.

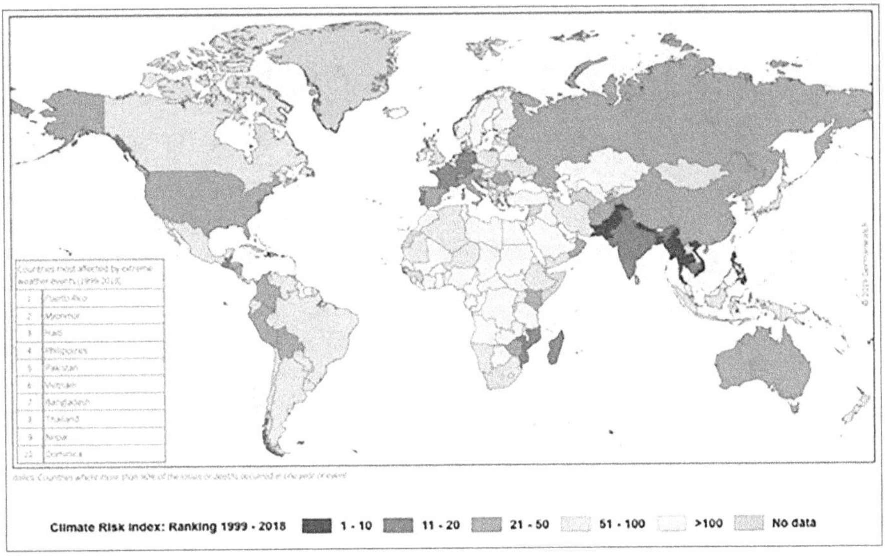

Figure 2.6. Map of climate risk index in different countries (Germanwatch, 2019).

Climate Change and Energy

The IPCC forecast that climate change will have far-reaching implications on the availability of renewable surface and groundwater resources – with knock-on effects on global food security systems that will increasingly push the most vulnerable regions and marginalized populations into even deeper poverty (IPCC, 2014). Germanwatch annually publishes a climate risk index based on fatalities and economic losses in countries due to extreme weather events (Figure 2.6 and Table 2.8).

The assessment of climate risks in 2020 shows that Germany and India are among the most affected countries worldwide, respectively ranking 3 and 5 out of 181 in total. While the annual assessments show high deviations such as in the case of Peru (ranked 5 in 2019 and 110 in 2020), the risk assessment over a 20-year period provides a more stable evaluation of impacts – India, France, Germany, Italy and Spain remain high-risk countries. The comparatively high values for fatalities in countries of the Global North are explained largely by a rise in heat wave events. Losses attributable to climate change in all analysed countries amount to USD 23,794 million PPP (Purchasing Power Parity) on annual average.

Technological innovations and green growth solutions can have an important impact on climate change and associated pressures such as water availability and pollution control. In the wastewater sector these can include decentralised sewerage and nature-based treatment systems (NBS). NBS are generally designed with lower operation and maintenance (O&M) requirements when compared with conventional solutions, and when coupled with decentralised approaches can minimize associated infrastructure and transport requirements. Aside from energy

Table 2.8. Climate risk impact (CRI) ranking in INNOQUA countries. *Source*: Germanwatch, 2019, 2020. PPP = Purchasing Power Parity.

	CRI Rank 2020	CRI Rank 20a Period (1998–2017)	CRI Score 20a	Fatalities		Fatalities Per 100,000		Losses in USD Million (PPP)		Losses Per Unit GDP	
				Avg	Rank	Avg	Rank	Avg	Rank	Avg	Rank
Ecuador	112	96	90.67	21	69	0.143	81	101.452	83	0.071	115
France	34	18	38.67	1,121	5	1.815	11	2,205.338	13	0.098	96
Germany	3	25	42.83	475	11	0.584	31	3,945.817	6	0.124	89
India	5	14	36.50	3,661	2	0.316	48	12,822.708	3	0.263	59
Ireland	81	125	111.67	2.1	133	0.048	130	173.495	69	0.087	104
Italy	21	28	46	1005	6	1,709	12	1,458.029	20	0.072	113
Peru	110	45	56.83	80.5	34	0.281	59	433.020	39	0.161	75
Romania	89	37	53.33	43.25	51	0.206	71	1,160.849	23	0.310	52
Spain	38	34	49.83	695	8	1,569	13	979.181	27	0.069	119
Tanzania	68	126	112	20.95	68	0.052	127	52.715	100	0.060	125
Turkey	85	121	110	29.35	63	0.041	140	461.532	37	0.036	140
UK	78	60	68	152.2	21	0.246	64	1,480.998	19	0.068	120

Table 2.9. Estimates for potential energy yields from human faeces (iteration based on population data from World Bank, 2018).

Countries	Biomethane Production[a] m^3/day	Calorific Value[b] kWh/day	
		Min	Max
World	**64,131,330**	**359,135,448**	**461,745,576**
Ecuador	144,277	807,953	1,038,797
France	565,622	3,167,481	4,072,476
Germany	700,139	3,920,780	5,041,003
India	11,422,853	63,967,979	82,244,544
Ireland	41,104	230,185	295,952
Italy	510,262	2,857,466	3,673,885
Peru	270,149	1,512,836	1,945,075
Romania	164,392	920,593	1,183,619
Spain	395,197	2,213,102	2,845,417
Tanzania	475,608	2,663,407	3,424,381
Turkey	695,190	3,893,064	5,005,368
UK	561,258	3,143,043	4,041,055
INNOQUA, Total	**15,946,052**	**89,297,889**	**114,811,572**

[a]On basis of 8,445 L/cap and day, in Taseli and Kilkis, 2016.
[b]On basis of 20–26 MJ/m^3 or 5.6–7.2 kWh/m^3; in Frost and Gilkinson, 2011, as reported in Andriani *et al.*, 2015.

savings (which represent short-term gains), sanitation systems can play a crucial role in achieving long-term greenhouse gas benefits through closed loop systems that both prevent or/and capture greenhouse gas emissions to convert resources into renewable energy (UNESCO WSSM, 2019). Estimates for biomethane yields from human faeces are provided in Table 2.9.

The assessment shows that daily cumulative energy resources of 15.9 Mm^3 in the form of biomethane could be generated from faeces in the assessed countries, which translates to 89,298 MWh/day. Other studies and calculations based on the median energy content per kg of COD for wastewater suggest theoretical energy yields as high as 562 TWh/year; this equates to 50 million tonnes of oil equivalent (MTOE) or 100×10^6 t CO_2 – which represents 1% of global CO_2 equivalent emissions (Capodaglio and Olsson, 2019). While this indicates the potential value of resources in wastewater it also indicates the immensely damaging impacts those resources can have if not captured. For example, septic tanks can act as greenhouse gas (GHG) point sources, and may be responsible for emissions as high as 27.1g CH_4 per capita and day (Diaz-Valbuena *et al.*, 2011). Campos *et al.* further report in (2016) that methane emissions in several wastewater treatment processes can exceed the amount of CO_2 avoided through the utilization of produced biogas and conclude that further development of alternative processes is needed.

2.3.4 INNOQUA as Responsive Action at Niche Level

INNOQUA was a research project and represented an investigation of novel niche technologies. Within a broad demonstration programme in 11 of the 12 partner countries, different system configurations of decentralised nature-based technologies were analysed under real environmental conditions for extended periods (up to two years) to identify wastewater treatment solutions for different geographical conditions and applications. The project investigations aimed to address public health and environmental pressures by:

(a) providing affordable sanitation for all, which is socially accepted and requires low expenditure on operations and maintenance (O&M) and
(b) bringing forward nature-based and more sustainable solutions with lower resource footprints.

Nature-based solutions do not require additional chemical inputs for treatment, are normally designed with low or no direct energy inputs, and usually incorporate recycling of treated water resources (and nutrients). INNOQUA investigated three different nature-based solutions:

1. Lumbrifilter – which harnesses the capacities of earthworms and micro-organisms to deliver primary and secondary treatment;
2. Daphniafilter – which harnesses the capacities of small crustaceans, micro-organisms and aquatic plants to deliver tertiary treatment; and
3. Bio-solar purification – which harnesses the capacities of micro-algae and bacteria to deliver tertiary treatment and partial disinfection through solar irradiation.

To meet the requirements for wastewater reuse in irrigation of edible crops, INNOQUA also investigated the potential for UV disinfection of wastewater treated first by the Lumbrifilter and/or Daphniafilter.

2.3.5 Assessing Possible Barriers at Socio-Technical Regime Level

As described in the assessment of global trends, one major reported barrier to sustainability transition and circular approaches in the wastewater sector is the lack of adaptive risk management approaches and regulations. Here, the state of adapted policies in analysed geographies is reviewed in regard to (a) progress on integrated water resources management and (b) regulatory measures supportive of overall integrated resources management in terms of recycling, reuse and recovery of water, nutrient and energy resources.

Integrated Water Resources Management

As described in the Agenda of the United Nations, integrated water resources management is a pre-requisite for the overall achievement of SDG 6. Table 2.10 presents the state of Integrated Water Resources Management (IWRM) implementation in the different countries under an overall IWRM score. This score is derived from four disciplines, based on the (a) enabling environment, (b) institutions and participation, (c) management instruments and (d) financing.

As presented in the assessment by the UN, progress towards integrated management can be observed in all studied locations to varying degrees, apart from India, where data were not available. The strongest achievements can be seen in France with a maximum score of 100 throughout all categories. Lower performances can be observed in the case of Peru and Ecuador, with overall scores of 30 and 42 and major difficulties in the financing field – a situation valid in many countries. In India the National Water Policy of 2012 promulgates the implementation of IWRM. In 2019 the Ministries for Water Resources, River Development & Ganga Rejuvenation and Ministry of Drinking Water and Sanitation were merged to form the new Ministry of Jal Shakti. The new ministry has subsequently mounted a national water mission to strengthen integrated water resource management (MoWR, 2012; MoJS, 2020).

Table 2.10. Progress on Integrated Water Resources Management (IWRM) (UN, 2018b; Ministry of Jal Shakti, 2020).

Countries	IWRM Scores	Enabling Environment	Institutions and Participation	Management Instruments	Financing
Ecuador	42	38	44	51	34
France	100	100	100	100	100
Germany	88	96	89	83	84
India	No submission, newly formed Ministry of Jal Shakti in 2019 envisions national water mission through IWRM				
Ireland	81	80	84	82	76
Italy	55	60	61	51	46
Peru	30	34	26	34	24
Romania	72	96	65	84	44

(Continued)

Table 2.10. Continued

Countries	IWRM Scores	Enabling Environment	Institutions and Participation	Management Instruments	Financing
Spain	82	93	81	90	66
Tanzania	50	57	55	40	50
Turkey	70	75	75	70	58
UK	77	83	82	76	66

Regulatory Measures Towards an Integrated Resource Management for Wastewater

Within the paradigm shift, wastewater is mainly considered as a resource hub for water, nutrients and energy. Given the increasing pressures due to water scarcity, land degradation, climate change and energy demand, an assessment has been undertaken to analyse the regime response and state of policy adaptation towards a circular approach in the wastewater sector. Table 2.11 gives an overview of existing regulations in the fields of (a) circular water management and permitted applications for treated wastewater, (b) nutrient recycling or recovery and (c) GHG mitigation or energy recovery from wastewater.

a. Regulations on Circular Water Management

Wastewater reuse regulations are available in around half the assessed countries, while being under development in Turkey and lacking in Germany, Ireland, Tanzania, India and the UK. In response to increasing water stress among Member States, the European Union recognized that limited awareness and the lack of a supportive and coherent regulatory framework were two major barriers to water reuse. As a response, the European Commission released guidelines for wastewater reuse in 2019 (European Commission, 2019). These harmonize the minimum requirements for irrigation in agriculture across the EU and particularly aim to promote reuse in Member States with rife water scarcity. Germany and Ireland have carried out assessments of the potential and requirement for wastewater reuse, and observed water scarcity in some regions (Umweltbundesamt, 2019; EPA, 2016a). However, given the low pressures on available water resources and constraints from emerging pollutants, water reuse does not play a significant role in these countries. In Tanzania wastewater reuse is reported as practiced but regulations are lacking, and the status of any related developments remains unclear (Kihila *et al.*, 2015). A wide number of reuse applications are permitted in Italy and Spain, ranging from irrigation to aquifer recharge, street cleaning and industrial applications.

While most countries have developed explicit reuse standards, in Ecuador the regulations instead relate to water quality criteria within receiving environments.

Given increases in indirect wastewater reuse, the shift in risk approach from standards at the point of use versus standards at the point of discharge is gaining increased attention. Despite having previously established quality criteria for a range of reuse applications, a regression can be observed in India (Schellenberg *et al.*, 2020). Recognizing the pressure of water scarcity, reuse of wastewater in India is explicitly targeted and envisioned in the new set of standards released in 2019. However, these discard the previous approach in which standards varied according to the receiving environment and included a dedicated set of standards for discharges on land. The current pollution measures focus instead on standards for discharge to receiving water bodies. It should be noted that, while water reuse standards support this practice in many countries – it is only rendered compulsory in a number of locations. For example, in some regions of the US, China and India wastewater reuse is mandated through Zero-Liquid-Discharge orders (Tong and Elimelech, 2016). Currently these orders apply mostly to industries, with one exception observed in the case of Bengaluru – which targets the municipal sector and commercial buildings (Evans *et al.*, 2009).

Table 2.11. Overview on regulations for reuse, recycling and recovery of water, nutrients and energy in the wastewater sector.

Countries	Wastewater Reuse Regulation[a]	Application Areas Considered in Regulation[a]	Nutrient Reuse/Recovery Regulations[b]	Energy Recovery Regulation[c]
Ecuador	Registro Oficial, No. 097-A	3 diverse sets for irrigation in agriculture	–	–
France	JORF num. 0153, 4 July 2014; EU regulation	8 diverse sets for irrigation in agriculture and green areas	EC Sewage Sludge Directive; ~80% of sewage sludge used in agriculture.	–
Germany	NA on country level, EU regulation	–	EC Sewage Sludge Directive; Sewage Sludge Ordnance mandates P recovery	Indirect measures
India	Diversified set NA; enforcement on reuse in some states	Reuse promoted in new fixed set of discharge standards. No variance according to reuse	Partially covered by NFSSM Policy, 2017 and Solid Waste Management Rules, 2016	–

(*Continued*)

Table 2.11. Continued

Countries	Wastewater Reuse Regulation[a]	Application Areas Considered in Regulation[a]	Nutrient Reuse/Recovery Regulations[b]	Energy Recovery Regulation[c]
Ireland	NA on country level; EU regulation	–	EC Sewage Sludge Directive; Code of Practice for Biosolids; ~80% of sewage sludge used in agriculture.	–
Italy	DM 185/2003, EU regulation	Sixteen, including various irrigation, urban, municipal and industrial uses	EC Sewage Sludge Directive; D. Lgs. 99/92 on sludge reuse; D. Lgs. 75/2010 on fertiliser; ~35% of sewage sludge used in agriculture	–
Peru	LMP by Ministry of Environment	Range unclear, include discharge and environmental quality standards. Possibility for Adaptation Plan for Environmental Management, in the case of non-compliance with established standards.	–	–
Romania	NA on country level; EU regulation	–	EC Sewage Sludge Directive; ~10% of sewage sludge used in agriculture; ~25% used in other sectors, and the rest landfilled	–

(Continued)

Table 2.11. Continued

Countries	Wastewater Reuse Regulation[a]	Application Areas Considered in Regulation[a]	Nutrient Reuse/Recovery Regulations[b]	Energy Recovery Regulation[c]
Spain	RD 1620/2007; EU regulation	Twenty-one, including various irrigation, urban, municipal, industrial and environmental uses	EC Sewage Sludge Directive; ~70% of sewage sludge used in agriculture.	–
Tanzania	–	–	–	–
Turkey	Under development	–	Soil Pollution Control Regulation	–
UK	–	–	EC Sewage Sludge Directive, Sludge Regulations 1989 No. 1263; ~98% of sewage sludge used in agriculture	Indirect measures
World	WHO, US EPA, FAO, ISO, EU, etc.	–	ISO/DIS19698 under development	–

[a]IMPEL, 2018; Ministrio del Ambiente, 2015; Schellenberg *et al.*, 2020; Ministerio de Vivienda, Construcción y Saneamiento, 2010; Ungureanu, 2018; Kihila *et al.*, 2015; Nas *et al.*, 2020; GWI, 2014; WHO, 2006; US EPA, 2012; FAO, 1992; ISO 23056, n.d.
[b]Hermann and Hermann, 2019; Hudcová *et al.*, 2019; Collivignarelli *et al.*, 2019; EUROSTAT, 2020; EPA, 2016a,b; UN Habitat, 2008; EC, 1986; BMJ, 2007; EC, 1999, 2003, 2008, 2018, 2019, 2020; ISO 16075, 2015; MoEFC, 2016; MoUD, 2017; UK Government, 1989.
[c]EC, 2001, 2009; BMU, 2017; UK Government, 2013; Defra, 2011.

b. Regulations on Nutrient Recycling and Recovery

Three major options are available to achieve closed-loop approaches for nutrients. These are (a) the reuse of treated nutrient-rich water, (b) reuse of nutrient concentrated biomass (sludge/biosolids or plant biomass such as algae) or (c) direct recovery of nutrients from wastewater media through physiochemical separation. Direct enforcement mechanisms exist only in the case of Switzerland and Germany, where treatment plants are mandated for retrofitting in order to recover phosphorus resources (Shaddel *et al.*, 2019). Elsewhere, a range of regulatory approaches is evident.

The reuse of nutrient-rich water strongly depends on the application in an environment where both water and nutrients are required. This should be supported by regulations covering both the irrigation and fertiliser value of the substrate. While some regions enable reuse for various applications, the characteristics of pollution control measures and their related limits can differ significantly. For example, in the case of irrigation water quality in Spain or France it can be observed that standards cover a small set of parameters including pathogens, turbidity, and suspended solids. In contrast, standards in Italy cover a broader range of parameters and include limits for BOD, COD, total phosphorus and total nitrogen. The new fixed set of standards in India requires the elimination of nutrients for all wastewater discharges, and in so-doing represents an active barrier to nutrient recycling.

Reuse regulations for sludge apply in most European countries where around 4 million tonnes out of 8 million tonnes (dry solids) of wastewater treatment sludges are re-used in agriculture (EC, 2020). The general regulatory approach evolved to cope with increasing quantities of sludge produced in conventional wastewater treatment plants. This led to the formulation of the Sewage Sludge Directive in 1986, which encouraged the reuse of sludges in agriculture and set out the framework within which this could be achieved. Following the explicit encouragement of reuse in the Urban Wastewater Treatment Directive in 1991, and subsequent limitations for landfilling under EC 1999/31, sludge recycling to land was further consolidated. However, the management of sewage sludge can still vary significantly between the Member States based on local economic, social, or technical contexts – as well as widely differing national regulations with limits set significantly below EC norms (Inglezakis *et al.*, 2014). As in the case of Ireland approximately 80% of produced sewage sludge was applied to agricultural land in 2016, in comparison to around 30% in Germany for the same year (Collivignarelli *et al.*, 2019; EUROSTAT, 2020). Collivignarelli *et al.* report further shifts in countries such as Germany, where quantities of sludge applied to agricultural land have consistently decreased, and under the new sewage sludge ordinance of 2017 the practice is prohibited for sludges derived from wastewater treatment plants treating Population Equivalents (PE) > 50,000.

The European Union follows a strong circular approach, manifested through the Waste Framework Directive in 2008 and subsequently the Circular Economy Action Plan (in 2015). However, while sewage sludge as such is excluded from the directive and is also excluded from new EU fertiliser regulations, derivatives such as struvite and phosphorus recovered from incinerated sewage sludge are permitted under these latest regulations (EC, 2003, 2008, 2018, 2019; Huygens *et al.*, 2019). The 2017 German ordinance mandates progressive phosphorus recovery at larger sewage treatment plants (BMJV, 2017; BMU, 2017).

Outside Europe the assessment shows that only Turkey provides a regulation for sewage sludge for land application, which is managed under the Turkish Soil Pollution Control Regulation, whereas technical requirements for compost are stated in the Turkish Solid Waste Control Regulation (UN Habitat, 2008). In India the first steps towards the regulation of faecal sludge have been recently formulated in the National FSSM Policy of 2017 and Solid Waste Management Rules of 2016, which envision a waste hierarchy under a circular approach, representing a move towards nutrient recycling – although concrete standards are yet to be developed. In contrast to Europe, the policy envisions a management strategy for untreated or processed media whereby the former will be landfilled, while the latter will be composted (MoUD, 2017; MoEFC, 2016). In the case of Ecuador, Peru, and Tanzania regulations for the management of septage, sewage sludge and nutrient recovery are lacking.

c. Climate Change and Energy Recovery

The assessment shows that as of now there are no direct mechanisms or regulations mandating energy recovery or GHG mitigation from wastewater processes. However, first signs for an adaptation and a possible transition are observed in the case of Europe. The Directives on the promotion of renewable energy in 2003 and 2009 in conjunction with the Directive on energy efficiency in 2012 stipulate targets of 20% share of energy from renewable sources and 20% savings in primary energy consumption in Member States (EC, 2001, 2009). Herein, biogas from wastewater treatment was included as a possible source of renewable energy and resultantly formed a path for further development. Several treatment plants in Europe achieve near or even complete energy self-sufficiency, and there are further aspirations to transform the wastewater sector from one of the highest energy consumers of the municipal sector into a fundamental pillar of the energy revolution (Gandiglio *et al.*, 2017; IEA, 2018; Powerstep, 2018). In Germany, for example, the national wastewater directive requires treatment plants under Section 3 (2a) to be installed and operated in a way which enables energy efficient operation. In the UK on the other hand the Renewables Order placed an obligation on electricity utilities to source an increasing percentage of their supply from renewable generators. Water utilities took advantage of the premium that this scheme created in the electricity market to invest in anaerobic digestion of their sewage sludges, with combustion of the biogas in combined heat and power (CHP) units to generate renewable electricity (Environment Agency, 2009). In both cases, while there is no direct mandate for the recovery of energy in the wastewater treatment sector, these indirect measures have led to the adaptation of sewage treatment plants (STPs) in Germany and the UK. IEA reports that in Germany 3,050 GWh/year have been produced from biogas in STPs. This represents the highest generation globally and biggest progress when compared to South Korea as second highest producer with 969 GWh/year (IEA Bioenergy, 2015). For context, sectoral energy consumption is reported at around 4,400 GWh/year in Germany indicating a possible near-term achievement in energy neutrality for the wastewater sector (Umweltbundesamt, 2009).

While the wastewater sector in Germany is estimated to consume the equivalent of the annual production from two power stations, the chemical energy potential in wastewater is estimated to be in the range of 12 large power stations (Powerstep, 2018). Unlocking this innate potential remains the focus of intense research activity, with EU research projects such as Powerstep, investigating innovations for energy-positive wastewater treatment at full scale. Apart from (off-site) GHG emissions arising from electrical consumption, it is reported that in 2010, methane emissions from the wastewater sector accounted for more than 500 Mt CO_2 equivalent and nitrous oxide emissions for 80 Mt of CO_2 equivalent (Tanmay Ram Kate and Sridevi, 2019). As a result, the sector represents one of the top five biggest sources of anthropogenic greenhouse gas emissions. Energy efficiency regulations target the optimization of processes and mainly orient towards reduced energy consumption. Regulations for direct GHG emission mitigation in wastewater treatment processes do not exist, although ambitious voluntary targets are in place, as with (for example) the UK water industry's target for net zero greenhouse gas emissions by 2030 (Water UK, 2020).

2.4 Discussion and the Way Forward

Integrated resources management, decentralised and nature-based wastewater treatment under circular approaches are considered as key elements in the achievement of SDG6 and the paradigm shift towards a sustainability transition in the wastewater sector. It is common for sectoral resource management approaches to consider just one aspect of a complex system in isolation and by this omit intricate interactions with other elements. Managed under anthropogenically-induced system structures, these sectoral approaches have been shown to incrementally fail when a holistic understanding of resource inventories and the complex dynamics of natural systems are considered. Furthermore, they can lead to conflicts through opaque handling among diverse sets of divided actors attempting to implement incoherent and fragmented regulatory frameworks to an essential resource such as water. Integrated resources management focuses on the system as a whole and through this can enable inclusive collaboration and management.

The paradigm shift towards a more sustainable development of the wastewater sector recognizes major benefits through recycling or recovery of resources derived from wastewater in the shape of water, nutrients and energy. While enabling more sustainable handling of resources, this approach also has the potential to open additional revenue streams and through this drive broader development and access to sanitation. This study shows that pressures for change can be observed and are increasing in all sectors. These range from lack of wastewater treatment with related contamination of the environment and public health impacts, to growing water scarcity, progressing to land degradation and food insecurity – which are all further exacerbated by ongoing climate change and rising demand for energy resources.

Water Scarcity and Water Reuse Regulation

Water scarcity is increasing in almost all countries. Overall, decreasing water availability forms a major driver and path behind the observed alignment between landscape and regime levels albeit that several of the studied geographies with lower or no stress factors such as Ireland, Germany or Romania do not have specific national regulations for wastewater reuse. The most progressive regulations in terms of water reuse in a wide range of applications can be observed for Spain or Italy, which are also among the most affected geographies threatened by water scarcity in Europe. On the other hand, while India and Peru belong to the most water-scarce nations globally, clear regulations and standards for reuse are lacking in those countries. Here a disparity in the progress towards appropriate risk management can be observed between the Global North and South, despite (often) controversial reuse realities on the ground with their associated risks. The magnitude of pressures and risks is often higher for the countries of the Global South, where financial constraints mean that the treatment coverage is usually lower. At the same time, growing water scarcity is leading to reported high levels of reuse for irrigation in agriculture in these countries. In this context the lack of guidelines or enforcement of pollution control measures (which normally require capital-intensive removal of nutrients) can form active barriers to the promotion of safe reuse practices or nutrient recycling. In some cases, regulations might also be considered to have regressed. In India, standards for discharge/reuse have been reduced from four to one set – requiring a degree of treatment that is not necessarily achievable or applicable to all intended end uses. Furthermore, some countries implement a BAT approach despite financial limitations. The reasons behind such counter-intuitive approaches are among others reported to include strong lobbying activities of national and international actors (in support of conventional systems) as well as vested interests of politicians (Never, 2016; Istenic *et al.*, 2014). Nature-based solutions on the other hand have historically been promoted among NGO's or green movements, and their importance has only recently started to receive attention in global agendas. However, given the strong power relations, the implementation of NBS remains sub-marginal at the regime level. Although the BAT approach to risk management is claimed to be technology-neutral, the opposite can be observed in practice (Schellenberg *et al.*, 2020).

While there are big differences in global guidelines, risk management approaches and diverse sets of discharge standards, another discussion around 'point of use' standards is gaining momentum. As in the case of Ecuador, water quality criteria apply for reuse scenarios. Given accelerating water scarcity, indirect reuse is commonly observed in peri-urban environments in locations with low wastewater treatment coverage – with widespread downstream use of contaminated water for irrigation. This means that point of use standards become highly significant. Such conditions may necessitate implementation of a bilateral approach to cope with the given risk reality (with separate standards for discharge and reuse). However, given the generally observed low capacities in monitoring, safe reuse practices may ultimately depend on awareness and responsibility of the end user.

Land Degradation and Nutrient Recycling

The current status of terrestrial systems reveals tremendous challenges relating to increasing land degradation, which has deep and far reaching impacts on the world's fundamental ability to secure food production in the long term, leading to a range of health issues and high vulnerability within marginalized populations. Although research studies report an immense degradation of soil quality through disaggregated consumption and production systems and associated nutrient stripping, the interlinked relevance of nutrient recycling and closed loops is not yet fully reflected in global agendas.

Under the scope of 'sustainable intensification', mineral fertilization has developed into a widely established practice with an immense and powerful global market, which is mostly controlled by transnational corporations in the Global North (Struik and Kuyper, 2017; Jorgenson and Kuykendall, 2008). This extensive overuse of mineral fertilisers is critically impacting water quality, while their production is reported to account for more than 1.2% of global GHG emissions (Kongshaug, 1998; Wood and Cowie, 2004). Although the green revolution and agricultural intensification initially led to an increase in food production, it is now reported that it undermines the resilience of agricultural ecosystems and their capacity to feed the global population in the long run (Struik and Kuyper, 2017). Current studies observe declining yields and adverse effects from decreasing soil fertility linked to the use of synthetic agricultural inputs, including both fertilisers and pesticides (Biswas *et al.*, 2014). There is a great need for more sustainable agricultural practices, which are reported to enhance food security and water resources, and also to mitigate GHG emissions (Geng *et al.*, 2019; Lin *et al.*, 2019; Struik and Kuyper, 2017). However, the structural globalization of investment and international trade in the world food economy has pushed forward industrial agronomic practices under inter-related networks and power relationships, as observed particularly in less developed geographies linked to foreign investment schemes (Jorgenson and Kuykendall, 2008). Although our review shows that nutrient recycling in the form of sewage sludge occurs widely in European countries, the prevalence of this practice varies hugely between EU Member States and in some cases increasingly strict regulations indicate negative trends in sludge recycling. As an alternative, phosphorus recovery is increasingly mandated, while the relevance of micronutrients in soil matrices for plant growth and (ultimately) for public health appears to be inadequately taken into account.

The highest vulnerabilities in terms of mature soils, harsh climatic conditions and related poverty patterns arise in the Global South – increasing difficulties in terms of mitigation of hunger or poverty eradication, or attainment of sustainability goals. However, this review further shows that regulations for nutrient recycling or sludge management are mostly absent in the countries of the Global South, despite the given pressures. Apart from biomass (sludge) reuse or nutrient recovery, recycling can also be achieved via nutrient-rich effluent and irrigation. Effluent quality and composition rely on measures formulated as standards. Despite the growing trend for reuse standards, especially in agricultural irrigation, comparatively strict limitations for nutrients and thus barriers towards their recycling can be observed, such as in the case of Italy, Israel or the new set of standards in India (Hanseok *et al.*, 2016; Schellenberg *et al.*, 2020). The reasons behind this may include the chronological development of the wastewater sector itself and complexity in the assessment, development, management and monitoring of various different risks with associated sets of standards in different receiving environments. Given the historically prevalent practice of discharge to water bodies, wastewater treatment technologies and treatment standards were typically developed in the context of risks to the aquatic environment. Given the low capacity for independent risk management, many countries formulate regulations based on pre-existing standards and thus often follow the BAT approach of the Global North. Following this approach can fail to take local conditions and risks into account, which can in turn result in the omission of appropriate alternative systems or feasible approaches.

Implemented recycling regulations can strongly depend on the wastewater collection approach or prevalent treatment processes. Innovative technologies as in the case of INNOQUA may enable both water and nutrient recycling. While algae-based treatment systems show good performances in the elimination of a wide range of pollutants, the formation of algae biomass can cause higher concentrations of suspended solids in the final effluent. However, the biomass can readily be used as a fertilizing substance in agriculture. While niches allow research and demonstration projects to develop innovative solutions that address current environmental pressures, the envisaged treatment objectives can differ from current practices. This can lead to non-compliance with current standards (as is frequently observed for nutrient limits in water intended for reuse in irrigation), and the absence of innovative and adaptive policies and standards-setting at the regime level would therefore eventually mean that these innovative systems may never gain a legal ground for implementation.

While prevailing approaches have been developed to address public health risks from pathogens, more recent constraints have evolved around emerging pollutants or heavy metals. Centralised collection of wastewater can result in the capture of a wider range of contaminants than might be expected in a decentralised system and thus increase the risks from agricultural application. In this case, the only suitable option may be resource recovery – suggesting financial investments that are not supportable in the Global South. This makes a stronger case and higher relevance for decentralization in less developed countries. While NBS such as constructed wetlands can have comparatively lower treatment performances for certain pollutants, studies indicate their potential to remove a wider spectrum of pollutants as compared to the current constellations of conventional systems, although further investigations are needed to assess elimination processes for all relevant contaminants under varied reuse scenarios (Salgot *et al.*, 2018; Markman *et al.*, 2007; Kumar *et al.*, 2015). Certain NBS have a higher land area requirement, which can make their installation difficult – especially in dense urban areas as found in the Global South. This aspect would benefit from further investigations in the field of urban planning, to explore the potential for integration of NBS within urban landscapes. NBS technologies with lower land area requirements should also be investigated.

There is a growing recognition that a complementary strategy including a wider set of technological options is necessary to allow more sustainable risk management and sectoral development. As in the case of IWRM, an integrated soil or nutrient resource management approach could enable a better assessment and understanding of resources and further build on more sustainable risk management strategies.

Climate Change and Energy Recovery

Shifting weather patterns are transforming the global environment and threaten aquatic and food production systems, while the increased demand on (bio)energy drives sectoral conflicts. Climate change is a global issue, and the need for global action and mitigation of GHG as incorporated in the Paris Agreement has been accepted by most countries. The highest climate change risks in the scope of this study were observed for Germany and India. Various strategies and directives were identified as having been implemented to address these risks, focussing largely on increasing the share of renewable energy or energy efficiency. Despite a major contribution to GHG emissions from the wastewater sector, direct mitigation strategies or regulations for the wastewater sector are overdue in all assessed geographies; the same applies for energy recovery. However, the first milestones can be observed in the case of Europe and especially in Germany and the UK. 'Sewage gas' has been recognized as a renewable energy source in Directive 2009/28/EC. In order to achieve targets for energy efficiency, STPs in Germany have been mandated to operate to state-of-the-art standards, which have indirectly led to the recovery of energy and the evolution of the sector towards energy neutrality. Currently, further ambitious research and demonstration projects strive towards a fundamental transformation by shifting this usually highly energy intensive sector towards becoming a key element of the energy revolution in Europe. While this can immensely advance the wastewater–energy nexus, further investigations for GHG mitigation are required to ensure sustainability at a wider view. These require a fundamental analysis of GHG emissions from different treatment processes and possible mitigation strategies. The energy footprint of treatment plants is often assessed in terms of energy consumption, which is typically higher for smaller conventional treatment plants (on the basis of PE load treated), but such comparisons may not consider a wider footprint such as overall construction and operation resources or benefits arising from local resource recovery.

Current risk management approaches, such as BAT, typically evaluate a limited number of technologies based on a limited set of parameters majorly focused on treatment performances for one receiving environment. However, with ever-increasing pressures and requirements for resource recycling under different receiving environments, these current evaluation matrices and parameters are obsolete and limit the identification and thus also the path for more sustainable practices or technologies. The continued adherence to this approach inevitably means that the additional benefits offered by innovative and more sustainable technologies may not be considered, making it ever-more difficult for innovations to gain ground and compete with conventional technologies.

There are various policy options to support more sustainable development – ranging from financial incentives to pricing mechanisms and different command and control instruments, sequenced towards gradual change. However, the evolution of an extended number of interlinked directives as in the case of Europe shows that adaptation of one element requires change in the whole apparatus, creating an inertia that can lead to inaction. The integration of energy and nutrient regulations represents a particular challenge in a sectoral approach, which rather focuses on water. Although progress towards the implementation of integrated water resource management can be observed, a nexus or integrated resource management approach is required to create a holistic and overall system function including all sectors. A concrete strategy for the implementation of such a concept in real practice is yet to be developed and demonstrated (Allouche *et al.*, 2015).

While pollution control measures give a mandatory range for discharge limits, the reuse, recycling or recovery in the fields of reviewed regulations remain mostly optional, unspecific in regard to the wastewater sector or indirect as in the case of targeted improvements for energy efficiency. Specific mandates paired with financial incentives and pricing mechanisms that directly address circular management in the wastewater sector could help to bring forward more sustainable development. In addition, there is a shared understanding that a gradual approach is more likely to foster an incremental transition rather than huge transformational shifts. This applies to both wastewater infrastructure and policy regulations.

Overall, a systematic consideration of the wastewater nexus and disruption of existing regimes is only observed to a limited and varying extent among the analysed sectors and geographies. Although environmental conditions at landscape level can drive adaptive policy measures, response times and action for delivering resilience in the Global South mismatch with the comparatively high pressures. Meanwhile, the Global North is facing lock-in mechanisms from cost-intensive centralised infrastructure. In contrast, the usually less developed sector in the Global South is assumed to provide huge potential for leapfrogging to more sustainable sectoral development. While innovative approaches and technologies are increasingly available and relevant to the Global South, regulatory frameworks that incorporate adaptive policies and encourage integrated cross-sector management require further progress to facilitate an adequate response to current pressures and pave the way for a sustainability transition.

This chapter has been peer-reviewed by: Dr. Eoghan Clifford, NUI Galway; Dr. David Tompkins, Aqua Enviro.

Conflict of Interest: The author declares that the research was conducted in the absence of any commercial or financial relationships that could be construed as a potential conflict of interest.

References

Adeel, M., Song, X., Wang, Y., Francis, D., Yang, Y., 2017, Environmental impact of estrogens on human, animal and plant life: A critical review, Environment International 99, 107–119, http://dx.doi.org/10.1016/j.envint.2016.12.010.

Al-Saidi, M. and Elagib, N. A., 2016, Towards understanding the integrative approach of the water, energy and food nexus, Science of the Total Environment, 574, 1131–1139, http://dx.doi.org/10.1016/j.scitotenv.2016.09.046.

Andriani, D., Wresta, A., Saepudin, A., Prawara, B., 2015, A review of recycling human excreta to energy through biogas generation: Indonesian case, Energy Procedia 68, 2019–225.

Angelakis, A. N. and Gikas, P., 2014, Water re-use: Overview of current practices and trends in the World with emphasis in EU. Water Utility Journal, 6: 67–78.

Alcalde-Sanz, L. and Gawlik, B. M., 2017, Minimum quality requirements for water re-use in agricultural irrigation and aquifer recharge, JRC Science for Policy Report, European Commission.

Allouche, J., Middleton, C., Gyawali, D., 2015, Technical Veil, Hidden Politics: Interrogating the Power Linkages behind the Nexus, Water Alternatives 8(1): 610–626.

Ban, Z. S., and Dave, G., 2004, Laboratory studies on recovery of n and p from human urine through struvite crystallisation and zeolite adsorption. Environmental Technology, 25(1), 111–121.

Baumann, D. D., 1983, Social acceptance of water re-use, Applied Geography, Vol. 3, Issue 1, 79–84, doi.org/10.1016/0143-6228(83)90007-3.

Beler-Baykal, B., Bayram, S., Akkaymak, E., and Cinar, S., 2004, Removal of ammonium from human urine through ion exchange with clinoptilolite and its recovery for further re-use. Water Science and Technology, 50(6), 149–156.

Benson, D., Gain, A. K., Rouillard, 2015, Water Governance in a Comparative Perspective: From IWRM to a 'Nexus' Approach?, Water Alternatives, 8(1): 756–733.

Bichai, F., Grindle, A. K. and Murthy, S. L., 2018, Addressing barriers in the water-recycling innovation system to reach water security in arid countries, Journal of Cleaner Production, Volume 171, 10, 97–109, https://doi.org/10.1016/j.jclepro.2016.07.062.

Biswas, S., Nasim Ali, Md., Goswani, R. and Chakraborty, S., 2014, Soil health sustainability and organic farming: A review, Journal of Food, Agriculture & Environment Vol. 12 (3&4), 237–243.

BMJ, 2007, BMJ readers choose the "sanitary revolution" as greatest medical advance since 1840, British Medical Journal, Editorial, 334:111–334.

doi: https://doi.org/10.1136/bmj.39097.611806.DB, retrieved at https://www.bmj.com/content/334/7585/111.2, accessed on 24th of April, 2020.

BMU, 2017, New Sewage Sludge Ordinance in force, Federal Ministry for the Environment, Nature Conservation and Nuclear Safety, retrieved at https://www.bmu.de/en/law/sewage-sludge-ordinance/ on June 24th, 2020.

BMJV, 2017, Verordnung zur Neuordnung der Klärschlammverwertung, Bundesgesetzblatt Jahrgang 2017, Teil I Nr. 65, Bundesministerium der Justiz und für Verbraucherschutz, Bonn.

Brundtland, G. H., 1987, Our Common Future, Report of the World Commission on Environment and Development.

Burek, P., Satoh, Y., Fischer, G., Kahil, M. T., Scherzer, A., Tramberend, S., Nava, L. F., Wada, Y., Eisner, S., Flörke, M., Hanasaki, N., Manuszewski, P., Cosgrove, B., Wiberg, D., 2014, Water Futures and Solution – Fast Track Initiative – Final Report ADA Project Number 2725-00/2014, IASA – International Institute for Applied Systems Analysis, Laxenburg, Austria, Working Paper-16-006.

Calloway, D. H., and Margen, S., 1971, Variation in endogenous nitrogen excretion and dietary nitrogen utilization as determinants of human protein requirement, The Journal of Nutrition, 101(2), 205–216.

Campos, J. L., Valenzuela-Heredia, D., Pedrouso, A., Val del Río, A., Belmonte, M., Mosquera-Corral, A., 2016, Greenhouse Gases Emissions from Wastewater Treatment Plants: Minimization, Treatment, and Prevention, Journal of Chemistry, Volume 2016, http://dx.doi.org/10.1155/2016/3796352.

Capodaglio, A. G., 2017, Integrated, Decentralized Wastewater Management for Resource Recovery in Rural and Peri-Urban Areas, Resources, 6, 22, doi: 10.3390/resources6020022.

Capodaglio, A. G., and Olsson, G., 2019, Energy Issues in Sustainable Urban Wastewater Management: Use, Demand, Reduction and Recovery in the Urban Water Cycle, Sustainability, 12, 266, doi: 10.3390/su12010266.

Collivignarelli, M. C., Abba, A., Frattarola, A., Miino, M. C., Padovani, S., Katsoyiannis, I., Torretta, V., 2019, Legislation for the Re-use of Biosolids on agricultural land in Europe: Overview, Sustainability, 11, 6015, doi:10.3390/su11216015.

Damania, Richard, Sébastien Desbureaux, Aude-Sophie Rodella, Jason Russ, and Esha Zaveri, 2019, Quality Unknown: The Invisible Water Crisis. Washington, DC: World Bank. doi: 10.1596/978-1-4648-1459-4. License: Creative Commons Attribution CC BY 3.0 IGO.

Daneshgar S., Callegari A., Capodaglio A. G., Vaccari D., 2018, The Potential Phosphorus Crisis: Resource Conservation and Possible Escape Technologies: A Review, *Resources* **2018**, *7*(2), 37; https://doi.org/10.3390/resources7020037.

Defra, 2011, Anaerobic digestion strategy and action plan – A commitment to increasing energy from waste through anaerobic digestion; Department for Environment, Food & Rural Affairs and Department of Energy & Climate Change, UK Government.

Diaz-Valbuena, L. R., Leverenz, H., Cappa, C. D., Tchobanoglous, G., Horwath, W. R., Darby, J. L., 2011, Methane, Carbon Dioxide, and Nitrous Oxide Emissions from Septic Tank Systems, Environmental Science and Technology, 45, 7, 2741–2747, https://doi.org/10.1021/es1036095.

Diem, K., and Lentner, C., 1970, Scientific tables: Excerpts from the 7th ed. Ciba-Geigy; seen in Rose *et al.*, 2015, The characterization of faces and urine: A review of literature to inform advanced treatment technology, Critical Reviews in Environmental Sciences and Technology, 45:1827–1879, Taylor and Francis, ISSN: 1064-3389 print/ 1547-6537 online, doi: 10.1080/10643389.2014.1000761.

Eastwood, M. A., Brydon, W. G., Baird, J., Elton, R., Helliwell, S., Smith, J., and Pritchard, J., 1984, Fecal weight and composition, serum lipids, and diet among subjects aged 18 to 80 years not seeking health care, The American Journal of Clinical Nutrition, 40(3), 628–634; seen in Rose *et al.*, 2015, The characterization of faces and urine: A review of literature to inform advanced treatment technology, Critical Reviews in Environmental Sciences and Technology, 45:1827–1879, Taylor and Francis, ISSN: 1064-3389 print/1547-6537 online, doi: 10.1080/10643389.2014.1000761.

EC, 1986, Sewage Sludge Directive, European Commission, European Union, Office Journal of the European Communities, Council Directive on the protection of the environment, an in particular of the soil, when sewage is used in agriculture (86/278/EEC); No. L 181/6.

EC, 1991, Urban Wastewater Treatment Directive, European Commission, European Union, Official Journal of European Communities.

EC, 1996, Council Directive 96/61/EC concerning integrated pollution prevention and control, Official Journal of the European Communities, No. L 257/26.

EC, 1999, Council Directive 1999/31/EC on the landfill of waste, Official Journal of the European Communities, European Commission.

EC, 2000, Directive 2000/60/EC of the European Parliament and of the Council establishing a framework for Community action in the field of water policy, L 327/1.

EC, 2001, Directive 2001/77/EC of the European Parliament and of the Council on the promotion of electricity produced from renewable energy sources in the internal electricity market, Official Journal of the European Communities, L283/33.

EC, 2003, Regulation (EC) No. 2003/2003 relating fertiliser, Official Journal of the European Union, European Commission.

EC, 2008, Directive 2008/98/EC of the European Parliament and of the council on waste and repealing certain Directives, Official Journal of the European Union, European Commission.

EC, 2009, Directive 2009/28/EC of the European Parliament and of the Council on the promotion of the use of energy from renewable sources and amending and subsequently repealing Directives 2001/77/EC and 2003/30/EC, Official Journal of the European Union, L140/6.

EC, 2018, Directive (EU) 2018/851 of the European Parliament and of the Council – amending Directive 2008/98/EC on waste, Official Journal of the European Union, European Commission.

EC, 2019, Regulation (EU) 2019/1009 of the European Parliament and of the Council, Official Journal of the European Union, European Commission.

EC, 2020, Sewage Sludge, European Commission, retrieved at https://ec.europa.eu/environment/waste/sludge/, accessed on June 26th, 2020.

Environment Agency, 2009. Renewable energy potential for the water industry, retrieved at https://assets.publishing.service.gov.uk/government/uploads/system/uploads/attachment_data/file/291637/scho1209brof-e-e.pdf, accessed on November 20th, 2020

EPA, 2016a, Water Re-use in the Context of the Circular Economy, Environmental Protection Agency Ireland, EPA Research Report No. 293 by Eoin Byrne, Kevin Fitzgibbon, Anca Minescu and Julia Blanke, EPA Research Programme 2014–2016.

EPA, 2016b, Identification and evaluation of phosphorus recovery technologies in an Irish context, EPA research Programme 2014–2020, Environmental Protection Agency, Ireland.

European Commission, 2019, ANNEXES to the Proposal for a Regulation of the European Parliament and of the Council on Minimum Requirements for Water Re-use, retrieved at https://data.consilium.europa.eu/doc/document/ST-9498-2018-ADD-1/en/pdf, accessed December 12, 2019.

EUROSTAT, 2020, Sewage sludge production and disposal from urban wastewater, Eurostat data browser, European Commission, retrieved at https://ec.europa.eu/eurostat/databrowser/view/ten00030/default/bar?lang=en, accessed on June 24th, 2020.

Evans, A. E. V., Varma, S., Krishnamurthy, A., 2009, Formal approaches to wastewater re-use in Bangalore, India, Conference Paper for 37th WEDC International Conference, Hanoi, Vietnam, 2014.

FAO Aquastat, n.d., Water Withdrawal by Sector, Aquastat databank, retrieved at http://www.fao.org/nr/water/aquastat/data/query/results.html, accessed on 15th of April, 2020.

FAO, 1992, Wastewater treatment and use in agriculture, Food and Agriculture organization of the United Nations, Rome, Italy.

FAO, 1996, Application of the pressure-state-response framework for the land quality indicators (LQI) programme, in Land Quality Indicators and Their Use in Sustainable Agriculture and Rural Development, Food and Agriculture Organization of the United Nations, 1997, Rome.

FAO, 2000, Land Resource Potential and Constraints at regional and Country Level, World Soil Resources Reports 90, based on the work of A.J. bot, F.O. Nachtergaele and A. Young, Land and water development Division, Food and Agriculture Organization of the United Nations, Rome, ISSN 0532-0488.

FAO, 2010, The wealth of waste – The economics of wastewater use in agriculture, FAO Water Reports 35, Food Agricultural Organization, Rome.

FAO, 2017, *The Future of Food and Agriculture – Trends and Challenges*, Food and Agriculture Organization of the United Nations, Rome.

FAO and IMWI, 2017, Water pollution from agriculture: a global review, Food and Agriculture Organization of the United Nations, Rome, 2017 and the International Water Management Institute on behalf of the Water Land and Ecosystems research program, Colombo, 2017.

Finger, M., Groenewegen, J., Künneke, R., 2005. The quest for coherence between institutions and technologies in infrastructures. Journal of Network Industries 6, 227–259.

Freeman, S., Booth, A. M., Sabbah, I., Tiller, R., Dierking, J., Klun, K., Rotter, A., Ben-David, E., Javidpour, J., Angel, D. L., 2020, Between source and sea: The role of wastewater treatment in reducing marine microplastics, Journal of Environmental Management 266, 110642; https://doi.org/10.1016/j.jenvman.2020.110642.

Frost, P. and Gilkinson, S., 2011, 27 Months Performance Summary for Anearobic Digestion of Dairy Cow Slurry at AFBI Hillsborough, seen in Andriani, D., Wresta, A., Saepudin, A., Prawara, B., 2015, A review of recycling human excreta to energy through biogas generation: Indonesian case, Energy Procedia 68, 2019–225.

Gandiglio, M., Lanzini, A., Soto, A., Leone, P., Santarelli, M., 2017, Enhancing the Energy Efficiency of Wastewater Treatment Plants through Co-digestion and Fuel Cell Systems, Frontiers Environmental Science, 5, 70, doi: 10.3389/fenvs.2017.00070.

Geels, F. W., 2002, Technological transitions as evolutionary reconfiguration processes: a multi-level perspective and a case study.; Research Policy 31, 1257–1274.

Geels, F. W., 2004, From sectoral systems of innovation to socio- technical systems: insights about dynamics and change from sociology and institutional theory. Research Policy 33: 897–920.

Geels, F., 2011, The multi-level perspective on sustainability transitions: response to seven criticisms, Environmental Innovation and Societal Transitions, 1 24–40, doi: 10.1016/j.eist.2011.02.002.

Gellings, C. W. and Parmenter, K. E., 2004, Energy Efficiency in fertilizer production and use, in Efficient Use and Conservation of Energy, [Eds. Clark W. Gellings, and Kornelis Blok], in Encyclopedia of Life Support Systms (EOLSS), developed under the auspices of the UNESCO, Eolss Publishers, Oxford, UK, [http://www.eolss.net]

Geng, Y., Cao, G., Wang, L. and Wang, Sh., 2019, Effect of equal chemical fertilizer substitutions with organic manure on yield, dry matter, and nitrogen uptake of spring maize and soil nitrogen distribution, PLoS One 14(7): e0219512, doi: 10.1371/journal.pone.0219512.

Germanwatch, 2019, Global Climate Risk Index 2019, Briefing Paper by Eckstein D., Hutfils M. L., Winges M., Germanwatch e.V., Bonn, Germany.

Germanwatch, 2020, Global Climate Risk Index 2020, Briefing Paper Eckstein D., Künzel V., Schäfer L., Winges M., Germanwatch e.V., Bonn, Germany.

Goldblith, S. A., and Wick, E. L., 1961, Analysis of human fecal components and study of methods for their recovery in space systems (ASO-TR-61-419). Washington, DC: NASA.

Guest, J. S., Barnard, J. L., Beck, M. B., Daigger, G. T., Hilger, H., Jackson, S. J., Karvazy, K., Kelly, L, Macpherson, L., Mihelcic, J. R., Pramanik, A., Raskin, L., Van Loosdrecht, M. C. M., Yeh, D., Love, N. G., 2009, A New Planning and Design Paradigm to Achieve Sustainable Resource Recovery from Wastewater, Environmental Science & Technology, Vol. 43, 6126–6130, 10.1021/es9010515.

GWI, 2014, Global Water Market 2014 – 10. Peru, Global Water Intel, retrieved at https://www.globalwaterintel.com/client_media/uploaded/20130220_GWM2014_Peru_sample_chapter.pdf, accessed no June 26th, 2020.

Hansen, J. A., and Tjell, J. C., 1979, Human excretion of heavy metals and other elements (pp. 1–15). Denmark: Technical University of Denmark. In Rose *et al.*, 2015, The characterization of faeces and urine: A review of literature to inform advanced treatment technology, Critical Reviews in Environmental Sciences and Technology, 45:1827–1879, Taylor and Francis, ISSN: 1064-3389 print/1547-6537 online, doi: 10.1080/10643389.2014.1000761.

Hanseok, J., Hakkwan, K., Taeil, J., 2016, Irrigation Water Quality Standards for Indirect Wastewater Re-use in Agriculture: A Contribution towards Sustainable

Wastewater Re-use in South Korea, Water, 8, 169, doi: 10.3390/w8040169.

Hermann, L. and Hermann, R., 2019, Report on regulations governing anaerobic digesters and nutrient recovery and re-use in EU member states. Wageningen, Wageningen Environmental Research, SYSTEMIC Project under EU Horizon 2020, doi: 10.18174/476673.

Hoekstra, A. Y. and Mekonnen, M. M., 2012, The water footprint of humanity, Proceedings of the National Academy of Sciences, USA 109(9), 3232–3237.

Hudcová, J., Vymazal, J., Rozkosný, M., 2019, Present restrictions of sewage sludge application in agriculture within the European Union, Soil and Water Research, 14(2), 104–120, https://doi.org/10.17221/36/2018-SWR.

Huygens, D., Saveyn, H. G. M., Tonini D., Eder P., Delgado Sancho, L., 2019, Technical proposals for selected new fertilising materials under the Fertilising Products Regulation (Regulation (EU) 2019/1009) – Process and quality criteria, and assessment of environmental and market impacts for precipitated phosphate salts & derivates, thermal oxidation materials & derivates and pyrolysis & gasification materials, EUR 29841 EN, Publications Office of the European Union, Luxembourg, ISBN 978-92-76-09888-1, doi: 10.2760/186684, JRC117856.

IEA Bioenergy, 2015, Sustainable biogas production in municipal wastewater treatment, International Energy Agency, ISBN 978-1-910154-22-9.

IEA, 2018, The energy sector should care about wastewater, International Energy Agency, retrieved at https://www.iea.org/commentaries/the-energy-sector-should-care-about-wastewater, accessed on 26th June, 2020.

IMPEL, 2018, Report on Urban Water Re-use – Integrated Water Approach and Urban Water Re-use Project, European Union Network for Implementation and Enforcement of Environmental Law, 2018/07, Version 2.0.

Inglezakis, V. J., Zorpas, A. A., Karagannidis, A., Samaras, P., Voukkali, I., Sklari, St., 2014, European Union Legislation on Sewage Sludge Management, Fresenius Environmental Bulleting, PSP Volume 23 – No. 2a 2014.

IPCC, 2014, Climate Change 2014: Impacts, Adaptation, and Vulnerability. Part A: Global and Sectoral Aspects. Contribution of Working Group II to the Fifth Assessment Reprot of the Intergovernmental Panel on Climate Change, [Field, C.B., V.R. Barros, D.J. Dokken, K.J. Mach, M.D. Mastrandrea, T.E. Bilir, M. Chatterjee, K.L. Ebi, Y.O. Estrada, R.C. Genova, B. Girma, E.S. Kissel, A.N. Levy, S. Mac Cracken, P. R. Mastrandrea, and L. L. White eds.)]. Cambridge University Press, Cambridge, United Kingdom and New York, NY, USA, 1132p.

ISO 16075, 2015, Guidelines for treated wastewater use for irrigation projects, International Organization for Standardization, Geneva, Switzerland.

ISO 23056, n.d., Water re-use in urban areas – Guidelines for decentralized/onsite water re-use system, International Organization for Standardization, Geneva, Switzerland, retrieved at https://www.iso.org/standard/74466.html?browse=tc, accessed on 25th, May, 2020.

Istenic, D., Bodik, I., Bulc, T., 2014, Status of decentralized wastewater treatment systems and barriers for implementation of nature-based systems in central and eastern Europe, Environmental Science and Pollution Research, Springer Verlag Berlin, doi: 10.1007/s11356-014-3747-1.

Jana, B. B., Rana, S., and Bag, S. K., 2012, Use of human urine in phytoplankton production as a tool for ecological sanitation, Water Science and Technology, 65(8), 1350–1356.

Jimenez, B. and Asano, T., 2008, Water Re-use – An International Survey of current practice, issues and needs, IWA Publishing, Scientific and Technical Report No. 20.

Jönsson, H., Stenström, T. A., Svensson, J., and Sundin, A. (1997). Source separated urine-nutrient and heavy metal content, water saving and faecal contamination. Water Science and Technology, 35(9), 145–152; adapted values seen in Rose, C., Parker, A., Jefferson, B., and Cartmell, E., 2015, The characterization of faces and urine: A review of literature to inform advanced treatment technology, Critical Reviews in Environmental Sciences and Technology, 45:1827–1879, Taylor and Francis, ISSN: 1064-3389 print/ 1547-6537 online, doi: 10.1080/10643389.2014.1000761.

Jones, D. L., Cross, P., Withers, P. J. A., DeLuca, T. H., Robinson, D. A., Richard S., Harris, I. M., Chadwick, D. R., Edwards-Jones, G., 2013, Nutrient stripping: the global disparity between food security and soil nutrient stocks, Journal of Applied Ecology, 50(4), 851–862, doi: 10.1111/1365-2664.12089.

Jorgenson, A. K. and Kuykendall, K. A, 2008, Globalization, Foreign Investment Dependence and Agriculture Production: Pesticide and Fertilizer Use in Less-developed Countries, 1990–2000, Social Forces, Volume 87, Issue 1, pages 529–560, Oxford University Press, https://doi.org/10.1353/sof.0.0064.

Kihila, J., Mtei, K. M., and Njau, K. N., 2015, A review of the challenges and opportunities for water re-use in irrigation with a focus on its prospects in Tanzania, International Journal of Environmental Engineering, Vol. 7, No. 2, doi: 10.1504/IJEE.2015.069817.

Kongshaug, G. 1998, Energy Consumption and Greenhouse Gas Emissions in Fertilizer Production, IFA Technical Conference, Marrakech, Morocco, 28 September-1 October, 1998, 18p.

Krantzberg, G., Tanik, A., Antunes do Carmo, J. S., Indarto, A., Ekda, A., 2010, Advances in Water Quality Control, Scientific Research Publishing Inc., USA.

Kumar, Ch., Ghosh, A. K., Sinha, S. and Sinha, R. K., 2015, Growing Scarcity of Water for Irrigation in World: Earthworms as Nature's Biofilter (Vermi-Filter) to purify Human wastewater for re-use in agriculture, saving groundwater and also removing deadly agrochemicals: Some Agronomic Studies on Rice (Oryza Sative) Crops in India, in Agricultural Research Updates, Volume 11, ISBN: 978-1-63482-968-7.

Larsen, T., Udert, K. M., Lienert, J., 2013, Source Separation and Decentralization for Wastewater Management, IWA Publishing, London, UK.

Lin, W., Lin, M., Zhou, H., Wu, H., Li, Z. and Lin, W., 2019, The effect of chemical and organic fertilizer usage on rhizospere soil in tea orchards, PLOS ONE 14(5): e0217018, doi: 10.1371/journal.pone.0217018.

Lüthi, Ch., Panesar, A., Schütze, Th., Norström, A, McConville, J., Parkinson, J., Saywell, D., Ingle, R., 2011, Sustainable Sanitation in Cities: A Framework for Action, Sustainable Sanitation Alliance (SuSanA) & International Forum on Urbanism (IFoU), Papiroz Publishing House, The Netherlands.

Markard, J. and Truffer, B., 2008, Technological innovation systems and the multilevel perspective: Towards and integrated framework, Research Policy, 37, 596–615, doi: 10.1016/j.respol.2008.01.004.

Markard, J., 2011, Transformation of infrastructures: sector characteristics and implications for fundamental change, Journal of Infrastructure Systems (ASCE) 17, 107–117.

Markard, J., Raven, R., Truffer, B., 2012, Sustainability transitions: An emerging field of research and its prospects, Research Policy 41, 955–967.

Markman, S., Guschina, I., Barnsley, S., Buchanan, K., Pascoe, D., Müller, C. T., 2007, Endocrine discurpting chemicals accumulate in earthworms exposed to sewage effluent, Chemosphere, Volume 70, Issue 1, 119–124, https://doi.org/10.1016/j.chemosphere.2007.06.045.

Meinzinger, F., and Oldenburg, M., 2009, Characteristics of source-separated household wastewater flows: A statistical assessment, Water Science and Technology, 59(9), 1785–1791.

Mekonnen, M. M. and Hoekstra, A. Y., 2011, National water footprint accounts: the green, blue and grey water footprint of production and consumption, Value of Water Research Report Series No. 50, UNESCO-IHE, Delft, Netherlands.

Meadows, D. H., Meadows, D. L., Randers, J., Behrens, W. W., 1972, The Limits to Growth, Club of Rome, A Potomac Associates Book.

Ministrio del Ambiente, 2015, Norma de Calidad Ambiental y de Descarga de Efluentes, Registro Official – Edición Especial No. 387, Ecuador.

Ministerio de Vivienda, Construcción y Saneamiento, 2010, Normas Legales Decreto Supremo No. 003-2010-MINAM, El Peruano, Peru.

Ministry of Jal Shakti, 2020, Implementation of National Water Mission, Department of Water Resources, River Development & Ganga Rejuvenation, Government of India, retrieved at http://mowr.gov.in/schemes-projects-programmes/schemes/implementation-of-national-water-mission, accessed on 29th May, 2020.

MoEFC, 2016, Solid Waste Management Rules, 2016, The Gazette of India: Extraordinary, Government of India.

MoJS, 2020, Objective of National Water Mission, Ministry of Jal Shakti, Government of India, retrieved at http://nwm.gov.in/?q=objective-national-water-mission, accessed on 2nd July, 2020.

MoUD, 2017, National Policy on Faecal Sludge and Septage Management (FSSM), Ministry of Urban Development, Government of India.

MoWR, 2012, National Water Policy 2012, Ministry of Water Resources, Government of India.

Nas, B., Uyanik, S., Aygün, A., Dogan, S., Erus, E., Batuhan Nas, K., Turgut, S., Cop, M., Dolu, T., 2020, Wastewater Re-use in Turkey: from present status to future potential, International Water Association, IWA Publishing, Water Supply, 20.1.

Never, B., 2016, Wastewater Systems and Energy Saving in Urban India – Governing the Water-Energy-Food Nexus Series, German Development Institute, ISSN 1860-0441, ISBN 978-3-96021-000-9.

Perraton, S. C., Blackwell, B. D., Fischer, A., Gaston, T. F. and Meyers, G. D., Systemic barriers to wastewater re-use in Australia: some jurisdictional examples, Australasian Journal of Environmental Management, doi: 10.1080/14486563.2014.955889.

Petrie, B., Barden, R., Kasprzyk-Hordern, B., 2015, A review on emerging contaminants in wastewaters and the environment: Current knowledge, understudied areas and recommendations for future monitoring, Water Research 72, 3–27, http://dx.doi.org/10.1016/j.watres.2014.08.053.

Powerstep, 2018, Policy Brief – The potential of the wastewater sector in the energy transition, Powerstep H2020 Project, European Union.

Putnam, D. F., 1971, Composition and concentrative properties of human urine (CR-1802). Washington, DC: NASA.

Rip, A., Kemp, R., 1998, Technological change. In: Rayner, S., Malone, E.L. (Eds.), Human Choice and Climate Change 2. Battelle Press, Columbus, Ohio, pp. 327–399.

Rose, C., Parker, A., Jefferson, B., and Cartmell, E., 2015, The characterization of faces and urine: A review of literature to inform advanced treatment technology, Critical Reviews in Environmental Sciences and Technology, 45:1827–1879, Taylor and Francis, ISSN: 1064-3389 print/ 1547-6537 online, doi: 10.1080/10643389.2014.1000761.

Sarma, J. S. and Sharma, P. D., 2010, Food Security – Indian Scenario, in Potassium Role and Benefits in Improving Nutrient Management for Food Production, Quality and Reduced Environmental Damages, Proceedings of the IPI-OUAT-IPNI International Symposium Bhubaneswar, Orissa, India, Volume I.

Sajid, M., Ilyas, M., Basheer, C., Tariq, M., Daud, M., Baig, N., Shehzad, F., 2015, Impact of nanoparticles on human and environment: review of toxicity factors, exposures, control strategies, and future prospects, Environ Sci Pollut Res (2015), 22, 4122–4143, doi: 10.1007/s11356-014-3994-1.

Salgot, M., Oron, G., Cirelly, G. L., Dalzios, N. R., Díaz, A., Angelakis, A. N., 2018, Criteria for Wastewater Treatment and Re-use under Water Scarcity, Handbook of Drought and Water Scarcity, Chapter 14.

Schellenberg, T., Subramanian, V., Ganeshan, G., Tompkins, D., Pradeep, R., 2020, Wastewater Discharge Standards in the Evolving Context of Urban Sustainability – The Case of India, Frontiers in Environmental Science, 8:30; doi: 10.3389/fenvs.2020.00030.

Schot, J., Geels, F. W., 2008, Strategic niche management and sustainable innovation journeys: theory, findings, research agenda, and policy, Technology Analysis & Strategic Management 20, 537–554, doi: 10.1080/095373 20802292651.

Schouw, N. L., Danteravanich, S., Mosbaek, H., and Tjell, J. C., 2002, Composition of human excreta—A case study from Southern Thailand. Science of the Total Environment, 286(1–3), 155–166.

Simpson G. B. and Jewitt G. P. W., 2019, The Development of the Water-Energy-Food Nexus as a Framework for Achieving Resource Security: A Review, Frontiers Environmental Science, Volume 7, Article 8, doi: 10.3389/fenvs.2019.00008.

Snyder, W., Cook, M., Tipton, I., Nasset, E., Karhausen, L., and Howells, G., 1975, Reference man: Anatomical, physiological and metabolic characteristics (Report of Task Group on Reference Man-International Commission on Radiological Protection), New York, NY: Pergamon Press.

Smith, A. and Stirling, A., 2010, The politics of social-ecological resilience and sustainable socio-technical transitions, Ecology and Society, 15(1), 11, http://www.ecologyandsociety.org/vol15/iss1/art11/.

Shaddel, A., Bakhtiary-Davijany, H., Kabbe, Ch., Dadgar, F., Osterhus, St. W., 2019, Sustainable Sewage Sludge Management: From Current Practices to Emerging Nutrient Recovery Technologies, Sustainability, 11, 3435, doi: 10.3390/su11123435.

Statista, 2020, Wassernutzung der deutschen Industrie nach Verwendungszweck 2016, veröffentlicht von A. Breitkopf in 2018, retrieved at https://de.statista.com/statistik/daten/studie/314739/umfrage/wassernutzung-der-deutschen-industrie-nach-verwendungszweck/, accessed on 18th August, 2020.

Struik, P. and Kuyper, T. W., 2017, Sustainable intensification in agriculture: the richer shade of green. A review, Agronomy for Sustainable Development 37, article number: 39, doi: 10.1007/s13593-017-0445-7.

Tan, Z. X., Lal, R., Wiebe, K. D., 2005, Global Soil Nutrient Depletion and Yield Reduction, Journal of Sustainable Agriculture, Vol. 26(1) 2005, The Haworth Press, doi: 10.1300/J064v26n01_10.

Tanmay Ram Kate and Sridevi H, 2019, Greenhouse Gas Emissions from Wastewater Treatment Plant. International Journal of Civil Engineering and Technology 10(9), pp. 81–89.

Taseli, B. K. and Kilkis, B., 2016, Ecological sanitation, organic animal farm and cogeneration: Closing the loop in achieving sustainable development-A concept study with on-site biogas fueled trigeneration retrofit in a 900-bed university hospital, Energy and Buildings 129, 201–119.

The Lancet, 2017, Estimates of global, regional, and national morbidity, mortality, and aetiologies of diarrhoeal diseases: a systematic analysis for the Global Burden of Disease Study 2015, GBD Diarrhoeal Diseases Collaborators, The Lancet Infectious Diseases, Volume 17, Issue 9, P90–948, doi: https://doi.org/10.1016/S1473-3099(17)30276-1, retrieved at https://www.thelancet.com/journals/laninf/article/PIIS1473-3099(17)30276-1/fulltext, accessed on 16th of April, 2020.

Tilley, E., Atwater, J., and Mavinic, D., 2008, Recovery of struvite from stored human urine. Environmental Technology, 29(7), 797–806; EAWAG, Zürich.

Tong, T. and Elimelech, M., 2016, The Global Rise of Zero Liquid Discharge for Wastewater Management: Drivers, Technologies, and Future Directions, Environmental Science & Technology, 50, 6846–6855, doi: 10.1021/acs.est.6b01000.

Tortajada, C., and Ong, C. N., 2016, Re-used water policies for potable use, International Journal of Water Resources Development, 32:4, 500–502, https://doi.org/10.1080/07900627.2016.1179177.

UK Government, 1989, The Sludge (Use in Agriculture) Regulations 1989, The National Archives, retrieved at https://www.legislation.gov.uk/uksi/1989/1263/made, accessed on 18th August, 2020.

UK Government, 2013, Guidance – Generating energy from waste, including anaerobic digestion, Department for Environment, Food & Rural Affairs, retrieved at https://www.gov.uk/guidance/generating-energy-from-waste-including-anaerobic-digestion, accessed on 18th August, 2020.

Umweltbundesamt, 2009, Energieeffizient Kommunaler Kläranlagen, retrieved at www.umweltbundesamt.de/uba-info-edien/mysql_medien.php?anfrage=Kennummer&Suchwort=3347, accessed on 26th June, 2020.

Umweltbundesamt, 2019, Wasserwiederverwendung, retrieved at https://www.umweltbundesamt.de/themen/wasser/wasser-bewirtschaften/wasserwiederverwendung, accessed on 29th May, 2020.

UN, 2015, A/RES/70/1 Transforming our world: the 2030 Agenda for Sustainable Development, General Assembly-Resolution adopted by the General Assembly on 25 September 2015, Seventieth session, Agenda items 15 and 116, United Nations.

UN WWAP, 2013, The United Nations World Water Development Report 2013. United Nations, indicated in FAO, 2017, Water pollution from agriculture: a global review.

UN WWAP, 2017, Wastewater – The Untapped Resource, The United Nations World Water Development Report 2017, UN Water.

UN WWAP, 2018, Nature-Based Solutions for Water, The United Nations World Water Development Report 2018.

UN, 2018a, Sustainable Development Goal 6 – Synthesis Report on Water and Sanitation, UN-Water, United Nations Publications, United Nations, Geneva, ISBN: 978-92-1-101370-2.

UN, 2018b, Progress on Integrated Water Resources Management, Global Baseline for SDG 6 Indicator 6.5.1: degree of IWRM implementation. United Nations Environment.

UNCCD, 2017, Global Land Outlook, Secretariat of the United Nations Convention to Combat Desertification, Platz der Vereinten Nationen 1, 53113 Bonn, Germany, www.unccd.int.

UNCCD, 2018, Poor land use costs countries 9 percent equivalent of their GDP, retrieved at https://www.unccd.int/news-events/poor-land-use-costs-countries-9-percent-equivalent-their-gdp-0, accessed on 31st of March, 2020.

UNESCO WSSM, 2019, Water Security and the Sustainable Development Goals, Global Water Issues Series 1, The United Nations Educational, Scientific and Cultural Organization (UNESCO), Paris, ISBN UNESCO 978-92-3-100323-3.

UNESCO WWAP, 2019, The United Nations World Water Development Report 2019: Leaving No One Behind, UNESCO World Water Assessment Programme, Paris.

UN Habitat, 2008, Global Atlas of Excreta, Wastewater Sludge, and Biosolids Management: Moving forward the sustainable and welcome uses of a global Resource, United Nations Human Settlements Programme.

Ungureanu, N., Vladut, V., Dinca, M., Zabava, B. S., 2018, Re-use of wastewater for irrigation, a sustainable practice in arid and semi-arid regions, Conference Paper: 7th International Conference on Thermal Equipment, Renewable Energy and Rural Development (TE-RE-RD)At: Drobeta Turnu Severin, Romania.

UN Water, n.d., Water Scarcity, retrieved at https://www.unwater.org/water-facts/scarcity/, accessed on 8th of April, 2020.

US EPA, 2004, Guidelines for Water Re-use, U.S. Environmental Protection Agency, Washington DC.

US EPA, 2012, Guidelines for Water Re-use, U.S. Environmental Protection Agency, Washington DC.

Vinneras, B., Palmquist, H., Balmer, P., and Jonsson, H., 2006, The characteristics of household wastewater and biodegradable solid waste—A proposal for new Swedish design values, Urban Water Journal, 3(1), 3–11.

Voulvoulis, N., 2018, Water re-use from a circular economy perspective and potential risks from an unregulated approach, Environmental Science & Health 2018, 2: 32–45, https://doi.org/10.1016/j.coesh.2018.01.005.

Water UK, 2020. Net Zero 2030 Routemap, retrieved at https://www.water.org.uk/routemap2030/, accessed on 20th November, 2020.

WEF, 2011, The Water-Food-Energy-Climate Nexus, World Economic Forum Water Initiative.

Wester, J., Timpano, K. R., Cek, D., Liebermann, D., Fieldstone, S. C., Broad, K., 2015, Psychological and social factors associated with wastewater re-use discomfort, Journal of Environmental Psychology, 42, 16–23, http://dx.doi.org/10.1016/j.jenvp.2015.01.003.

WHO, 2006, WHO Guidelines for the safe use of wastewater, excreta and greywater, World Health Organization, Volume IV, Geneva, Switzerland.

WHO, 2019, Sanitation – Key Facts, Progress on drinking water, sanitation and hygiene, Joint Monitoring Programme 2017 update and SDG baselines, World Health Organization, retrieved at https://www.who.int/news-room/fact-sheets/detail/sanitation, accessed on 1st June, 2020.

Wignarajah, K., Pisharody, S., and Fisher, J. W., 2003, Investigating the partitioning of inorganic elements consumed by humans between the various fractions of human wastes: An alternative approach (2003-01-2371), Moffett Field, CA: SAE International.

Wood, S. and Cowie, A., 2004, A review of greenhouse gas emissions factors for fertiliser production, Cooperative Research Centre for Greenhouse Accounting, For IEA Bioenergy Task 38.

WRI, 2013, New Study Raises Question: What Don't We Know About Water Scarcity, by Betsy Otto, World Resources Institute, retrieved at https://www.wri.org/blog/2013/05/new-study-raises-question-what-don-t-we-know-about-water-scarcity, accessed on 16th of April, 2020.

WRI, 2015, Ranking the World's Most Water-Stressed Countries in 2040, by Andrew Maddocks, Robert Samuel Young and Paul Reig, World Resources Institute, Washington DC, retrieved at https://www.wri.org/blog/2015/08/ranking-world-s-most-water-stressed-countries-2040 accessed on 17th of April, 2020.

WRI, 2018, Water Use and Stress, by Hannah Ritchie and Max Roser, first published in 2015, retrieved at https://ourworldindata.org/water-use-stress, accessed on 15th of April, 2020.

World Bank, n.d., World Development Indicators, World Bank databank, retrieved at https://databank.worldbank.org/home.aspx, accessed on 15th of April, 2020.

World Bank, 2018, Population, World Development Indicators, World Bank databank, retrieved at https://databank.worldbank.org/reports.aspx?source=2&series=SP.POP.TOTL&country=#, accessed on 27th of April, 2020.

World Bank, 2020, Total Agricultural Land and Irrigated Land Area, World Development Indicators, World Bank, retrieved at https://databank.worldbank.org/reports.aspx?source=2&series=AG.LND.AGRI.ZS&country=# accessed on 11th May, 2020.

World Data Lab, n.d., Average water availability per person, data retrieved from graphics provided at https://worldwater.io/?utm_source=google&utm_medium=search&utm_campaign=WaterscarcityData&campaignid=6444167483&adgroupid=77198318295&adid=389909590979&gclid=EAIaIQobChMI2bbD6-nB6AIVkjUrCh13xwWPEAAYASAAEgJuAvD_BwE, accessed 30th of March, 2020.

DOI: 10.1561/9781680837810.ch3

Chapter 3

Nature-based Wastewater Treatment – Overview and Current Common Systems

By Evelyn Herrera and Alexander Meneses-Jácome

The current dominant approach of the 'networked city' reinforced by local governments, practitioners and multilateral agencies has failed to close the sanitation gap in countries of the Global South. Nature-based wastewater treatment technologies are increasingly considered a resilient and adaptable alternative with lower investment costs that could complement current centralised approaches, minimising resource consumption and reducing the growing pressures on freshwater resources through enabling recycling and reuse of treated wastewater. This chapter focuses on nature-based wastewater treatment systems with an emphasis on those installed in countries of the Global South: septic tanks, waste stabilisation ponds (WSPs), constructed wetlands and vermifilters. While the performance of nature-based solutions can be comparable to conventional wastewater treatment systems, two or more solutions usually have to be combined in order to meet stringent discharge standards. In particular, the reduction of nutrients by individual technologies can be limited, although nutrient removal may not be a requirement where the effluents have value for irrigation purposes. Nature-based solutions can share some of the same issues as conventional centralised approaches, such as overloading, underloading, lack of de-sludging, improper operation and maintenance and lack of dedicated financial/human resource. Further adoption and robust implementation of nature-based solutions in the Global South will require appropriate governmental/institutional/communal interventions, including realistic and enforceable discharge and water reuse standards, and exchange among practitioners from the Global South and technology providers from the Global North to provide better solutions adapted to the realities and contexts of the Global South.

3.1 Introduction: The Sanitation Challenge in the Global South

Although the proportion of the Global population with basic sanitation services increased from 54% to 73% between 2000 and 2017, 4.2 billion people (over half of the global population) lack safe sanitation,[1] 90% of whom live in three regions of the Global South: Central and Southern Asia, Sub-Saharan Africa and Eastern and Southern-Eastern Asia (UNICEF and WHO, 2019). To achieve the goal of global coverage of basic sanitation services by 2030, these regions would need to double their efforts and increase the coverage rate at a minimum of 1 to 2 percent points per annum (UNICEF and WHO, 2019). Even in areas where the population connected to a sewerage system is relatively high (e.g., >65% in Latin America and the Caribbean), less than 40% of the wastewater collected is treated and the efficiency of the treatment can be limited due to poorly maintained and partially non-functional wastewater treatment plants (Vairavamoorthy and Nolasco, 2020; WaterAid, 2019). The lack of treatment is the consequence of past trends which were focused on sewage collection only and in combination with the common practice of discharging untreated sewage into water bodies has resulted in acute health, environmental and economic risks in Latin America and the Caribbean (Looker, 1998; Martin-Hurtado and Nolasco, 2016; Nolasco *et al.*, 2020).

Additionally to the underserved challenge, countries of the Global South need to cope with a rapidly urbanising world. It is projected that by 2050, around 68%

1. Used of improved facilities (flush/pour flush to piped sewer system, septic tanks or pit latrines; ventilated improved pit latrines, composting toilets or pit latrines with slabs) which are not shared with other households and where excreta are safely disposed in situ or transported and treated off-site (WHO and UNICEF, 2017).

of the world's population will live in urban areas (UN, 2019). Most of this growth will occur in medium-sized cities and urban slums of the Global South. These are commonly characterised by unreliable sanitation services due to deficient infrastructure, fragmented legislation, insufficient resource allocation and unstable governance (Ramos-Mejía *et al.*, 2018). Furthermore, in recent years, the construction and planning of sanitation systems in the Global South have followed the concept of 'networked' cities. This static and uniform approach is often confined to more modern and developed parts of urban areas. This creates islands of prosperity in vast underserved areas where the distribution and access to water and sanitation is already uneven (Gutterer and Reuter, 2016; Monstadt and Schramm, 2017). High connection costs to energy-intensive and large-scale wastewater treatment technologies leave behind lower income groups that in the long term are forced to pay higher rates for service provision (Kooy and Bakker, 2008; Monstadt and Schramm, 2017). Moreover, the shortage of skilled human resources contributes to poor operation of more 'advanced' wastewater treatment systems (Edokpayi *et al.*, 2016). This situation, combined with the presence of weak or non-existent utilities and related institutions, will hinder the recovery of investment in maintenance and operation and thus systems are left in states of disrepair (Luth *et al.*, 2011). This, in turn, creates distrust among low income urban populations who have no other option than to rely on such services.

In recent years, the International Development Bank, European Investment Bank and World Bank, among other multilateral agencies have approved loans to different countries in the Global South to expand or construct their centralised wastewater management infrastructure. Wastewater treatment and distribution services are normally sub-components of development programmes with budgets between 10% and 80% of the total cost of the project. Some examples of such projects are summarised in Appendix 1. These developments are welcome but it should also be noted that centralised urban planning approaches that rely on 'advanced' technologies can ignore options that might be more sustainable for many Global South communities, hindering the creation of an inclusive sanitation infrastructure. Indeed, care must be taken that such projects do not lead to fragmented solutions that depend on ongoing donor or sector interest. Centralised solutions, often imported from the Global North, are now increasingly seen as unsustainable in many applications there (Tchobanoglous *et al.*, 2003; WWAP (UNESCO World Water Assessment Programme), 2019).

Nature-based solutions to wastewater treatment can provide resilient and adaptable alternatives to established technology with lower life cycle costs. They can allow the recovery and reuse of valuable resources present in wastewater, and contribute to meeting environmental and social inclusion targets demanded by the fast and dynamic developing urban areas – as well as making cost-effective contributions to the achievement of cross-cutting sustainable development goals (Andersson *et al.*, 2016b).

This study seeks to provide an overview of nature-based wastewater treatment systems installed in countries of the Global South and discusses treatment performance, technical considerations associated with their design, operation and maintenance. Environmental impacts and investment costs are also analysed against the background of the classical approach of a 'networked city'.

3.2 Natured-based Technologies for Wastewater Treatment

Nature-based treatment systems or natural treatment systems for wastewater can be defined as technologies that mimic natural processes such as the interaction of soil, microorganisms and/or plants, and utilise natural components (e.g., gravity forces for sedimentation) for removal of multiple pollutants. Natural systems are also seen as technologies that require low or no external energy sources to maintain the major treatment processes and contain no movable parts (Boano *et al.*, 2020; Crites *et al.*, 2014; Kumar and Asolekar, 2016). Nature-based treatment systems aim to protect and sustainably manage and restore natural or modified ecosystems by addressing societal challenges in an efficient manner while providing improved human well-being and biodiversity (IUCN, 2016). This study explores several of the technologies that fall within these definitions: septic tanks, waste stabilization ponds (WSP), constructed wetlands, vermifiltration and small anaerobic systems (specifically the anaerobic baffled reactor or 'ABR').

3.2.1 Septic Tanks

Septic tanks are one of the simplest and most widespread means of collecting and providing basic treatment of wastewater. Septic tanks do not require significant infrastructure, are easy to construct and are able to (partially) treat wastewater close to the source. They are among the most common on-site sanitation systems in many urban and rural areas of countries of the Global South (Table 3.1). In fact, in countries such as Vietnam the use of septic tanks is institutionalised, meaning that houses and apartment buildings must construct a septic tank regardless of whether their effluent is discharged into soil, surface water or sewer lines (Nguyen *et al.*, 2007).

A well-designed septic tank is a watertight chamber made of concrete, fiberglass, PVC or other plastic, typically comprising two or three compartments used for the treatment of household as well as community grey and/or black wastewater.

Table 3.1. Septic tank coverage in various countries and cities of the Global South (Williams and Overbo, 2015).

Countries/cities	Septic System Coverage (%)
Egypt	49 (rural) 7 (urban)
Nigeria	46
Accra, Ghana	40
Bangkok, Thailand	25
Colombo, Sri Lanka	33
Dakar, Senegal	58
Dar es Salaam, Tanzania	15
Hanoi, Vietnam	63
Ho Chi Minh City, Vietnam	79
Jakarta, Indonesia	39
Karachi, Pakistan	50
Kathmandu, Nepal	70
Kuala Lumpur, Malaysia	20
New Delhi, India	40
Phnom Penh, Cambodia	37
Yamoussoukro, Côte d'Ivoire	90

It is a preliminary treatment unit which combines sedimentation and anaerobic digestion to mainly remove suspended solids and organic carbon. The retention time of a septic tank typically varies between one or two days (Tilley *et al.*, 2008). It usually consists of three zones:

1. The sludge zone where the settled solids are anaerobically digested;
2. The clear zone which retains the primary treated wastewater; and
3. The scum layer which includes, among others, oil and grease (Fig. 3.1).

While the effluent (liquid portion) typically overflows to soak-pits or soak away fields where it percolates into the surrounding soil, the settled solids, also known as faecal sludge, must be removed by a de-sludging process on a regular basis (e.g., every 1–3 years), to ensure proper functioning of the system (Mehta *et al.*, 2019). In countries of the Global South, the faecal sludge has caused several environmental and public health problems due to the lack of safe treatment and its direct disposal into the environment (Cairns-Smith *et al.*, 2014; Strande *et al.*, 2014). Hence, the

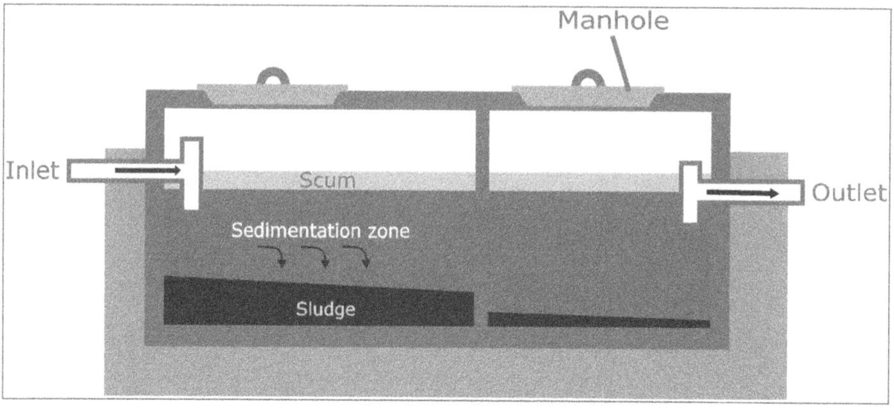

Figure 3.1. Section view of a septic tank. Adapted from Ministry of Water – United Republic of Tanzania, 2018.

demand for faecal sludge management services (collection, transport, treatment, safe reuse and disposal of faecal sludge from on-site systems) is rapidly increasing. Faecal sludge management (FSM) is out of the scope of this study, but various business models and processing techniques can be applied (Strande *et al.*, 2014).

3.2.1.1 Treatment performance of septic tanks in the Global South

Septic tanks when properly designed and regularly de-sludged can be considered as a low cost and low impact sanitation solution with removal efficiencies between 46% and 68% for BOD_5 and 30–81% for TSS (Bitton, 2005). However, studies in countries of the Global South have demonstrated that septic tanks often perform poorly, with removal percentages ranging from 4% to 30% for BOD_5 and 22–49% for TSS (Hegg *et al.*, 2018; Mehta *et al.*, 2019; Nguyen *et al.*, 2007; Williams and Overbo, 2015). There can also be misuse of the term 'septic tank' as it can be applied to latrines or soak away pits. These structures may not follow any technical design guidelines and result in contamination of water sources which may be falsely attributed to septic tanks.

3.2.1.2 Main causes of malfunction of septic tanks in the Global South

The most common causes of failure of septic tanks include poor design and construction, overloading, inappropriate installation and lack of maintenance. In the Kathmandu Valley of Nepal, most of the so-called 'septic tanks' are tanks with only one chamber or large pits lined with brick walls and covered with concrete. Similarly in Muaeng Klong Luang (Thailand), 32% of 'septic tanks' are bottomless concrete ring tanks which allow the infiltration of the untreated effluent into groundwater (Williams and Overbo, 2015) leading to contamination of water sources. In Dar es Salaam, Tanzania a study revealed that the quality of water from wells in a shallow aquifer located in an area with a high population density where septic tanks were in close proximity had similar characteristics to septic tank effluent: faecal coliform levels ranged between 10^5 and 10^6 CFU/ml (Gondwe *et al.*, 1997), 1–2 log units above the microbial discharge standards for treated municipal wastewater of Tanzania (TBS, 2005).

Consistent septic tank de-sludging is rare among households and usually occurs when the tanks are visibly clogged, overflowing or damaged. In some areas of Vietnam, septic tanks have not been de-sludged since their construction (Harada *et al.*, 2008) while in the Philippines times between emptying are often a decade (Harder *et al.*, 2013). Williams and Overbo (2015) found that irregular or non-existent de-sludging was caused by a lack of information and education of septic tank owners on the operation and maintenance of the system, as well as the lack of sufficient and affordable de-sludging services. In India, the towns of Mira-Bhayandar (population 815,000) and Wardha (population 125,000) are served by only one vacuum suction service, meaning that only a limited number of septic tanks can be emptied per year (Mehta *et al.*, 2019).

3.2.2 Anaerobic Baffled Reactor

The anaerobic baffled reactor (ABR) can be described as an improved septic tank, developed in the 1980s at Stanford University (Hoffmann *et al.*, 2011; Reynaud, 2014). It consists of a series of chambers in which the water flows up through settled sludge, increasing the contact time between organic pollutants and biomass, thereby reducing total solids and organic matter concentrations in the wastewater (Figure 3.2). Anaerobic digestion and solids' retention are the two principle treatment mechanisms within the ABR. Anaerobic digestion occurs when organic pollutants in the wastewater (suspended or dissolved) enter into contact with the anaerobic bacteria located in the sludge settled at the bottom of each chamber. Solids' retention occurs through physical settlement in each reactor chamber (Reynaud, 2014). In some cases, the final chamber is filled with coarse granular material which supports a microbial biofilm – further contributing to treatment. Where this 'biofilm' module is added the system is usually known as combined ABR (BORDA, 2017). ABRs are simple to operate under a wide range of hydraulic and organic loading rates and are known to be robust to shock-load events. They can also generate biogas for various uses (Reynaud and Buckley, 2015).

China could be the country with the largest number of ABR implemented at full-scale. Up to 2015 the Ministry of Agriculture recorded around 200,000 decentralised wastewater treatment systems with ABR technology (Cheng *et al.*, 2017). The German NGO Bremen Overseas Research and Development Association (BORDA) has defined the decentralised wastewater systems with the term

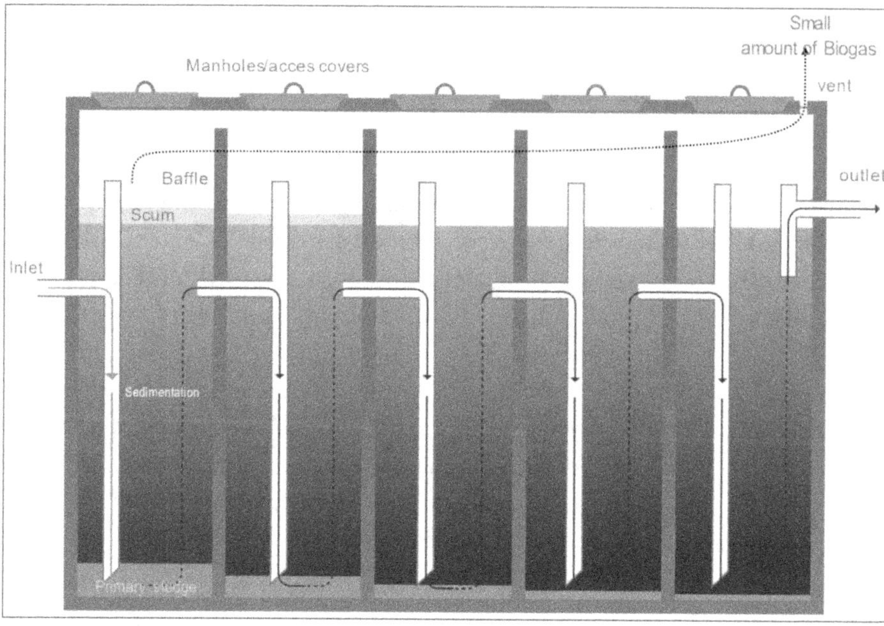

Figure 3.2. Anaerobic Baffled Reactor and its main treatment mechanisms. Adapted from MoW, 2018.

'DEWATS' and made the ABR its core technology due to its minimal maintenance, low (or zero) electricity requirements, lack of movable parts and suitability for low income communities and densely populated urban areas (Reynaud, 2014). As a result of the Settlement Sanitation Development Acceleration Program in Indonesia, launched in 2010, the majority of the 2,700 ABRs installed by BORDA have been there.

3.2.2.1 Treatment performance of ABRs in countries of the Global South

ABRs have typical removal efficiencies for organic matter of between 40% BOD_5 and 60% BOD_5 (Hoffmann *et al.*, 2011) and between 70% COD and 90% COD at laboratory scale (Reynaud, 2014). Anaerobic treatment processes are not specifically designed to remove nutrients and in certain areas are limited to 1 or 2 log reduction for faecal coliforms (Ligy, 2018; Mahenge and Malabeja, 2018). In fact, Kerstens *et al.*, 2012 found that in the municipalities of Yogyakarta, Surakarta and Blitar from Java, Indonesia, the concentration of faecal coliforms in the effluent is tremendously high (between 10^8 and 10^{18} coliforms/100 ml). If ABRs are not combined with other treatment systems to enable nutrient removal or wastewater reuse, their discharges may increase the risk of eutrophication in receiving water bodies. Chemical oxygen demand and nutrient concentrations measured at the effluent of ABR or combined ABR systems installed in India, Indonesia and South Africa are presented in Table 3.2. Similar to the pilot study published by Dama *et al.* (2002), the effluents from ABR systems have COD values ranging between 70 and 400 mg/l. Although COD removal in the ABR can reach efficiencies as high as 90% under controlled conditions (Dama *et al.*, 2002). Reynaud (2014) found that COD reduction is non-optimal in ABRs implemented at full-scale in some countries of the Global South. According to Reynaud and Buckley (2015), a high amount of biodegradable organics can remain untreated in an ABR, leading to high BOD_5 concentrations in the effluent. Tchobanoglous *et al.* (2003) reported a BOD_5/COD ratio for wastewater after biological treatment of 0.1 to 0.3, which is significantly lower than the average BOD_5/COD ratio of 0.46 reported by Reynaud and Buckley (2015) and Kerstens *et al.*, 2012 for ABRs.

3.2.2.2 Main causes of malfunction of ABR

Studies have not shown clear and consistent reasons for poor ABR performance. Charles *et al.* (2003) observed that increased effluent concentrations of viruses and pathogens from small scale wastewater treatment systems are usually associated with insufficient or absent operation and maintenance. However, Reynaud (2014) reported that operational problems such as irregular de-sludging of the systems, lack of operator training and training of users did not have any significant influence on effluent quality from the ABR systems. Instead rain-water intrusion, organic-underloading and salinity were found to be the main factors limiting ABR performance (the relationship between elevated raw water salinity and treatment performance is still unclear; however, high salinity was associated with poorer treatment performance of the ABRs built close to the coast).

3.2.3 Waste Stabilisation Ponds

Waste stabilisation ponds (WSPs) were the first nature-based treatment technology implemented in both the Global North and Global South to treat municipal and domestic wastewater (Ho and Goethals, 2020). They are considered a good option for removing solids, organic matter and pathogens if sufficient land is available. WSPs can also remove persistent pollutants such as carbamazepine, diclofenac and trimethoprim and paracetamol with up to 90% efficacy (K'oreje *et al.*, 2018; Macedo *et al.*, 2011; Taherkhani *et al.*, 2018).

In general, anaerobic, facultative and maturation ponds in series are the most common combination for treating municipal and domestic wastewater (Fig. 3.3). Anaerobic ponds are typically 2–5 m deep. They are designed to remove BOD primarily through the sedimentation of suspended organic matter and their subsequent anaerobic digestion at the bottom of the pond. The biodegradable part

Table 3.2. Design, operation and treatment performance of DEWATS systems installed in India, Indonesia and South Africa. Adapted from Kerstens *et al.* (2012) and Reynaud and Buckley (2015).

Countries	Number of People	Types	HRT(h)	$Q(m^3/d)$	COD (mg/l) In	COD (mg/l) Out	Removal Efficiency COD (%)	BOD_{5out} (mg/l)	$NH_4\text{-}N_{out}$ (mg/l)	$PO_4\text{-}P_{out}$ (mg/l)
India, BWC	575	SSS	73	16.5	513	**320**	37.6	n.a.	123	–
Indonesia, GB	195	SSS	27	16.6	393	**127**	67.7	69	**76**	6
Indonesia, MM	251	SSS	10	22.9	436	77	82.3	42	**49**	11
South Africa, NLM	420	SSS	–	35.9	–	**406**	–	n.a.	**61**	9
Indonesia, ST	450	SSS	13	36.4	349	**108**	69.1	83	**50**	4
Indonesia	–	SSS	–	–	–	**122**	–	49.7	**46**	3.8
Indonesia	–	CSC	–	–	–	**131**	–	50	**57.4**	4.8
Tanzania[a]	–	SSS	–	–	–	**86**	–	n.a.	36	–

In: Inlet, Out: outlet, SSS: Shallow sewer system, CSC: Community Sanitation Centres with toilets, showers and at times laundry areas, [a]BORDA, 2019a,b, BWC: Beedi Workers Colony, GB: Gambiran, MM: Minomartani, NLM: Newlands Mashu, ST: Santan, n.a. Not available. Values in **bold** do not comply with national discharge standards.

Tanzania standards for Municipal and Industrial effluent discharges into water bodies are less than: 60 mg/l COD, 30 mg/l BOD_5, 100 mg/l TSS, n.a. TN; n.a. NH_4-N , 6 mg/l TP, n.a. PO_4-P (TBS, 2005).

India National Green Tribunal norms for effluent discharge from sewage treatment plants in mega and metropolitan cities are less than: 50 mg/l COD, 10 mg/l BOD_5, 20 mg/l TSS, 10 mg/l TN, n.a. NH_4-N, 1 mg/l TP, n.a. PO_4-P (NGT, 2019).

Indonesia National discharge standards of domestic wastewater effluent on water stream are less than: 100 mg/l COD, 30 mg/l BOD_5, 30 mg/l TSS, n.a. TN, 10 mg/l NH_4-N, n.a. TP, n.a. PO_4-P (Ministry of Environment and Forestry of Indonesia, 2016).

South Africa: Wastewater values applicable to discharge of wastewater into a water resource are less than: 75 mg/l COD, n.a. BOD_5, 25 mg/l TSS, n.a. TN, 6 mg/l NH_4-N, n.a. TP, 10 mg/l PO_4-P (Department of Water Affairs-South Africa, 2013).

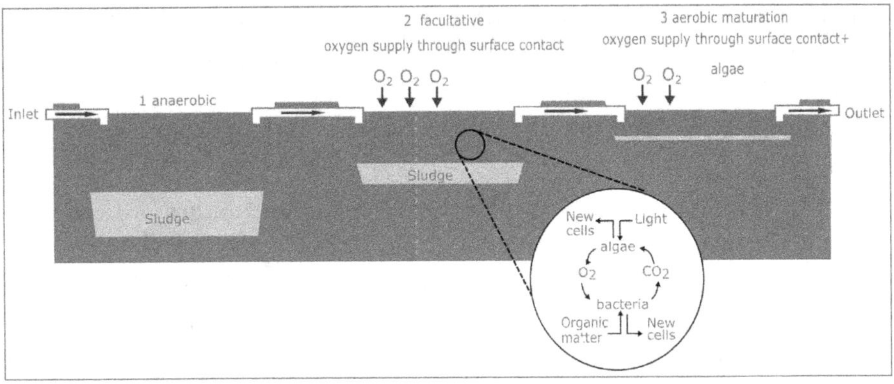

Figure 3.3. Design and principles of anaerobic, facultative and maturation ponds. Adapted from Tilley *et al.* (2008).

of the settled sludge is transformed into carbon dioxide, methane and other compounds while the non-biodegradable fraction or inert part remains at the bottom (Sperling, 2007). Since there is a regular accumulation of digested solids, anaerobic ponds should be de-sludged once every 1–3 years (Mara, 2003). With proper design and maintenance anaerobic ponds are able to remove more than 60% of BOD at 20°C (Mara, 2003; Ronteltap *et al.*, 2014).

Facultative ponds are relatively shallow, typically between 1 m and 1.8 m deep. They can be of two types: primary facultative ponds which receive the effluent of a preliminary treatment and secondary facultative ponds which receive pre-settled wastewater usually corresponding to the effluent from the anaerobic ponds. They are designed to reduce BOD concentration at a surface loading of between 100 and 400 kg/ha/day, which allows the development of healthy algae populations (Mara, 2003). Non-settleable organic matter in suspension together with the dissolved organic matter is aerobically digested through a mutualistic arrangement between algae and bacteria (Sperling, 2007). The oxygen required by the bacteria for the oxidation of organic matter is supplied by the algae (naturally found in the pond) as a result of their photosynthetic activity. The retention period in facultative ponds varies between 5 and 30 days depending on the surface organic loading and the type of climate (temperate, warm tropic, etc.) where the pond is implemented (Mara, 2003; Muga and Mihelcic, 2008).

Maturation ponds are typically 1–1.5 m deep and correspond to the polishing step of the treatment. Their main function is pathogen reduction through UV radiation. Maturation ponds can have removal efficiencies of more than 99.9% for *E. coli* and 4–6 log units for faecal coliforms (Sperling, 2007). If the anaerobic ponds and/or primary facultative ponds work properly, the sludge accumulation in secondary facultative ponds and maturation ponds is much lower, thus their de-sludging is infrequent or not required during their design life (Mara, 2003).

3.2.3.1 Treatment performance of WSPs in the Global South

WSPs have been widely deployed in the Global South due to their low cost, simple construction and basic operation and maintenance. Currently in Tanzania, WSPs represent the main full-scale technology available for treating domestic and municipal wastewater (The United Republic of Tanzania: National Audit Office, 2018). In Kenya out of 39 publicly operated wastewater treatment plants 27 are WSPs (Wang *et al.*, 2014) and in Latin America 38% of the installed technologies for wastewater treatment are WSPs (Chernicharo *et al.*, 2015). The National River Conservation Directorate (2003) of India stated that 'only waste stabilization ponds, which are eco-friendly and simple to operate, will be mainly supported to treat wastewater in India' (although recent centralised facilities have tended to implement more intensive engineered solutions typical of those found in the Global North).

WSPs in the Global South have in general a poor performance. Table 3.3 summarises some of the most recent literature reports on the efficiency of wastewater treatment plants employing pond technology. Although Mara (2003) established that WSPs are able to achieve removal efficiencies of more than 90% for BOD, TSS and ammonia, in the Global South the removals are normally below this threshold. Observed percentage removals vary from 27% in Dar es Salaam, Tanzania to 90% in Shirere, Kenya for BOD, and between 31% in Ponta Negra, Brazil and 90% in Shirere, Kenya for TSS. A limited number of studies report nitrogen and phosphorus removals; however, some authors have reported removal efficiencies of ammonium, TN and TP between 40 and 70%, 13 and 60% and 5 and 9%, respectively in countries such as Tanzania, Brazil and Palestine.

WSPs are considered especially efficient in the removal of pathogens. Mara (2003) stated that well operated, designed and maintained WSPs can reduce *E. coli* and faecal coliforms by 99.99% or 4 log units. Few of the analysed systems matched this potential (Table 3.3), which would not meet WHO guidelines for unrestricted irrigation (4–6 log reduction) (WHO, 2006). In general lower than expected performance has led to concentrations of BOD, TSS in the effluent that exceed the national discharge standards of Tanzania, Botswana, Kenya, India, Egypt and Palestine. WSPs situated in countries with regulations for ammonium, nitrates, total nitrogen and/or total phosphorus generally do not meet the national guidelines for municipal effluents into water bodies or drains or discharge standards for treated effluent.

Table 3.3. A summary of design/operation and removal efficiencies of WSPs treating municipal wastewater in countries of the Global South. All concentrations measured in the effluent.

	Removal Efficiencies															Design and Operation			
Cities/Towns Countries	**COD (mg/l)**	**%**	**BOD (mg/l)**	**%**	**TSS (mg/l)**	**%**	**NH_4-N (mg/l)**	**%**	**NO_3-N (mg/l)**	**%**	**TN (mg/l)**	**%**	**TP (mg/l)**	**%**	**Faecal Coliforms (FC) (MPN/100 ml)**	**Log Unit**	**Flow (m^3/d)**	**HRT (Days)**	**References**
Mwanza, Tanzania	**215**	63	–	–	**105**	66	–	–	–	–	38	33	–	–	4.9E+03	3.8	5,952	12	Zacharia *et al.*, 2019
Dar es Salaam, Tanzania	**285**	45.5	**184**	27	–		6.58	44.7	13.1a		58.4	13.6	**22.6**	6	6.5E+02	3.7	840	19.9	Mbwele *et al.*, 2004
Palapye, Botswana	**156**	72	**110**	46	**305**[a]	–	–	–	**203**[a]	–	–	–	**15**[a]	–	–	–	4,507	7.1	Gopolang and Letshwenyo, 2018
Shirere, Kenya	112	89.6	**75**	90	**50**	90.1	–	–	–	–	–	–	–	–	–	–	11,877	25–30	K'oreje *et al.*, 2018
Jodhpur, India	**124.7**	53.4	**26.6**	84.1	**91.8**	79.7	–	–	–	–	**61**[b]	61.7	–	–	1.1E+08	2.5	17,000	10	Goyal and Mohan, 2013
Rishikesh, India	**82.8**	78.8	**31.4**	83.5	**46.8**	84.6	–	–	–	–	–	–	–	–	2.5E+04	2.3	6,000	11	Tyagi *et al.*, 2011
Ponta Negra, Brazil	259	55	82	73	241	31	–	–	–	–	–	–	–	–	–	–	4,742	31	Macedo *et al.*, 2011
Ceara, Brazil	176	75.7	81	81	115	60	10.8	70	–		–	–	7.5	5	3.7E+03	3.8	28,418	61	Da Silva *et al.*, 2011
Catacama, Honduras	–	–	**100**	64.7	92	64	–	–	–	–	–	–	–	–	4.1E+05	1.6	–	18	Oakley *et al.*, 2000

(Continued)

Table 3.3. Continued

Cities/Towns Countries	Removal Efficiencies															Design and Operation			References
	COD (mg/l)	%	BOD (mg/l)	%	TSS (mg/l)	%	NH_4-N (mg/l)	%	NO_3-N (mg/l)	%	TN (mg/l)	%	TP (mg/l)	%	Faecal Coliforms (FC) (MPN/100 ml)	Log Unit	Flow (m^3/d)	HRT (Days)	
El–Mofti, Egypt	**289**	48.9	**145.3**	50.6	**157.8**	44.3	–	–	0.4	57	–	–	–	–	1.6E+06	4.6	225	–	El-Deeb Ghazy *et al.*, 2008
Talitha Kumi, Palestine	**387.6**	41	**289.5**	38	**1392**	39	**32.5**	42	9.7	46	–	–	22.6	9	–	–	38	15.6	Al-Sàed, 2007

[a]Concentration higher in the effluent than in the influent.
[b]TKN.

Values in bold don't comply with national discharge standards.

Tanzania standards for municipal and industrial effluent discharges into water bodies are less than: 60 mg/l COD, 30 mg/l BOD_5, 100 mg/l TSS, n.a. TN, n.a. NH_4-N, 20 mg/l NO_3-N, 6 mg/l TP (TBS, 2005).

Kenya National Environment Standards for discharge of effluents into water bodies are less than: 50 mg/l COD, 30 mg/l BOD_5, 30 mg/l TSS, 100 mg/l TN, 0.5 mg/l NH_4-N, 10 mg/l NO_3-N, n.a. TP, 1,000 MPN/100 ml FC for use in irrigation (NEMA, 2006).

India National Green Tribunal norms for effluent discharge from sewage treatment plants are less than: 50 mg/l COD, 10 mg/l BOD_5, 20 mg/l TSS, 10 mg/l TN, n.a. NH_4-N, n.a. NO_3-N, 1 mg/l TP, 230 MPN/100 ml FC (NGT, 2019).

Brazilian standards for discharge of effluents from sewage treatment plants into water bodies are less than: n.a. COD, 120 mg/l or 60% removal efficiency BOD_5, 20% removal efficiency TSS, n.a. TN, n.a. NO_3-N, 20 mg/l NH_4-N, n.a. TP, n.a. FC (CONAMA, 2011).

Honduras discharge standards of effluents into water bodies are less than: 200 mg/l COD, 50 mg/l BOD_5, 100 mg/l TSS, n.a. TN, 20 mg/l NH_4-N, n.a. NO_3-N, n.a. TP, 5,000 MPN/100 ml FC (CTN, 1996).

Egyptian discharge standards for treated effluent to drains, lakes or ponds, are less than: 80 mg/l COD, 60 mg/l BOD_5, 50 mg/l TSS, n.a. TN, n.a. NH_4-N, 50 mg/l NO_3-N, n.a. TP, 5,000 MPN/100 ml TC (NRWC, 1995).

Palestine guidelines for high quality treated wastewater for reuse are less than: 50 mg/l COD, 20 mg/l BOD_5, 30 mg/l TSS, n.a. TN, 5 mg/l NH_4-N, 20 mg/l NO_3-N, n.a. TP, 200 MPN/100 ml FC (PSI, 2012).

3.2.3.2 Main causes of malfunction of WSPs

A mismatch between design and expected load appears to be a common problem for WSPs installed in the Global South. Al-Sàed (2007) observed that the surface organic load in the Thalita Kumi WSP was >200% higher than the design. Pescod (1992) found something similar in Jordan where the As Samra Pond exceeded its design organic and hydraulic load by 57% and 25%, respectively. The assessment of 95 WSPs in Ceará, Brazil indicated, on the contrary, that all ponds were under-loaded as a result of a low household connection to the sewer network (Da Silva *et al.*, 2011). Maintenance problems are also a factor that can explain poor performance of WSPs. The lack of regular de-sludging allows settlement and accumulation of solids which decreases the effective volume and the hydraulic retention time (HRT) and therefore, the overall performance of the system. In Botswana, the volume reductions of the anaerobic and facultative ponds ranged between 60% and 90% as a result of the operation of WSPs for more than 20 years without de-sludging (Gopolang and Letshwenyo, 2018). By 2018, in Dar es Salam (Tanzania) there were no records of the last time WSPs were de-sludged mainly due to low priority given to sanitation service provision, therefore insufficient funds allocated to the maintenance, expansion and rehabilitation of the sewerage infrastructure (usually less than 10% of the revenues collected from sanitation services) (The United Republic of Tanzania: National Audit Office, 2018). Table 3.4 shows the designed and calculated flow rate and HRT for WSPs that have been studied.

Table 3.4. Differences between the design and measured flow rate and hydraulic retention times in WSPs.

	Flow Rate (m^3/d)		Hydraulic Retention Time (HRT) (d)		
Countries	Design	Measured	Design	Measured	References
Mwanza, Tanzania	5,000	5,952	–	12	Zacharia *et al.*, 2019
Palapye, Botswana	2,779[a]	4,507	20	7.1	Gopolang and Letshwenyo, 2018
Ile-Ife, Nigeria	1,171	21,600	–	–	Oke *et al.*, 2009
Ponta Negra, Brazil	8,208	4,742	–	31	Macedo *et al.*, 2011
Talitha Kumi, Palestine	–	38	AP: 2 FP: 11 MP: 11	AP: 1.2 FP: 4.5 MP: 4.5	Al-Sàed, 2007

[a]Assuming 80 L/cap/d.
AP: Anaerobic pond; FP: Facultative pond; MP: Maturation pond.

The challenges faced with the operation and maintenance of WSPs together with the increasing implementation of other nature-based technologies such as constructed wetlands has reduced their popularity, thus, the number of wastewater treatment plants utilising pond technology has not changed significantly in the last decades (Ho and Goethals, 2020). Instead, this low-cost technology has been upgraded through combination with other technologies or been replaced entirely in order to comply with stricter discharge standards and the increasing demand of safer reclaimed water. Furthermore, increased land prices have impacted the cost-effectiveness of WSPs in some cases. For example, despite high investment requirement (USD 150 million) and higher operating costs of advanced electromechanical treatment systems As Samra, Jordan is replacing WSPs for these technologies in order to improve the effluent quality and expand the range of wastewater reuse applications to industrial cooling, groundwater recharge and municipal use (Ammary, 2007).

3.2.4 Constructed Wetlands

After WSPs, constructed wetlands were among the first nature-based treatments implemented worldwide at full-scale as a result of their low construction, maintenance and energy costs – and simple operation (Li *et al.*, 2017; Rahi *et al.*, 2020). Wetland technology has been adapted to treat industrial sewage, storm water and domestic wastewater (Zhang *et al.*, 2015). Treatment performance can be sufficient to produce effluents that can be reused for restricted irrigation in agriculture (Chandrakanth *et al.*, 2016; Colares *et al.*, 2019; Nivala *et al.*, 2019; Rahi *et al.*, 2020; Sehar *et al.*, 2016) and a recent study of full-scale wetlands in Mexico reveals exceptional capacity of wetlands to deal with some types of emerging pollutants (Herrera-Cárdenas *et al.*, 2016).

The mechanisms of pollutant removal in constructed wetlands comprise several biological transformations and physicochemical processes including sedimentation, filtration, precipitation, adsorption, microbial and plant uptake (Boano *et al.*, 2020) (Fig. 3.4). Macrophytes act as supporting media for microbial attachment which promotes several biological and chemical reactions, including oxygen allocation, nitrification and denitrification. Selection of support media is important, as this will support the active biofilm and may act as an adsorbent for some pollutants. Kumar and Dutta (2019) describe the key components and controlling factors critical to achieving optimum performance in constructed wetlands.

3.2.4.1 Treatment performance of constructed wetlands

Vertical flow constructed wetlands (VFCWs) and horizontal flow constructed wetlands (HFCWs) are the most common types of wetland technology implemented in

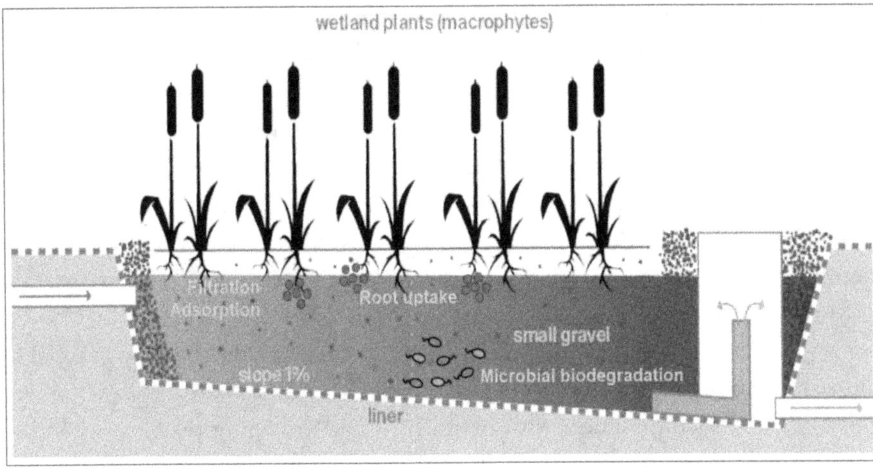

Figure 3.4. General mechanisms for the removal of pollutants in CWs, adapted from Gorito *et al.* (2017); Tilley *et al.* (2008).

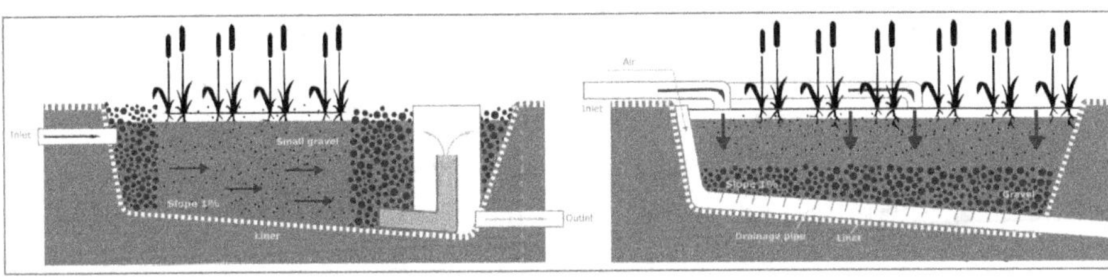

Figure 3.5. Flow path for Horizontal Flow Constructed Wetland (HFCW) and Vertical flow constructed Wetlands (VFCW). Adapted from Gorito *et al.* (2017) and Tilley *et al.* (2008).

the Global South (Fig. 3.5). Although at laboratory scale, both show high removal efficiencies for TSS (81% for HFCW, 89% for VFCW), their performance is limited for COD (56.3% HFCW, 60% VFCW) and BOD (66.2% for HFCW, 74.6% for VFCW) (Chandrakanth *et al.*, 2016). It has been demonstrated that VFCWs can achieve better removals when compared to HFCWs as a result of intermittent loading of the system which increases the transfer of oxygen creating a better oxidising environment for the degradation of organic matter (Zhang *et al.*, 2015). Total Nitrogen removal rates in both systems remain close to 50% due to their inability to provide appropriate sequential aerobic and anoxic conditions. In response to this situation, wetland technologies can be combined, e.g., HFCW-VFCW or VFCW-HFCW, to leverage the strengths of the separate systems. Hybrid constructed wetlands have been demonstrated to have better removal efficiencies up to 80% of the organic matter (BOD_5) independently of the HFCW-VFCW or VFCW-HFCW combination (Brix *et al.*, 2011; Singh *et al.*, 2009; Taylor *et al.*, 2015).

Phosphorus removal in constructed wetlands is a result of two principle mechanisms: plant uptake and harvesting (as examined by authors including

He *et al.*, 2018) media sorption/precipitation (as examined by authors including Colares *et al.*, 2019). However, despite these parallel mechanisms, constructed wetlands have a relatively low efficiency in phosphorus removal (Kumar and Asolekar, 2016; Vasconcellos *et al.*, 2019; Zhang *et al.*, 2015), therefore to comply with national and international discharge standards, it is usually necessary to use a final polishing step such as chemical precipitation (Colares *et al.*, 2019).

In countries of the Global South constructed wetlands have been utilised as a secondary treatment technology – commonly after septic tanks (Decezaro *et al.*, 2018) or ABR systems (CSE, 2020; Singh *et al.*, 2009; Taylor *et al.*, 2015) – to treat domestic or municipal wastewater, often at small scale (Zhang *et al.*, 2015). Numerous authors report combined treatment efficiencies where constructed wetlands are used in conjunction with established and emerging primary/secondary treatment technologies (Colares *et al.*, 2019; Decezaro *et al.*, 2018; Tomar and Suthar, 2011). An overview of the efficacy of some of these combined systems is presented in Table 3.5. Zhang *et al.* (2014) compared the removal efficiencies of constructed wetlands in the Global South with similar systems in Europe, and despite differences in design and operation, most treatment wetlands in the Global South were found to perform with similar efficiency to those in Europe.

Constructed wetlands are reliable and robust systems able to maintain consistent treatment performances over time. In Iraq, although wastewater production and wastewater characteristics change between summer and winter (e.g., daily flow rate increases from 90 L/capita in winter to 150 L/capita in summer), the treatment performance of constructed wetlands has not been found to change significantly from season to season (Rahi *et al.*, 2020). In India the removal efficiency of CWs at pilot scale and laboratory scale has been shown to be independent of the climate variability caused by monsoon seasons (Ramesh *et al.*, 2017). Additionally, a HFCW installed in the Arrudas Wastewater Treatment Plant in Belo Horizonte, Brazil has demonstrated consistent treatment efficiency for a study period of 10 years with average removals of 55% BOD_5, 64% COD, 66% TSS 66% and 54% Total Phosphorus (Vasconcellos *et al.*, 2019). Such data are encouraging, although it should be noted that this chapter explores only a small sample of the papers which examine constructed wetland performance in countries of the Global South.

3.2.4.2 Main causes of malfunction of constructed wetlands

Leakages, substrate clogging (and consequent ponding), insufficient vegetation, improper design and lack of operation and maintenance are the most common causes of failure identified in constructed wetlands. Saturation and clogging of the filter material have been classified as the worst operational problems by Gokalp *et al.* (2014) and Kimwaga *et al.* (2013), exacerbated by incomplete

Table 3.5. A summary of treatment efficiency of constructed wetlands in countries of the Global South (All concentration values in mg/L were measured in the effluent of the last constructed wetland).

Cities/Towns/Countries Treatment Systems	Types of Wastewater	Removal Efficiency (%)										Design and Operation		References
		BOD_5		COD		TSS		TN		TP				
		mg/l	%	mg/l	%	mg/l	%	mg/l	%	mg/l	%	Area (m^2)	Flow (m^3/d)	
Tiruphati, India. HFCW	Municipal	**125.1**	66.2	**209.8**	56.3	**95**	81	–	–	–	–	0.18	0.173	Chandrakanth *et al.*, 2016
Tiruphati, India. VFCW	Municipal	**92.9**	74.9	**192**	60	**55**	89	–	–	–	–	0.3	0.288	Chandrakanth *et al.*, 2016
Kampala, Uganda. VFCW (6 CW in series)	Municipal	–	–	–	–	–	–	**16.1**	72.5	2.6	83.2	Each CW: 0.26 Total: 1.6	0.064	Kyambadde *et al.*, 2004
Full-scale CW														
Belo Horizonte, Brazil. HFCW	Domestic	26	55	51	64	15	66	–	–	–	–	75	7.5	Vasconcellos *et al.*, 2019
Rio Grande do Sul, Brazil. ST + VFCW	Municipal	52	67	117	63	44	51	–	–	–	–	24.5	1.5	Decezaro *et al.*, 2018
Mansarde-Rancée, Martinique, French West Indies. VFCW+UV lamp	Domestic	20	–	125	–	30	–	8	–	–	–	680	204	Global Wetland Technology, 2020
Rainbow Drive, Bangalore . ABR+CW	Municipal	–	81	–	–	–	75	–	73	–	58	ABR: 16CW: 278.7	250	CSE, 2020
Bithoor, Kampur, India. HSFCW	Municipal	**13.3**	68.5	**66.6**	91.6	**78**	94.3	–	–	–	–	1184.5	610	CSE, 2020
Gebze, Turkey. ABR and hybrid constructed wetlands 3HFCW + 2 VFCW	Domestic	9.5	81	31.0	83	10.0	90	17.0	39	2.6	53	Total HFCW: 486 Total VFCW: 250	50–60	Taylor *et al.*, 2015

(Continued)

Table 3.5. Continued

Cities/Towns/Countries Treatment Systems	Types of Wastewater	Removal Efficiency (%)										Design and Operation		References
		BOD_5		COD		TSS		TN		TP				
		mg/l	%	mg/l	%	mg/l	%	mg/l	%	mg/l	%	Area (m^2)	Flow (m^3/d)	
Thimi Municipality, Nepal. ABR and hybrid CW: 2HFCW+2VFCW	Domestic	**173.3**	HFCW57.5 VFCW: 44.9 Hybrid 70 Total[a] 90.1	**318.6**	HFCW 51.4 VFCW: 45.7 Total[a] 90.0	37.8	HFCW 69.3 VFCW: 57.6 Total[a] 95.9	–	–	17.1	HFCW 27.3 VFCW: 0 Total 26.1	Total HFCW: 150 Total VFCW: 150	10	Singh *et al.*, 2009
Kho Phi Phi, Thailand. VFCW+HFCW+Free Water surface Flow CW+Disinfection Pond	Municipal	**56**[b] 25[c]	VFCW68.2 HFCW 42.1 Hybrid 81.6	–	–	16[c]	–	33[c,d]	VFCW 20.7 HFCW 4.6 Hybrid 24.3	4.5	–	Total VFCW: 2300 Total HFCW: 750	400	Brix *et al.*, 2011
Hachenoua, Mayotte. VFCW	Domestic	33	91	71	91	26	91	–	–	–	–	128	55	Molle *et al.*, 2015
Bois d'Opale, French Guiana. VFCW	Domestic	16.2[e]	80.5	75	68	54.5[e]	62.2	–	–	–	–	–	300–400 PE	Molle *et al.*, 2015

[a]Removal efficiency of treatment system including pre-treatment.
[b]Value measured in the outlet of the HFCW.
[c]Value measured in the outlet of the treatment plant.
[d]TKN.
[e]Average values between dry and rainy season.

Values in **bold** do not comply with national discharge standards.

India National Green Tribunal norms for effluent discharge from sewage treatment plants are less than: 50 mg/l COD, 10 mg/l BOD_5, 20 mg/l TSS, 10 mg/l TN, 1 mg/l TP, 230 MPN/100 ml FC (NGT, 2019).

Brazilian standards for discharge of effluents from sewage treatment plants into water bodies are less than: n.a. COD,120 mg/l or 60% removal efficiency BOD_5, 20% removal efficiency TSS, n.a. TN, n.a. NO_3-N, 20 mg/l NH_4-N, n.a. TP, n.a. FC (CONAMA, 2011).

Nepalese standards for wastewater from combined wastewater treatment plant into inland surface water are less than: 50 mg/l BOD_5, 250 mg/l COD, 50 mg/l TSS, n.a. TN, n.a. TP, n.a. FC (WEPA, 2016).

Ugandan standards for discharge of effluent or wastewater are less than: 100 mg/l COD, 50 mg/l BOD_5, 100 mg/l TSS, 10 mg/l TN, 10 mg/l TP, 10,000 counts/100 ml TC (NEMA, 1999).

French regulations for discharge standards PE 20–20,000 (The French outermost regions under tropical climate still have to comply with both French and EU regulations) are less than: 60% reduction for COD, 60% reduction for BOD5, 50% reduction for TSS, 10 mg/l (Legifrance, 2009).

Turkish regulations for discharge standards are less than: 125 mg/l COD, 25 mg/l BOD_5, 60 mg/l TSS, 15 mg/l TN, 10 mg/l, 2 mg/l TP (Taylor *et al.*, 2015).

vegetation establishment and inappropriate vegetation management (BORDA, 2019b; Gokalp *et al.*, 2014). Lack of maintenance and inadequate operation are cross-cutting issues among all nature-based wastewater treatment technologies, but Gokalp *et al.* (2014) found that 96% of the CWs in the province of Kayseri, Turkey had not received any kind of maintenance after their construction which together with substrate clogging, leakages, and their implementation in flood-prone areas had transformed constructed wetlands into swamp-like systems.

3.2.5 Vermifiltration

Vermifiltration (also known as lumbrifiltration) was first advocated in 1992 by Prof. Jose Tohá at the University of Chile (Samal *et al.*, 2017). It has been recognised as a novel, odourless, sludge-free and low cost technology for wastewater treatment (Choudhary and Medok, 2017; Kumar and Ghosh, 2019; Singh *et al.*, 2017). Vermifiltration is a liquid state vermi-conversion process used to treat domestic, municipal and industrial wastewater (Samal *et al.*, 2017) that is related to, but distinct from vermi-composting. It involves the synergistic and symbiotic action of earthworms and microorganisms, and has been shown to be a process that requires low investment and operation costs with high removal efficiencies of organic matter. When designed and operated well these systems do not clog and produce limited excess sludge (Singh *et al.*, 2017).

Vermifilters comprise an earthworm population in a filter bed which can, for example, comprise soil, sand and gravel (Singh *et al.*, 2017). The filter bed is normally stratified, with an upper organic layer hosting the earthworms, while lower mineral layers support a microbial biofilm. The earthworms enhance the biological reactions within the filter media through their ingestion and digestion of solids introduced in wastewater (Singh *et al.*, 2019).

Through burrowing and channelling/tunnelling earthworms intensify aerobic degradation processes, increasing the surface area of the bedding and maintaining the overall hydraulic conductivity of the system. These actions prevent clogging, increase the reduction of organic matter and aid in overall nutrient removal (Luth *et al.*, 2011; Singh *et al.*, 2017, 2019; Sinha *et al.*, 2008). For instance, it has been demonstrated that earthworms in a vermifilter are able to remove small loadings of BOD (200–400 mg/l) within a 30–40 minute HRT and high BOD loads (in excess of 10,000 mg/l), such as those found in food-processing industries, when the HRT is increased to 4–10 h (Sinha *et al.*, 2008).

3.2.5.1 Treatment performance of vermifilters

Vermifilters treating domestic or municipal wastewater in the Global South have demonstrated removal efficiencies up to 95.5% of BOD_5 (Gallegos Valqui, 2019),

Table 3.6. Treatment performance of vermifiltration. All concentrations measured in the effluent of the vermifilter.

Cities/ Towns, Countries	Types of Waste-Water	Removal Efficiencies: COD		BOD_5		TSS		TN		NH_4-N		NO_3-N		PO_4-P		Total Coliforms		Design and Operation Parameters				References
		mg/l	%	mg/l	%	mg/l	%	mg/l	%	mg/l	%	mg/l	%	mg/l	%	CFU/ 100 ml	Log unit	Flow (m^3/d)	HLR ($m^3/m^2/d$)	HRT (h)	Earthworm Species	
Langos La Nube, Ecuador	Domestic	**228**	51.7	70	84.4	–	–	–	–	5.4	68.2	4.1	80	5.2	73	–	–	102[b]	–	6	*E. fetida*	Coronel, 2015
Manchay, Peru	Primary settled domestic wastewater	69.4	90.8	6.8	95.5	–	–	–	–	–	–	–	–	–	–	4.9E+03	2.21	184[b]	–	4.5	*E. fetida*	Gallegos Valqui, 2019
Moche, Peru	Domestic	92.2	72.4	27.9	83.9	–	–	–	–	–	–	–	–	–	–	–	–	50[b]	–	–	*E. fetida*	Mitma, 2017
Orellana, Ecuador	Primary settled domestic wastewater	174	70.1	23	89	40	82.1	–	–	–	–	–	–	–	–	1.0E+05	–	70[b]	–	1.44	*E. fetida*	Reyes *et al.*, 2016
Santa Cruz do Sul, Brasil	Domestic	144	92	58.5	87	~0	99	less than 20	91	less than 20	92	–	–	–	–	–	–	0.02	0.5	–	*E. andrei*	Rothmund and Becker, 2018
São Carlos, Brasil	Domestic	169	76	34	87	15	92	232	11.1	135	38.6	79[a]	–	–	–	–	–	0.02–0.08	0.5–1.6	–	*E. andrei*	Peña *et al.*, 2019
São Carlos, Brasil	Domestic	233	68	46	82	29	86	234	10.3	128	41.8	80[a]	–	–	–	–	–	0.01–0.04	0.25–0.8	–	*E. andrei*	Peña *et al.*, 2019

(*Continued*)

Table 3.6. Continued

Cities/ Towns, Countries	Types of Waste-Water	Removal Efficiencies: COD mg/l	COD %	BOD$_5$ mg/l	BOD$_5$ %	TSS mg/l	TSS %	TN mg/l	TN %	NH$_4$-N mg/l	NH$_4$-N %	NO$_3$-N mg/l	NO$_3$-N %	PO$_4$-P mg/l	PO$_4$-P %	Total Coliforms CFU/ 100 ml	Total Coliforms Log unit	Design and Operation Parameters: Flow (m^3/d)	HLR (m^3/m^2/d)	HRT (h)	Earthworm Species	References
Roorke, India	Synthetic	118.6	74	19.4	92	–	–	–	–	–	–	–	–	–	–	**8.3E+05**	2.60	0.3	1	–	*E. fetida*	Arora *et al.*, 2014
Roorke, India	Domestic	–	–	**28**	88	**162**	78	–	–	4.4	85.6	51.1[a]	–	34.5[a]	–	**3.0E+03**	3.12	–	2.5	–	*E. fetida*	Kumar and Asolekar, 2016
Roorke, India	Domestic	–	–	**72**	70	**378**	67	–	–	8	73.8	28.4[a]	–	20[a]	–	**3.9E+05**	0.98	–	2.5	–	*Eudrilus eugeniae*	Kumar and Asolekar, 2016
Mexico	Domestic	–		**140.5**	–	20		22	–	–	–	–	–	–	–	–	–	0.75	–	–	Unknown	Cplantae, 2019

[a]Concentration in the effluent increased after the treatment.

[b] mL/min.

E. fetida = *Eisenia fetida*, *E. andrei* = *Eisenia andrei*.

HLR: Hydraulic loading rate.

HRT: Hydraulic retention time.

Values in **bold** do not comply with national discharge standards.

India National Green Tribunal norms for effluent discharge from sewage treatment plants are less than: 50 mg/l COD, 10 mg/l BOD5, 20 mg/l TSS, 10 mg/l TN, 1 mg/l TP, 230 MPN/100 ml FC (NGT, 2019).

Brazilian standards for discharge of effluents from sewage treatment plants into water bodies are less than: n.a. COD,120 mg/l or 60% removal efficiency BOD$_5$, 20% removal efficiency TSS, n.a. TN, n.a. NO$_3$-N, 20 mg/l NH$_4$-N, n.a. TP, n.a. FC (CONAMA, 2011).

Ecuador standards for discharge of effluent or wastewater are less than: 200 mg/l COD, 100 mg/l BOD5, 100 mg/l TSS, 30 mg/l TKN, 30 mg/l NH4-N, 10 mg/l TP, 2,000 counts/100 ml FC (Ministerio del Ambiente, 2015).

Peru regulations for discharge standards of effluent from sewage treatment plants are less than: 200 mg/l COD, 100 mlg/l BOD5, 150 ml/l TSS, 10,000 MPN/100 ml TC (Minister of Natural resources and Environment, 2008).

Mexican discharge standards of treated wastewater into rivers for the protection of aquatic life are less than: 60 mg/l BOD5, 60 mg/l TSS, 25 mg/l TN, 10 mg/l TP, 1,000–2,000 MPN/100 ml FC (Note: Mexico has several standards for reuse or discharge into different water bodies, therefore depending on the end use or disposal of the effluent the treatment system might meet some of the other limits that were not presented in this description) (SEMARNAT, 1998).

92% of COD (Rothmund and Becker, 2018) and 92–99% of TSS (Peña *et al.*, 2019; Rothmund and Becker, 2018). Peña *et al.* (2019) highlighted that vermifilters treating raw wastewater function as both a primary and secondary treatment system able to reach similar or higher removals than the combination of septic tanks and biological reactors. Vermifilters have been shown to remove more than 80% of NH_4-N and Kumar and Asolekar (2016) and Peña *et al.* (2019) have reported enhanced nitrification as a result of the earthworms in the filter-beds. However, removals of TN and PO_4-P are generally limited (Table 3.6). While the discharge of nutrient-rich effluent could have severe impact in water bodies, some authors have seen it as an opportunity to reuse such effluents for irrigation as in Manchay, Peru where the vermifilter effluent has been recommended for the irrigation of a forest area of approximately 6 ha (Gallegos Valqui, 2019).

The efficiency of vermifiltration depends on several factors, among them the amount of earthworms in the vermi-bed and their species. Gallegos Valqui (2019) demonstrated that a change from 5,900 earthworms/m^2 to 24,605 earthworms/m^2 increased the removal efficiency of BOD_5 from 45% to 95.5% within one month (March – April 2019). In fact, Choudhary and Medok (2017) recommend to start the vermifiltration of wastewater with 15,000–20,000 worms/m^3 of soil in order to have higher removal efficiencies. In vermifiltration, earthworm species *Eisenia fetida (E. fetida)* have been identified as the optimal candidate to decompose organic matter in wastewater (Singh *et al.*, 2019). They have their maximum removal efficiency in a temperature range between 20°C and 25°C (Sinha *et al.*, 2008). Nevertheless, in tropical and subtropical regions of the Global South, wastewater temperatures can reach values above 30°C which could decrease the treatment performance of vermifilters that use *E. fetida* species (Rahi *et al.*, 2020). Thus, other species like *Eudrilus eugeniae* which are native to regions such as West Africa (Table 3.7) have been identified as a possible alternative species to be used in the vermifiltration process. As an example, Kumar and Asolekar (2016) compared at laboratory scale the capacity of *E. eugeniae* and *E. fetida* in Roorkee, India to treat domestic wastewater at a hydraulic loading rate (HLR) of 2.5 $m^3/m^2/d$. Removal efficiencies were significantly higher for filters with *E. fetida* (88% for BOD_5, 78% for TSS, 85% for NH_4-N and 3.12 log reduction for total coliforms) than those with *E. eugeniae* (70% for BOD_5, 67% for TSS, 66% for NH_4-N and 0.98 log reduction for total coliforms); furthermore, *E. eugeniae* populations were not stable during the research. It should also be noted that in the INNOQUA project, earthworm species native to colder countries (in this case Ireland) were used – both *Eisenia hortensis* (syn. *Dendrobaena venata*) and *E. fetida* were successfully trialled in vermifilters treating municipal wastewater at temperatures between 5°C and 15°C (data not shown).

Table 3.7. Earthworms species characteristics and distribution in the Global South (Patel and Gajera, 2018).

Earthworm Species	Weight Adult Worm (g)	Temperature Tolerance (°C)	Moisture Tolerance (%)	Active Phase	Distribution
Eudrilus eugeniae	1.5–2.5	18–35	20–40	Throughout the year	Tropical Africa and South Africa
Eisenia fetida	0.3–0.7	15–30	20–40	Throughout the year	Temperate regions of Europe, north America and India
Perionyx excavatus	0.8–1.2	8–30	30–50	Throughout the year	Tropical countries
Dichogaster bolaui/ Dicogaster affinis	0.04–0.07	20–28	20–30	July–October	Tropical countries
Drawida barwelli	0.2–0.5	20–30	40–50	August–November	Tropical countries, shade essential for establishment
Lampito mauritii	0.8–1.5	18–28	20–40	June–August	Plains of Indian peninsula

In the Global South there is still limited information about the performance of this technology in full scale applications though its potential as a 'plug and play' addition to existing systems has recently encouraged the commercialisation of the technology in Chile, Mexico, Venezuela and India (Sinha *et al.*, 2014).

3.2.5.2 Main causes of malfunctioning of vermifilters

HLR is a critical factor for vermifiltration (Sinha *et al.*, 2008). Depending on soil characteristics, high HLR may cause saturation of the vermibed, reducing concentrations of oxygen in the filter media, and thus increasing mortality of earthworms (Xing *et al.*, 2010). Higher HLR also leads to the infiltration of higher volumes of wastewater in the media, which reduces the contact time needed for the biochemical reactions, thus reducing the treatment efficiency of the vermifilter (Singh *et al.*, 2017). Gallegos Valqui (2019) and Paico Revilla (2017) also highlighted that the treatment efficiency depends on the height of the active layer or vermibed which should be kept consistent, hence filter material for the vermibed has to be added periodically. As with other wastewater treatment systems, poor maintenance will result in poor performance and this is a common challenge in the Global South.

3.3 Nature-based Treatment Performances and Their Compliance with National Standards

Many of the nature-based solutions analysed in this study would be categorised as small-scale sanitation systems, defined by Klinger *et al.* (2020) as wastewater treatment systems serving between 10 and 10,000 households or treating 5,000 to 700,000 L of wastewater per day. However as can be seen from Figure 3.6 the volume treated by each technology referenced in this study is quite broad.

Klinger *et al.* (2020) performed a country-wide study to establish, among others, the treatment performance of established, conventional and nature-based, small-scale wastewater treatment systems in India. Although, small-scale systems have been implemented in different contexts, nature-based solutions (ABR and CW) are mostly built in low-income residential areas while highly automated and mechanised processes (sequencing batch reactors and activated sludge systems) are mainly found in middle to high-income residential areas, institutions and commercial areas. Factors such as water accessibility, type of water supply, diet and fibre intake influence per capita wastewater production and pollution load (Henze *et al.*, 2008; Reymond *et al.*, 2014; Rose *et al.*, 2015; Wolter, 2018). Appendix 2. summarises wastewater production and wastewater characterisation for different countries. Municipal wastewater (mixture of wastewater from different sources including households, restaurants, small businesses, etc.) tends to have lower organic and solids concentration than domestic wastewater. Besides, wastewater from areas with intermittent water supply (e.g., public water tap) is more concentrated than areas with permanent water supply usually associated with higher water consumption, which produce a more diluted wastewater (Klinger *et al.*, 2020; Wolter, 2018).

The type of wastewater to be treated by the sanitation systems influences the removal rate of the different technologies and their capacity to provide effluents that comply with the national discharge standards of the country. For example, nature-based systems installed in low-income residential areas and public toilets treat more concentrated wastewater and are exposed to higher changes in the inlet concentration than mechanised systems installed in commercial, middle-high-income and institutional areas (Klinger *et al.*, 2020) (Fig. 3.7).

According to the results of Klinger *et al.*, 2020, the removal efficiencies of nature-based solutions for BOD, COD and TSS, and in some cases effluent concentrations, are comparable to mechanised treatment systems such as activated sludge processes and sequencing batch reactors (Fig. 3.8). Contrarily, TN removal efficiencies of ABR and CW technologies are slightly lower than mechanised systems as a result of an insufficient nitrification capacity and the concentrated wastewater treated by those systems. In other respects, the variable inlet concentration seems

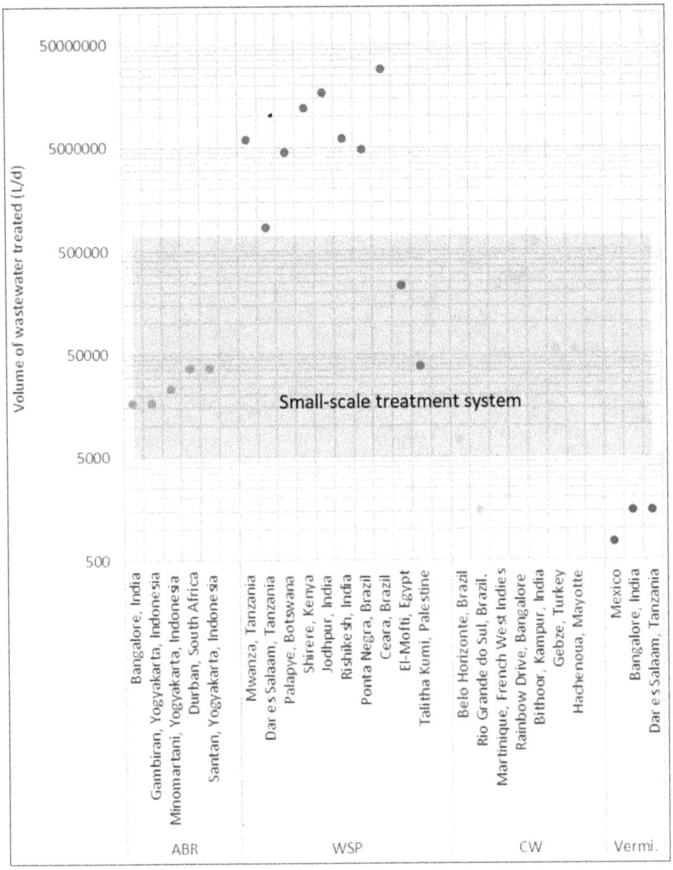

Figure 3.6. Volumes treated by nature-based solutions referenced in this study.

ABR: Bangalore, India; Gambiran, Yogyakarta, Indonesia; Minomartani, Yogyakarta, Indonesia; Durban, South Africa; Santan, Yogyakarta, Indonesia (Reynaud and Buckley, 2015).

WSP: Mwanza, Tanzania (Zacharia *et al.*, 2019); Dar es Salaam, Tanzania (Mbwele *et al.*, 2004); Palapye, Botswana (Gopolang and Letshwenyo, 2018); Shirere, Kenya (K'oreje *et al.*, 2018); Jodhpur, India (Goyal and Mohan, 2013); Rishikesh, India (Tyagi *et al.*, 2011); Ponta Negra, Brazil (Macedo *et al.*, 2011); Ceara, Brazil (Da Silva *et al.*, 2011); El-Mofti, Egypt (El-Deeb Ghazy *et al.*, 2008); Talitha Kumi, Palestine (Al-Sàed, 2007).

CW: Belo Horizonte, Brazil (Vasconcellos *et al.*, 2019); Rio Grande do Sul, Brazil. (Decezaro *et al.*, 2018); Martinique, French West Indies (Global Wetland Technology, 2020); Rainbow Drive, Bangalore (CSE, 2020); Bithoor, Kampur, India (CSE, 2020); Gebze, Turkey (Taylor *et al.*, 2015); Hachenoua, Mayotte (Molle *et al.*, 2015).

Vermi(filtration): Mexico (Cplantae, 2019), Bangalore, India (INNOQUA, 2018); Dar es Salaam, Tanzania (INNOQUA, 2018).

not to affect the treatment performance of nature-based solutions. Besides, their implementation might avoid operational problems such as intermittent functioning – characteristic of mechanised systems in India which aim to reduce Operation and Maintenance (O&M) costs and or noise nuisances (Klinger *et al.*, 2020). Since other factors as a proper design, financial resources for the operation of the system,

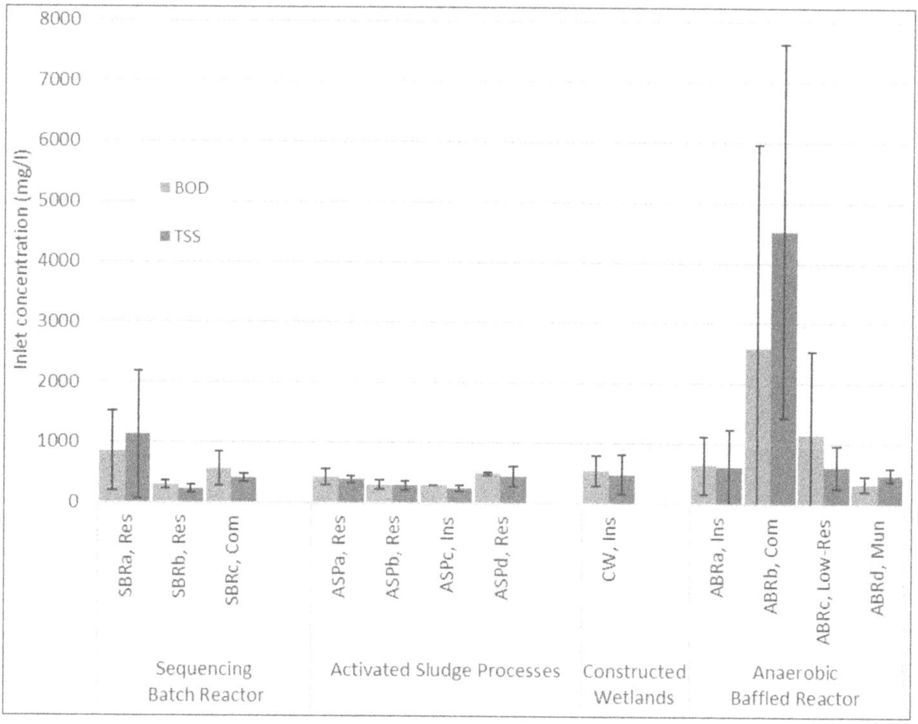

Figure 3.7. Average BOD and TSS influent concentration of small-scale sanitation systems in India. Error bars indicate standard deviations. Approximate values taken from Klinger *et al.*, 2020.

Res: Sanitation systems installed in middle-income or high-income residential areas, usually conformed by multistoried buildings.

Ins/Com: Systems installed in institutions such as schools, hospitals or offices as well as commercial centres, hotels and restaurants.

Low-Res: Sanitation systems installed in formal and informal low-income settlements.

Mun: Refers to systems treating a mixture of wastewater coming from households, restaurants and small businesses which are connected to the treatment plant. These systems are managed by municipalities.

SBRa: Primary sedimentation tank – Sequencing Batch Reactor – Secondary sedimentation Tank–Chlorination.

SBRb: Sequencing Batch Reactor – Activated Carbon Filter – Chlorination.

SBRc: Sequencing Batch Reactor – Secondary Sedimentation Tank – Ultrafiltration – Reverse Osmosis.

ASPa: Extended Aeration/Activated Sludge (suspended growth) Processes – Secondary Sedimentation Tank – Pressure sand filter – Activated Carbon Filter – Ultrafiltration.

ASPb: Extended Aeration/Activated Sludge (suspended growth) Processes – Secondary Sedimentation Tank – Pressure sand filter – Activated Carbon Filter – Chlorination.

ASPc, ASPd: Extended Aeration/Activated Sludge (suspended growth) Processes – Secondary Sedimentation Tank – Pressure sand filter – Activated Carbon Filter.

CW: Primary sedimentation tank – Horizontal Flow Constructed Wetland.

ABRa: Primary sedimentation tank – Anaerobic Baffled Reactor – Horizontal Flow Constructed Wetland – Polishing pond.

ABRb: Primary sedimentation tank – Anaerobic Baffled Reactor – Horizontal Flow Constructed Wetland.

ABRc, ABRd: Primary sedimentation tank – Anaerobic Baffled Reactor – Anaerobic Filter – Horizontal Flow Constructed Wetland.

correct O&M, etc., also influence the treatment performance of the different technologies, the equivalence in removal rates of nature-based and mechanised systems presented by Klinger *et al.* (2020) might not be transferrable to other countries and should be carefully analysed in other contexts.

For several years countries of the Global South have tried to protect human health and the environment through stringent wastewater discharge standards (Table 3.8). However, discharge standards can often be inappropriate and unattainable (Schellenberg *et al.*, 2020) and can hinder the development of more sustainable alternatives that could mitigate challenges presented in growing cities. With the exception of some countries (Mexico, Brazil and Jordan for example) several countries of the Global South lack standards for the design of small-scale systems or for water reuse. When comparing among the countries, it is noticeable that the limits have a high variability and usually are more stringent than the EU Urban Wastewater Treatment Directive requirements. For example, India has very significant challenges with provision of sanitation and wastewater treatment but

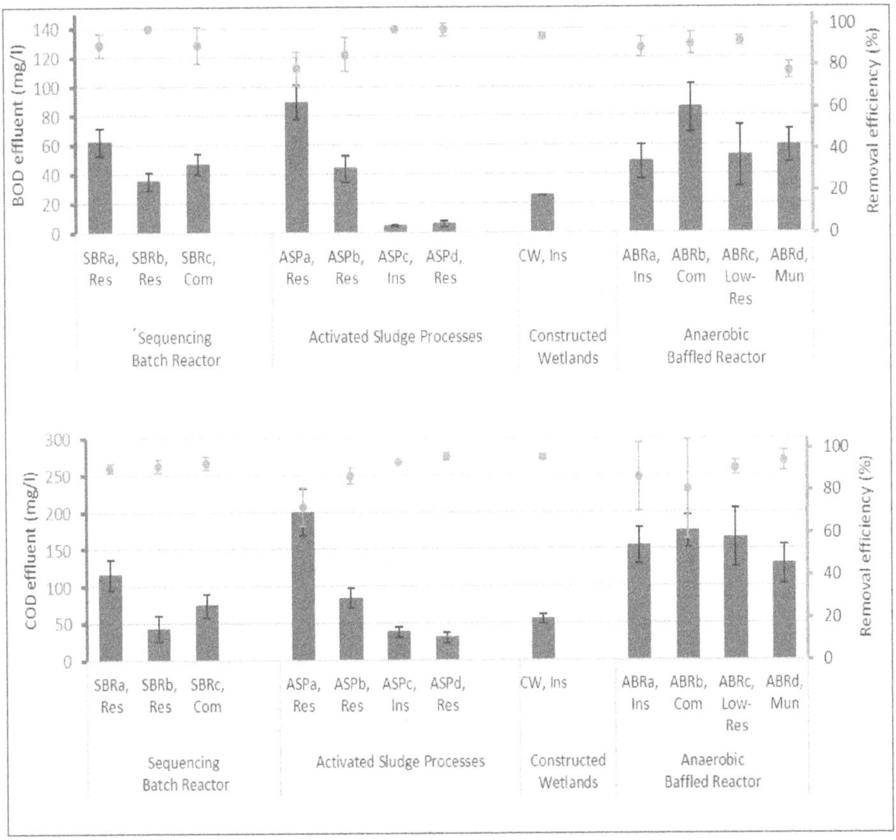

Figure 3.8. See next page for key to abbreviations used.

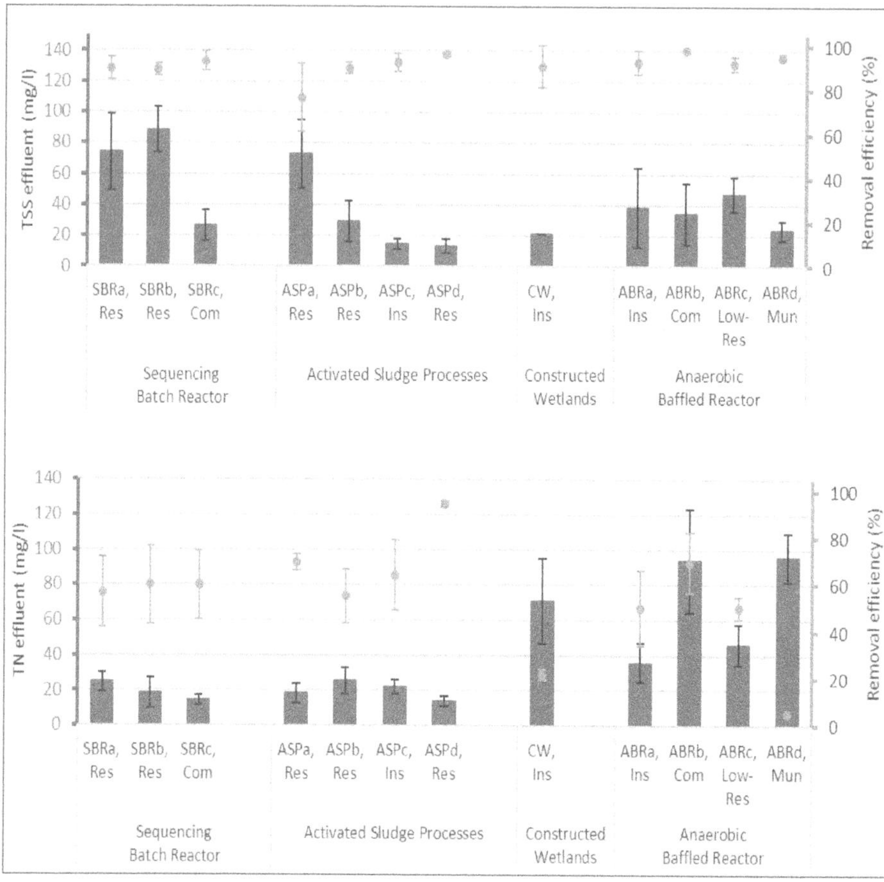

Figure 3.8 (Continued). BOD, COD, TSS and TN removal efficiency and treated water quality of conventional and nature-based technologies implemented at small-scale in India and Nepal. Error bars indicate standard deviation. Approximate values taken from Klinger *et al.*, 2020. See Fig. 3.7 for explanation of abbreviations.

its discharge standards are strict in comparison to other countries of the Global South (and indeed Global North). Schellenberg *et al.* (2020) analysed the causes and consequences of such standards and concluded that resilient, economically and environmentally sustainable technologies should be included in future standards to better ensure enforceable, achievable and realistic discharge limits in line with what can be sustainably achieved.

Where individual nature-based treatment systems are incapable of meeting local discharge standards a combination of treatment modules is often required – which in turn can imply higher capital, operation and maintenance costs that directly or indirectly promote the reuse and discharge of untreated water (Kramer *et al.*, 2003). In Tanzania for example, the Ministry of Water analysed the approximate

Table 3.8. Discharge standards of different countries in the Global South and Germany (as an EU example).

Countries	Discharge to	pH	BOD_5 (mg/l)	COD (mg/l)	TSS (mg/l)	TDS (mg/l)	NH_4-N (mg/l)	PO_4-P (mg/l)	References
European Union	Receiving waters for systems > 2000 PE	–	25 mg/l or 70–90% reduction	125 mg/l or 75%	35 mg/l or 90% reduction	–	–	–	Council of the European Community, 1991
Germany	Water bodies[a]	–	40	150	–	–	–	–	Bundesminister für Umwelt Naturschutz und Reaktorsicherheit, 2004
Brasil	Water bodies[b]	6 to 9	120	330	–	–	20	–	Colares *et al.*, 2019
Cambodia	Public water area and sewer	5 to 9	80	100	80	2000	7	2	Royal Government of Cambodia, 2009
India 2019	Water bodies	6.5 to 9	10	50	20				Schellenberg *et al.*, 2020
Indonesia	Water bodies	5 to 9	100	–	100	–	–	–	State Minister of Environment, 2003
Jordan	Irrigation Class A: Cooked vegetables, parks, playgrounds and sides of roads within city limits	6 to 9	30	100	50	–	–	–	Nivala *et al.*, 2019
	Irrigation Class B: Fruit trees, sides of roads outside city limits, and landscape.	6 to 9	200	500	150	–	–	–	

(*Continued*)

Table 3.8. Continued

Countries	Discharge to	pH	BOD_5 (mg/l)	COD (mg/l)	TSS (mg/l)	TDS (mg/l)	NH_4-N (mg/l)	PO_4-P (mg/l)	References
	Irrigation Class C: Field crops, industrial crops, and forest trees.	6 to 9	300	500	–	–	–	–	
Lao	Wastewater discharge standard from urban area	6 to 9.5	50	150	–	2000	40	–	Lao People's Democratic Republic, 1999
Philippines	Inland waters	6.5 to 9	50	100	70	–	–	–	WEPA, 1990
South Africa	Inland waters	5.5 to 9	–	75	–	–	4.5	3	Department of Water Affairs-South Africa, 2013
	Agricultural areas when discharge is <5000 m^3 ww/d	6 to 9	n.a.	400	–	–	–	–	
Tanzania	Water bodies	6 to 8.5	30	60	100	–	–	–	TBS, 2005
Vietnam	Water bodies not used for domestic water supply	5 to 9	50	–	100	1000	10	10	Minister of Natural resources and Environment, 2008

[a]For small systems treating less than 60 kg BOD/d.
[b]For systems treating less than 200 m^3 of wastewater per day, discharge standards based on the treatment capacity of the wastewater treatment plant.

CAPEX and OPEX of a combination of different small scale wastewater treatment systems (Table 3.9). According to the analysis, primary treatment, ABR, advanced tertiary treatment and UV would be the only combination able to meet their national discharge standards (30 mg/l BOD_5, 60 mg/l COD, 100 mg/l TSS (TBS, 2005)). By comparing System 3 (primary treatment, ABR, advanced tertiary treatment and UV) to System 2 (primary treatment, Combined ABR and CW), it is observed that to increase the removal efficiency of COD by only 3 percentage points implies a doubling of the CAPEX and OPEX. Delivering sanitation that is sustainable in its truest sense and at a meaningful pace will require flexible approaches to discharge standards – both in terms of developing and also regulating any limits.

Table 3.9. Effluent characteristics, capital and operation costs of small-scale systems in Tanzania. Values in parenthesis correspond to removal efficiencies (%). Modified from MoW, 2018.

	Inlet	Primary Treatment, ABR	Primary Treatment, Combined ABR and CW	Primary Treatment, Combined ABR, Advanced Tertiary Treatment and UV*
TSS (mg/l)	350	250 (28.6)	30 (91.4)	25 (92.8)
COD (mg/l)	1400	350 (82)	100 (93)	60 (96)
PO_4-P (mg/l)	15	15 (0)	10 (33.3)	6 (60)
NH_4-N (mg/l)	20	87 (–)	70 (–)	10 (50)
Faecal coliforms (counts/100 ml)	10^{12}	10^7	10^5	10^3
CAPEX (USD/cap)		20	80–120	150–200
OPEX (USD/cap/year)		5	5–10	8–15

*UV: Ultraviolet.

3.4 Nature-based Solutions in the Context of Resource Recovery and Reuse: Some Examples

The current global outlook in terms of water availability is characterised by a growing demand for food and growing pressure on freshwater resources (Andersson *et al.*, 2016a). According to Mateo-Sagasta *et al.* (2015) the nitrogen contained in the 900 million m^3 of wastewater produced every day in the world would be enough to replace 25% of the nitrogen currently used in agriculture, while the volume would be sufficient to irrigate 15% of the farmland in the world. Compared with energy-intensive and centralised wastewater plants, nature-based solutions can be installed at small-scale, supporting local water and nutrient recovery initiatives. Despite barriers such as lack of regulations for wastewater reuse in several countries of the Global South (CONPES, 2020), whenever possible, nature-based solutions are designed to allow reuse of treated wastewater for landscape or crop irrigation (Table 3.10).

Table 3.10. Examples of wastewater reuse in countries of the Global South.

Countries	Technology	Types of Wastewater	System Capacity	Comments	References
Morocco	HFCW	Municipal	10 m^3/d	Reuse of water for irrigation of a garden	Global Wetland Technology, 2020
Lima, Peru	VFCW	Greywater	2000 Population Equivalent (PE)	Irrigation of green public areas	Global Wetland Technology, 2020
Miraflores, Peru	CW	Grey and blackwater	70 PE	Irrigation of green public areas	Global Wetland Technology, 2020
Peru	VFCW	Greywater	160 PE	Reuse for toilet flushing and irrigation in green areas	Global Wetland Technology, 2020
Peru	CW	Domestic	55 PE	Irrigation for agricultural area	Global Wetland Technology, 2020
India	Vermifiltration	Domestic	1.5 m^3/d	Irrigation of a community garden that will provide fruit and vegetables to a nearby school	INNOQUA, 2018
Tanzania	Vermifiltration	Domestic	1.5 m^3/d	Irrigation of a banana plantation located near the treatment plant	INNOQUA, 2018

The limited removal of nutrients that can be characteristic of nature-based wastewater treatment solutions can be presented as an opportunity rather than a challenge in the Global South if water reuse is intended. However, the desirability of nutrient-rich effluents from wastewater treatment can sometimes impact negatively on the performance of those treatment systems. In Moshi, Tanzania, a constructed wetland has been connected to a WSP (consisting of an anaerobic pond, two facultative ponds arranged in parallel and six maturation ponds arranged in series) to partially treat the effluent of the second maturation pond.

While the effluent of the WSP is used for irrigating ~113,000 m^2 of paddy fields and sometimes other crops such as tomatoes, spinach and amaranths, the WSP-CW effluent is directed to a fish pond and a pilot paddy field (8,094 m^2) located downstream of the treatment system and owned by Moshi Urban Water Supply and Sewerage Authority (MUWSA) (Kihila *et al.*, 2014; Kimwaga *et al.*, 2013). According to Kihila *et al.*, 2014 in periods of high demand (planting and growing season) for irrigation water, farmers normally block the outlet of the wetland to prevent water to flow to the fish pond. The accumulated and partially treated wastewater is then diverted to the paddy fields in order to have more available water for irrigation. This practice impacts the performance of the constructed wetland reducing its removal efficiency for parameters such as faecal coliforms, whose high levels increase the health risk for farmers and final consumers of the irrigated crops. Nevertheless, due to the reliability of treated effluent flows, farmers close to the treatment plant not only avoid the use of artificial fertilisers but also grow twice as much as local farmers who practice rain-fed agriculture (Kihila *et al.*, 2014). Similarly, in India, the use of untreated or partially treated wastewater for rice and vegetable production is a common practice in urban and peri-urban areas (Starkl *et al.*, 2015). High level decision makers, governmental stakeholders and researchers are aware of the health problems this practice might generate to farmers and end-consumers – and increasingly acknowledge the potential for nature-based wastewater treatment to deliver water suitable for irrigation (Starkl *et al.*, 2015).

3.5 Comparison of Technologies, Cost, Land Requirement, Energy Consumption and GHG

Usually the selection of wastewater treatment technologies is limited to assessment of construction, operation and maintenance costs. Nevertheless, there is an increasing need to transform the paradigm from choosing the most economical technology to choosing the most sustainable one. A sustainable wastewater treatment technology can be defined as economically affordable, protective of the environment, technically and institutionally consistent and socially acceptable (Capodaglio *et al.*, 2017; Molinos-Senante *et al.*, 2014). This section presents a comparison of direct and indirect environmental impacts and economic costs of conventional wastewater treatment systems (conventional activated sludge) and natural wastewater treatment systems (constructed wetlands, WSPs and ABR systems). It is important to highlight that even though the data that are presented in the following sections draw from various scenarios, the overall results may inform more holistic comparisons between nature-based and conventional wastewater treatment systems in the Global South.

3.5.1 Capital, Operation and Maintenance Costs

The overall cost of a sewage treatment plant depends on numerous factors, including time frame, location, regulations, technology, size of population served, expertise required and climatic conditions (Sato *et al.*, 2007). Table 3.11 summarises CAPEX and OPEX of several conventional and nature-based treatment systems in selected countries of the Global South. In India, the CAPEX and O&M costs of the Activated Sludge Process (ASP) are almost 5 and 11 times, respectively, higher than of WSP for the same volume treated. Sato *et al.* (2007) report annual O&M costs of ~47USD per m^3/d for ASP treatment, but ~4USD per m^3/d for WSP treatment (Sato *et al.*, 2007). In other countries the cost of construction of natural treatment systems such as vermifilters (35–60 USD per capita in Chile (Sinha *et al.*, 2014)) or constructed wetlands (133–373 USD/m^3 in China (Liu *et al.*, 2009)) compares favourably to costs for an ASP (150–300 USD per capita in Chile (Sinha *et al.*, 2014)) or to a 'conventional' wastewater treatment system (200–533 USD/m^3 in China (Liu *et al.*, 2009)). However, when different nature-based systems are combined to deliver treated wastewater compliant with local discharge standards, capital costs may be significantly higher than for a conventional system delivering equivalent levels of treatment due to their higher land area requirement – although annual O&M are likely to be much lower in the nature-based systems as a result of low energy requirements and simple, low-cost maintenance activities such as harvesting of vegetation and insect/pest control (Singh *et al.* (2019) state O&M of 4.4 USD/m^3 per year for a combined settlement, ABR and CW system treating 35 m^3 per day).

Land use and land prices are among the limiting factors for the implementation of wastewater treatment technologies and in particular nature-based wastewater treatment systems. Nature-based systems have higher unit land use[2] (1–20 m^2 per capita) than conventional treatment systems (0.11–2.50 m^2 per capita) (Su *et al.*, 2019). Thus their applicability might be limited in high density urban areas, normally characterised by high land prices. For example, Suriyachan *et al.* (2012) highlight land prices in central Bangkok (Thailand) of up to 1250USD per m^2, while land on the city fringe might cost as little as 125 USD per m^2. In India the price of land in rural areas is assumed to be around 2USD per m^2, while in towns surrounding Delhi prices are upwards of 10 times this price (Sato *et al.*, 2007). Hence the land for a 1000 m^3/day WSP (with a land requirement of 30 m^2/m^3/d) might cost from 60,000 to 600,000 USD depending on whether it was installed in a rural or urban location in India.

2. Unit land use is the ratio of wastewater treatment plant area to number of inhabitants served by the system.

Table 3.11. Investment, operation and maintenance costs of conventional and nature-based wastewater treatment systems in countries of the Global South. O&M = Operation and Maintenance; WSP = Waste Stabilization Pond; ASP = Activated Sludge Process; ABR = Anaerobic Baffled Reactor; CW = Constructed Wetland.

Technology	Treatment Volumes	Units	Capital Cost	Units	Annual O&M Cost	Units	Countries	Land Price Included?	References
WSP	2150	m^3/d	40.7[a]	USD/m^3/d	4.3[b]	USD/m^3/d	India	No	Sato *et al.*, 2007
ASP	2150	m^3/d	186	USD/m^3/d	47	USD/m^3/d	India	n.s.	Tare *et al.* 2003 (see Sato *et al.*, 2007)
Vermifilter	>1000	m^3/d	172–210	USD/m^3/d[c]	14.6[d]	USD/m^3/d	India	No	Cshankar, 2019
Settler+ABR+ CW	35	m^3/d	730[c]	USD/m^3/d	4.5	USD/m^3/d	India	n.s.	Singh *et al.*, 2019
Conventional wastewater treatment systems	–	–	200–533[e]	USD/m^3	34–73[e]	USD/m^3	China	No	Liu *et al.*, 2009
CW	–	–	133–373[e]	USD/m^3	2.4–9.7[e]	USD/m^3	China	No	Liu *et al.*, 2009
ASP	–	–	150–300	USD/person	43.8	USD/m^3	Chile	No	Sinha *et al.*, 2014
Vermifilter	–	–	35–60	USD/person	18.3	USD/m^3	Chile	No	Sinha *et al.*, 2014
Settler+Combined ABR	200	households	240	USD/household	–	–	Bangkok	n.s.	Kerstens *et al.*, 2012
	100	households	460–490	USD/household	–	–	Bangkok	n.s.	Kerstens *et al.*, 2012

[a] $y = 474x^{-0.32}$ where y is the capital cost per unit volume (USD/m^3/d) and x the treatment volume (m^3/d).
[b] $y = 995x^{-0.71}$ where y is the annual O&M cost per unit volume (USD/m^3/d) and x the treatment volume (m^3/d).
[c] 1USD = 75.5 INR (July 2020).
[d] Assumes two operators.
[e] 1US = 7.5 RMB (December 2007).
n.s. = Not stated.

3.5.2 Direct and Indirect Environmental Impacts of Conventional and Nature-based Wastewater Treatment Systems Based on Life Cycle Assessment

Any comparison of natural wastewater treatment systems with conventional wastewater treatment systems in countries of the Global South should not be limited only to treatment performance, compliance with the environmental regulatory standards, and CAPEX/OPEX/O&M. The use of holistic comparison factors such as geographical locations, socioeconomic conditions and global environmental impacts have also gained attention recently. In this view, Life Cycle Assessment (LCA) – one of the most widely used tools to assess defined environmental impacts related to a product, service or process – has also been applied to evaluate the environmental response and improvement opportunities of diverse wastewater treatment typologies (Corominas *et al.*, 2013). Nevertheless, its application to domestic wastewater treatment with nature-based solutions in the Global South is still insignificant in comparison to countries of the Global North (Gallego-Schmid and Tarpani, 2019). Life cycle assessments can use different system boundaries and impact categories as well as different functional units (such as PE, cubic metre(s) of treated wastewater, number of households, etc.), and their interpretation is often restricted by limitations in the quality of inventoried processes and their associated datasets, which are particularly scarce for low-income countries (Gallego-Schmid and Tarpani, 2019). Thus, some caution must be taken when comparing sanitation systems on the basis of LCAs, especially in the Global South.

Comparison between conventional, and usually centralised, wastewater treatment systems and nature-based sanitation solutions with an LCA approach is challenging. For example, it is observed that the majority of impacts from conventional wastewater treatment plants occur during the operation phase (accounting from 85% to 97% of the total value in all impact indicators evaluated), while for natural wastewater treatment systems the construction phase seems to be the most significant element (Garfí *et al.*, 2017; Machado *et al.*, 2006). Consequently, any LCA comparison that excludes the construction phase (Corominas *et al.*, 2013), would penalise nature-based systems when compared with centralised approaches. Nevertheless, such comparative studies can be useful for indicative purposes, and help to promote a more representative and appropriate use of LCA in the sanitation sector of the Global South; a number are summarised in Table 3.12.

Conventional wastewater treatment systems such as activated sludge processes are high energy consumers during operation and are thus relatively large emitters of greenhouse gases (Machado *et al.*, 2006; Su *et al.*, 2019). Between 57% and 84% of the Global Warming Potential (GWP) of an ASP can result from the operation phase (Su *et al.*, 2019). These facilities are also relatively large indirect emitters of greenhouse gases, given the high demand of external electricity to drive their

Table 3.12. Impact categories for some conventional and nature-based wastewater treatment systems in the Global South. PE = Population Equivalent.

	Goals and Scope				Impact Categories										
Technologies*	**Phases Included****	**Boundaries*****	**Functional Units**	**Impact Assessment Methods**	**GWP**	**AP**	**EP**	**ODP**	**ADP**	**FAETP**	**HTP**	**MAETP**	**TETP**	**References**
WSP	Op	(1)(2) (+)(ST)	100 m^3/d	CML baseline 2000	less than 1E+05[a,b]	1.2E+03	1.2E+04	2.2E−01	1.2E+03	1.1E+04	6.8E+04	3.3E+07	3.8E+02	Su *et al.*, 2019
CW	Op	(1)(2)	100 m^3/d	CML baseline 2000	−2E+05[a,b]	1.2E+03	1.2E+04	2.2E−01	1.2E+03	1.2E+04	7.2E+04	3.5E+07	4.3E+02	
CW (VFCW + HFCW) Filter material: gravel	Op, Cons, Dem	(1)(2) (ST)(SD)	1 PE, 10 years	CML 2000	−2.9E+04[c]	31.1	3.4E+03	1.7E−03	48.2	–	–	–	–	Machado *et al.*, 2007
AS	Op, Cons, Dem	(1)(2) (ST)(SD)	1 PE, 10 years	CML 2000	4.0E+04[c]	434	3.4E+03	3.0E−03	323	–	–	–	–	
CW (VFCW + HFCW) Filter material: lightweight expanded clay aggregate	Op, Cons	(1)(2)	1 PE, 15 years	ReCiPe (Midpoint)	0.1[a,c,d]	–	–	–	–	–	–	–	–	Lopsik, 2013
AS	Op, Cons	(1)(2)	1 PE, 15 years	ReCiPe (Midpoint)	0.04[a,c,d]	–	–	–	–	–	–	–	–	

(*Continued*)

Table 3.12. Continued

	Goals and scope				Impact categories									
Technologies*	Phases included**	Boundaries***	Functional units	Impact Assessment methods	GWP	AP	EP	ODP	ADP	FAETP	HTP	MAETP	TETP	References
Settler + ABR + CW	Op, Cons	(1)(2)	1 PE, 20 years	CML 2000	164.3[c,e]	0.6	1.5	5.7E-06	–	–	–	–	–	Sapkota, 2016
Settler + CW + Collection Tank	Op, Cons	(1)(2)	1 PE, 20 years	CML 2000	52.7[c,e]	0.17	0.02	2.0E-6	–	–	–	–	–	
Settler + CW	Op, Cons	(1)(2)	1 PE, 20 years	CML 2000	127.3[c,e]	0.4	0.05	4.5E-6	–	–	–	–	–	

[a]Approximation.
[b]Software GaBi.
[c]Software SimaPro.
[d]Only normalised values are reported. It is also assumed that climate change equals to global warming potential. Other impact categories were not included in this table due to possible differences in the units of measurement in regard to other studies.
[e]Global Warming Potential 100.

GWP:Global Warming Potential (kg CO_{2-}eq).
AP:Acidification Potential (kg SO_2-eq).
EP: Eutrophication Potential (kg $PO_4{}^{3-}$-eq).
ODP: Ozone Layer Depletion Potential (kg R-11-eq).
ADP: Abiotic Depletion Potential (kg Sb-eq).
FAETP: Freshwater Aquatic Ecotoxicity Potential (kg DCB-eq).
HTP: Human Toxicity Potential (kg DCB-eq).
MAETP: Marine Aquatic Ecotoxicity Potential (kg DCB-eq).
POCP: Photochem Ozone Creation Potential (kg C_2H_4-eq).
TETP: Terrestric Ecotoxicity Potential (kg DCB-eq).

*CW: Constructed wetlands; WSP: Waste stabilisation ponds; VFCW: Vertical flow constructed wetland; HFCW: Horizontal flow constructed wetland; AS: Activated sludge.
**Op: operation; Cons: construction; Dem: demolition.
***(1): primary treatment; (2): secondary treatment; (3): tertiary treatment, (ST): sludge treatment; (SD): sludge disposal; (+): advanced treatment.

operation (Garfí *et al.*, 2017). Furthermore, as electricity consumption per cubic meter of treated wastewater increases in inverse proportion to the size of the conventional wastewater treatment plant (Lorenzo-Toja *et al.*, 2015), the smaller the size of the community, district or location, the lower the environmental performance of such systems and therefore the more appropriate natural wastewater treatment systems become (Garfí *et al.*, 2017). By contrast, natural wastewater treatment systems such as CW can have a CO_2 fixation capacity, meaning that their contribution to GWP during operation can be negative (depending on the type of filter material used (Lopsik, 2013)) and whether N_2O and CH_4 emissions are properly considered in environmental budgets for CW units (Fuchs *et al.*, 2011).

On the other hand, Lopsik (2013) determined that activated sludge processes have a better environmental performance than CW if filter materials which require large energy inputs for their production (such as lightweight expanded clay aggregate (LECA)), are used. However, if LECA is replaced by sand and gravel, the environmental impacts can be reduced by around 16% depending on the impact assessment method applied. Furthermore, by using sand and gravel, the Global Warming Potential of Constructed Wetlands can become negative as CO_2 uptake by the plants is greater than the CO_2 emissions during the construction and operation phases.

The selection of construction materials will inevitably have an impact in both conventional and nature-based systems (Pandey, 2015; Sapkota, 2016), and must also be considered carefully – both in terms of material type and quantity used. For example, reduced use of construction materials in CW systems can result in a GWP that is up to five times less than for an equivalently sized ASP system (Garfí *et al.*, 2017; Machado *et al.*, 2006). Overall, such comparisons should be treated with caution – it is essential that system boundaries are set appropriately and that high-quality LCA inventories are used. Developments outside the sanitation sector can also have a huge impact on GWP within it – for example, if aeration in ASP systems is powered by renewable energy, then the overall lifetime impact of the process will reduce very significantly, and possibly to the disadvantage of nature-based alternatives. Objectivity is essential.

WSPs have also been subjected to comparative LCA. In such studies it has been observed that greenhouse gas production as methane (CH_4) gas and indirect emissions of nitrous oxide (N_2O) are usually higher than their CO_2 sequestration capacity, making WSP and small facultative lagoons contributors to GWP impact over their lifetime. WSP are also associated with a lack of nutrient removal; consequently, eutrophication potential (EP) is the other relevant operational impact attributed and evaluated for these natural remediation systems (Godin *et al.*, 2012; Machado *et al.*, 2007).

It must also be noted that LCA studies on wastewater treatment often consider only a limited number of impact categories (particularly GWP and EP)

(Gallego-Schmid and Tarpani, 2019). LCA studies for nature-based wastewater treatments, at small-scale, follow the same trend, and the scarce available studies show a deficiency in assessment of factors such as land use (Fuchs *et al.*, 2011; Singh *et al.*, 2019). Since these technologies are extensive land users, a major effort to appraise this aspect must be imperative, in order to avoid the overestimation of their environmental advantages. Other aspects, such as the capacity of different wastewater technologies to deal with different types of emergent pollutants, pharmaceuticals and pathogens must also be considered to develop a detailed picture of toxicity impacts or other harms (Parra-Saldivar *et al.*, 2020).

Finally, while the vision for decentralised wastewater treatment claims a major potential for water and nutrient reuse on-site or in their immediate neighbourhood, the number of LCA studies considering water reuse or reclamation of urban or domestic effluents are relatively few and dominated by studies from the Global North (Arias *et al.*, 2020; Pasqualino *et al.*, 2011; Singh *et al.*, 2019).

3.6 Discussion and the Way Forward

Growth of cities in the Global South is currently overwhelming the pace at which centralised wastewater treatment systems can be extended or implemented (Reymond *et al.*, 2020). Conventional wastewater treatment systems are unable, as the only standard solution, to close the sanitation gap and provide sustainable sanitation for the underserved communities of the Global South. Government and institutions responsible for the construction, operation, maintenance and expansion of wastewater treatment systems usually do not have the resources to manage wastewater adequately, and rely on donor agencies to finance required infrastructure. In many cases these agencies may inherently support and often follow the conventional 'networked city' characterised by energy-intensive, centralised treat and dispose technologies.

The limited implementation of nature-based wastewater treatment systems can be related, in part, to the emphasis of aid programmes towards technologies with commercial spin off for the donors and the 'translocation of northern designs' to the realities and cultures of the Global South (Denny, 1997). Interviews with representatives of multilateral agencies and a deeper analysis of their development programmes on wastewater management would be necessary to analyse and understand preferences towards imported technologies and Global North designs for wastewater treatment. Nonetheless, an initial analysis of approved loans for sanitation programmes in some countries of the Global South seems to indicate a trend towards support of more technological, centralised approaches (Appendix 1).

Since the design, operation, construction and maintenance of on-site wastewater treatment systems is scattered among several actors who are not necessarily aligned with the interests of local governments and institutions, local authorities and water utilities may prefer conventional wastewater treatment systems because their management is distributed among larger bodies that are easier to regulate, monitor and influence. Nevertheless, the number of on-site sanitation systems keeps growing, and regulations covering the design and operation of nature-based solutions (particularly de-sludging of septic tanks) are increasingly common (Mehta *et al.*, 2019).

Nature-based systems can be low cost, with simple operation and construction, and are part of the solution for provision of sanitation in urban and rural areas of the Global South. Septic tanks are a key component of city-wide inclusive sanitation which together with appropriate faecal sludge management can complement conventional centralised wastewater treatment systems. However, there are large disparities between septic tank designs and their proper implementation in the field. For instance, design guidelines are not followed and in several cities septic tanks are constructed as underground pits that are only partially lined or not lined. Their effluents are a significant environmental problem, especially where infiltration into soil contributes to the pollution of underground freshwater reservoirs – potentially hindering the uptake of other 'low-tech' (but robust) systems in the Global South.

De-sludging and safe faecal sludge management are critical factors for the proper functioning of several on-site systems, and for the protection of the environment and public health. Nevertheless, at the household level, unaffordable emptying fees or non-existent emptying services hinder the proper management of faecal sludge from such systems. Inadequate communication or user training can also be an issue. Householders may not be sufficiently informed regarding the functioning of septic tanks even if they have resources for system maintenance and access to appropriate service providers.

Treatment performance is also an important issue. Nature-based solutions such as vermifiltration are able to reach removal efficiencies of more than 80% for organic matter and solids – removal rates that are comparable to conventional activated sludge technologies. Nevertheless, as with conventional systems, the performance of nature-based processes can be affected by operational factors such as overloading or underloading. Hence, WSPs, which are widely deployed in the Global South, often have HRTs different from their design, leading to effluents with high concentrations of organic matter and solids. Failure to account for the ecosystem services offered by WSPs, the introduction of (in some cases) inappropriately stringent effluent standards, and high land prices, are all leading to the replacement of WSP systems with more advanced technologies which are also likely to fail unless loadings are correctly understood and managed, and the whole treatment system is adequately operated and maintained.

According to Wang *et al.* (2012), standards which aim to reduce the environmental consequences of effluents discharged into water bodies actually reduce the sustainability of wastewater treatment systems by increasing the number of required treatment modules with their associated construction and operational impacts. Sustainability gains must therefore be sought in aspects such as resource recovery. This can have its own consequences. For example, the removal of nutrients can often be a side effect of the treatment of organic pollutants rather than the treatment goal of nature-based systems (Klinger *et al.*, 2020). On the other hand, nutrient-rich effluent is a highly valuable irrigation resource (for landscape and agriculture) in water-scarce areas, or where local reuse opportunities are available – provided that pathogen removal can be ensured (Colares *et al.*, 2019). Such challenges highlight the importance of local context and the necessity to consider wider system impacts before selecting between conventional and nature-based sanitation solutions. Indeed, in many situations it is likely that both will be needed to close the sanitation gap.

Despite the numerous market barriers, nature-based technologies such as constructed wetlands, anaerobic baffled reactors and vermifilters have all found their niches in the treatment of domestic or municipal wastewater within small communities, residential buildings and peri-urban areas – and countries such as India and Indonesia have institutionalised the implementation of small-scale wastewater treatment systems for larger buildings (Reymond *et al.*, 2020) and sanitation centres (GoI, 2014). In México, Chile and Venezuela, several nature-based solutions have been developed and commercialised in the absence of regulations for discharge standards from small-scale wastewater treatment systems.

In order to ensure further penetration of these systems as sustainable options for sanitation coverage in the Global South, it is necessary to create appropriate governance arrangements that prevent the malfunctioning or non-functioning of nature-based systems and the resulting discharge of untreated or poorly treated wastewater. The creation of incentives for the implementation of sustainable wastewater treatment systems and the establishment of appropriate discharge standards for water reuse and, where reuse is not possible, for discharges from small systems, should be considered. Furthermore, cooperation and exchanges among practitioners in the Global South and technology providers (usually from the Global North) should be strengthened. This information exchange will lead to solutions that are better adapted to local contexts and will help the uptake of nature-based solutions as complementary alternatives to conventional wastewater treatment systems in both the Global South and North.

This chapter has been peer-reviewed by: Dr. Eoghan Clifford, NUI Galway; Dr. David Tompkins, Aqua Enviro.

Conflict of Interest: The authors declare that the research was conducted in the absence of any commercial or financial relationships that could be construed as a potential conflict of interest.

Appendix 1. Sanitation Projects Financed by International Developments Banks in Countries of the Global South

Countries	Names of the Project	Year of Approval	Objectives	Financing	References
Ecuador	Guayaquil Wastewater Management	2016	To provide infrastructure to treat 100% of domestic wastewater in the southern and northeastern of the city of Guayaquil through 35,000 new intra-household sanitation connections the construction of two wastewater treatment plants (chemical enhance primary treatment followed by aerated stabilisation ponds).	Total project cost: US $461 millionTotal financing (IBRD): US $336.1 million Wastewater treatment: US $378.3 million	World Bank Group, 2019
Kenya	Kenya – Nairobi Metropolitan Services Improvement Project	2012	To provide large-scale metropolitan infrastructure in the areas of solid waste, transport and sewerage services. It includes construction and rehabilitation of truck sewer lines, sewerage treatment plants (waste stabilisation ponds of 10,500 m^3/d and 6,522 m^3/d, 20,736 m^3/day).	Total project cost: US $330 million Total financing (IDA): US $300 million Wastewater collection systems and treatment: US $53 million	World Bank Group, 2020a,b; AWSB, 2015; NCWSC, 2019
Vietnam	Danang Sustainable City Development Project (SCDP)	2013	To expand access of city residents to public services and transport in selected areas of Da Nang city. Construction of sewer line and new wastewater treatment unit of 40,000 m^3/day and 20,000 m^3/day using sequencing batch reactor (SBR) technology.	Total project cost: US $272.20 million Wastewater drainage system: US $53 million	World Bank Group, 2020a

(Continued)

Continued

Countries	Names of the Project	Year of Approval	Objectives	Financing	References
Palestine	Hebron Regional Wastewater Management Project – Phase 1	2015	To reduce the environmental pollution from wastewater produced in the municipality. It includes the construction of a wastewater treatment plant of 15,000 m^3/d and facilities for the reuse of treated effluents and biosolids.	US $61.65 million	World Bank and Agence Francaise de Developpement (afd), 2014
Vietnam	Can Tho drainage and wastewater treatment project*	From 2003 to 2014	To collect and treat wastewater in Ninh Kieu and Cai Rang. To construct and improve interceptor sewers and construct a wastewater treatment (30.000 m^3/day). In 2015 the first phase of the project was finalised. A wastewater treatment plant that can treat 6,000 m^3/day (US $6.5 million) was constructed.	German Development Bank (KfW): €18.7 million	IAC Vietnam and SINH THAI CICE, 2015, (Vietnam Investement Review, 2020)
China	Ningbo Sustainable Urbanisation Project	2016	To improve the use of urban public space, improve urban mobility and reduce flood risk. It includes the construction of 377.2 km sewers and the reconstruction of 1.06 km sewers.	Flood risk management component: US 317,106.6	Zhejiang RenXin HuanKeYuan Co Ltd, 2019
Tanzania	Second Tanzania Water Sector Support Project	2017	To improve access to water supply and sanitation services in in Dar-es-Salaam. The third component of the programme will support the construction of a wastewater treatment plant with a capacity of 100,000 m^3/day and a truck sewer and network.	Total project cost: US $230 million Wastewater treatment plant and FSM: US $65 million	The United Republic of Tanzania – Ministry of Water, 2016

IBRD: International Bank for Reconstruction and Development, IDA: International Development Association.

Appendix 2. Municipal and Domestic Wastewater Characteristics for Different Countries

This Appendix has been adapted from Wolter, 2018.

Locations	WW Production (l/cap/d)[a]	COD (mg/l)	BOD_5 (mg/l)	TSS (mg/l)	TKN (mg/l)	TP (mg/l)	Type of Wastewater (Comments)	References
Industrialised countries								
Germany	169	821[c]	365	527	81[d]	13.8	Domestic	Henze *et al.*, 2008
US	200[b]	1,000[c]	445	445	82.2[d]	13.7	Domestic	UNEP and WHO, 1997, Henze *et al.*, 2008
Middle east and North Africa								
Egypt	310	329[e]	110	176	35.5	1.6	Domestic	Henze *et al.*, 2008; Tchobanoglous *et al.*, 2003
Iraq	117.5	281.3	145.5	74.7	91.5[f]	3.9[g]	Municipal (samples from January 2018 to July 2018)	Rahi *et al.*, 2020
Palestine[h]	21	4167[i]	2,381	2,952	74	7	Domestic	Tchobanoglous *et al.*, 2003
Jordan	90	1,830	770	900	150[d]	25	Domestic (Amman, Jordan)	Pescod, 1992
Iran	–	811	527	459	149	12	Municipal (composite samples from January 2003 to December 2003)	Miranzadeh, 2005
Eastern Europe and Central Asia								
Turkey	156	494	247	349	71	8	Domestic	FAO, 2012; Tchobanoglous *et al.*, 2003
Sub-Saharan Africa								
South Africa	86	907	443	–	105	14	Domestic	Reynaud, 2014
Uganda	68	2,037	904	706	162	7	Domestic	Tchobanoglous *et al.*, 2003
Zambia	–	–	940	662	72[j]	–	Municipal	UNEP – IETC, 2000
	74	851	–	–	–	–	Domestic (in-house water supply, cistern flush toilet)	Laramee *et al.*, 2018
	13	2,708	–	–	–	–	Domestic (public water tap and pour flush toilet)	Laramee *et al.*, 2018

(*Continued*)

Continued

Locations	WW Production (l/cap/d)[a]	COD (mg/l)	BOD_5 (mg/l)	TSS (mg/l)	TKN (mg/l)	TP (mg/l)	Type of Wastewater (Comments)	References
	17	4,766	–	–	–	–	Domestic (private yard tap and pour flush toilet)	Laramee *et al.*, 2018
South Asia								
India	31	1,879	1,097	–	161	23	Domestic (low-income areas of Bengaluru)	CPCB, 2009; Reynaud, 2014; Tchobanoglous *et al.*, 2003
Pakistan	170	200	130	455	35	7[g]	Domestic (grab samples every 10 days from January 2015 to December 2015)	Ali *et al.*, 2018
East Asia and Pacific								
Indonesia	82	902	366	–	76	8	Domestic	Reynaud, 2014
Thailand	213	3,81.2	217.8	–	54[d]	8.7	Municipal (urban and peri-urban areas of Bangkok)	Tsuzuki *et al.*, 2013
Latin America								
Argentina	331	352	152	195	45[d]	5.4	Municipal (152.400 dwellers in average)	Bachur and Ferrer, 2013
Brazil	122	882	504	504	90.2	6.6	Municipal	Campos and Von Sperling, 1996; Tchobanoglous *et al.*, 2003
Colombia	211	–	362	323	41[d]	5	Municipal	Arias and Brown, 2009; FAO, 2012

[a]Average values from Wolter, 2018.
[b]Average values (UNEP and WHO, 1997).
[c]COD/BOD = 2.25 (partially diluted wastewater (Henze *et al.*, 2008)).
[d]TN.
[e]COD/BOD = 3 (highly diluted wastewater (Henze *et al.*, 2008)).
[f]Only NH_4-N and NO_3.
[g]PO_4-P.
[h]West Bank and Gaza Strip,
[i]COD/BOD = 1.75 (concentrated wastewater (Henze *et al.*, 2008)).
[j]NH_4-N.

References

Afini, B. J. (1989). DBO per capita. *DAE*, *48*(156). http://revistadae.com.br/artigos/artigo_edicao_156_n_87.pdf

Al-Sàed, R. (2007). Sustainability of natural and mechanized aerated ponds for domestic and municipal wastewater treatment in palestine. *Water International, 32*(2), 310–324. https://doi.org/10.1080/02508060708692209

Alamgir, A., Khan, M. A., and Shaukat, S. S. (2016). Algal growth and waste stabilization ponds performance efficiency in a sub-tropical climate. *Pakistan Journal of Botany, 48*(1), 377–385.

Ali, M., Rousseau, D. P. L., and Ahmed, S. (2018). A full-scale comparison of two hybrid constructed wetlands treating domestic wastewater in Pakistan. *Journal of Environmental Management, 210*, 349–358. https://doi.org/10.1016/j.jenvman.2018.01.040

Ammary, B. Y. (2007). Wastewater reuse in Jordan: Present status and future plans. *Desalination, 211*(1–3), 164–176. https://doi.org/10.1016/j.desal.2006.02.091

Andersson, K., Rosemarin, A., Lamizana, B., Kvarnström, E., McConville, J., Seidu, R., Dickin, S., and Trimmer, C. (2016a). *Sanitation, Wastewater Management and Sustainability: from Waste Disposal to Resource Recovery* (C. Trimmer (ed.)). UNEP & SEI.

Andersson, K., Rosemarin, A., Lamizana, B., Kvarnström, E., McConville, J., Seidu, R., Dickin, S., and Trimmer, C. (2016b). *Sanitation, Wastewater Management and Sustainability: from Waste Disposal to Resource Recovery.*

Arias, M. E., and Brown, M. T. (2009). Feasibility of using constructed treatment wetlands for municipal wastewater treatment in the Bogotá Savannah, Colombia. *Ecological Engineering, 35*(7), 1070–1078. https://doi.org/10.1016/j.ecoleng.2009.03.017

Arias, A., Rama, M., González-García, S., Feijoo, G., and Moreira, M. T. (2020). Environmental analysis of servicing centralised and decentralised wastewater treatment for population living in neighbourhoods. *Journal of Water Process Engineering, 37*(April), 101469. https://doi.org/10.1016/j.jwpe.2020.101469

Arora, S., Rajpal, A., Bhargava, R., Pruthi, V., Bhatia, A., and Kazmi, A. A. (2014). Bioresource Technology Antibacterial and enzymatic activity of microbial community during wastewater treatment by pilot scale vermifiltration system. *BIORESOURCE TECHNOLOGY, 166*, 132–141. https://doi.org/10.1016/j.biortech.2014.05.041

AWSB. (2012). *Nairobi Metropolitan Services Improvement Project (NAMSIP): Construction of Sewerage System for Ruiru Municipality – Environmental Impact Assessment Report* (Issue November).

AWSB. (2015). *Environmental and Social Impact Assessment Study Report for the Proposed Sewerage System in Juja and Thika South* (Issue August).

Bachur, J., and Ferrer, V. (2013). Caracterización de líquidos cloacales y su relación con variales socioeconómicas de la población en la provincia de Santa Fé.

In *Congreso Interamericano de Ingeniería Sanitaria y Ambiental.* http://forum.susana.org/media/kunena/attachments/10250/CongresoAIDISBachur-Ferrer.pdf

Benetto, E., Nguyen, D., Lohmann, T., Schmitt, B., and Schosseler, P. (2009). Life cycle assessment of ecological sanitation system for small-scale wastewater treatment. *Science of the Total Environment, 407*(5), 1506–1516. https://doi.org/10.1016/j.scitotenv.2008.11.016

Bitton, G. (2005). *Wastewater Microbiology.* John Wiley & Sons.

Boano, F., Caruso, A., Costamagna, E., Ridolfi, L., Fiore, S., Demichelis, F., Galvão, A., Pisoeiro, J., Rizzo, A., and Masi, F. (2020). A review of nature-based solutions for greywater treatment: Applications, hydraulic design, and environmental benefits. *Science of the Total Environment, 711*, 134731. https://doi.org/10.1016/j.scitotenv.2019.134731

BORDA. (2017). *DEWATS Implementation by BORDA.*

BORDA. (2019a). *Monitoring & Evaluation Country Report 2018/2019: Tanzania. Unpublished Document.*

BORDA. (2019b). Monitoring & Evaluation Country Report 2018/2019: Sustainable performance assessment of BORDA DEWATS in Tanzania. In *BORDA e.V.*

Brix, H., Koottatep, T., Fryd, O., and Laugesen, C. H. (2011). The flower and the butterfly constructed wetland system at Koh Phi Phi-System design and lessons learned during implementation and operation. *Ecological Engineering, 37*(5), 729–735. https://doi.org/10.1016/j.ecoleng.2010.06.035

Bundesminister für Umwelt Naturschutz und Reaktorsicherheit. (2004). *Verordnung über Anforderungen an das Einleiten von Abwasser in Gewässer (Abwasserverordnung – AbwV)* (Issue L). http://www.gesetze-im-internet.de/abwv/index.html#BJNR056610997BJNE003008311%5Cnhttp://www.gesetze-im-internet.de/abwv/anhang_22_30.html

Cairns-Smith, S., Hill, H., and Nazarenko, E. (2014). Urban sanitation: Why a portfolio of solutions is needed. *The Boston Consulting Group, Boston, USA., December*, 1–30. https://www.bcg.com/documents/file178928.pdf%0Ahttps://www.bcgperspectives.com/Images/December_2014_Sanitation_WORKING_PAPER_FINAL.pdf

Campos, H. M., and Von Sperling, M. (1996). Estimation of domestic wastewater characteristics in a developing country based on socio-economic variables. *Water Science and Technology, 34*(3-4-4 pt 2), 71–77. https://doi.org/10.1016/0273-1223(96)00558-6

Capodaglio, A. G., Callegari, A., Cecconet, D., and Molognoni, D. (2017). Sustainability of decentralized wastewater treatment technologies. *Water Practice and Technology, 12*(2), 463–477. https://doi.org/10.2166/wpt.2017.055

Chandrakanth, G., Srimurali, M., and VivekVardhan, C. M. (2016). A study on domestic wastewater treatment by pilot-scale constructed wetlands. *International Journal of ChemTech Research*, *9*(6), 376–383.

Cheng, S., Zhao, M., Mang, H. P., Zhou, X., and Li, Z. (2017). Development and application of biogas project for domestic sewage treatment in rural China: Opportunities and challenges. *Journal of Water Sanitation and Hygiene for Development*, *7*(4), 576–588. https://doi.org/10.2166/washdev.2017.011

Chernicharo, C. A. L., van Lier, J. B., Noyola, A., and Bressani Ribeiro, T. (2015). Anaerobic sewage treatment: state of the art, constraints and challenges. *Reviews in Environmental Science and Biotechnology*, *14*(4), 649–679. https://doi.org/10.1007/s11157-015-9377-3

Choudhary, M. P., and Medok, P. (2017). Vermifiltration: An Efficient Method of Wastewater Treatment. *International Journal of Modern Trends in Engineering & Research*, *4*(10), 189–195. https://doi.org/10.21884/ijmter.2017.4326.5vvqk

Colares, G. S., da Silva, F. P., de Souza Celente, G., de Loreto, A. C., Lutterbeck, C. A., Machado, Ê. L., and Kist, L. T. (2019). Combined system for the treatment and reuse of urban wastewater: The efficiency of anaerobic reactors þ hybrid constructed wetlands þ ozonation. *Water Science and Technology*, *80*(2), 254–264. https://doi.org/10.2166/wst.2019.270

CONAMA. (2011). *Resolução N° 430, De 13 De Maio De 2011*. http://www.mma.gov.br/port/conama/legiabre.cfm?codlegi=646

CONPES. (2020). *CONPES 4004: Economía circular en la gestión de los servicios de agua potable y manejo de aguas residuales*.

Corominas, L., Foley, J., Guest, J. S., Hospido, A., Larsen, H. F., Morera, S., and Shaw, A. (2013). Life cycle assessment applied to wastewater treatment: State of the art. *Water Research*, *47*(15), 5480–5492. https://doi.org/10.1016/j.watres.2013.06.049

Coronel P., N. P. (2015). *Diseño e implementación a escala piloto de un biofiltro Tohá en la Espoch para la depuración de aguas residuales domésticas procedientes de la comunidad Lagos la Nube*. Escuela Superior Politécnica de Chimborazo.

Council of the European Community. (1991). Urban Wastewater Treatment Directive. *Https://Eur-Lex.Europa.Eu/Legal-Content/EN/TXT/?Uri=CELEX%3A01991L0271-20140101*. https://eur-lex.europa.eu/legal-content/EN/TXT/?uri=CELEX%3A01991L0271-20140101

CPCB. (2009). *Status of water supply, wastewater generation and treatment in class-i cities & class-ii towns of India*. https://doi.org/http://cpcb.nic.in/upload/NewItems/NewItem_153_Foreword.pdf

Cplantae. (2019). *Laboratory results WormPod*. https://www.facebook.com/CPlantae/Photos/a.129479047200716/1263270040488272.

Crites, R. W., Middlebrooks, E. J., Bastian, R. K., and Sherwood, C. R. (2014). *Natural Wastewater Treatment Systems* (2nd Editio). CRC Press. https://doi.org/https://doi.org/10.1201/b16637

CSE. (2020). *MOUNT*. Centre for Science & Environment. https://www.cseindia.org/mount/home

Cshankar. (2019). *Soil Bio Technology type vermifiltration systems developed at IIT Bombay*. Https://Forum.Susana.Org/205-Vermifilters-or-Vermi-Digesters-for-Wastewater-Treatment/23494-Soil-Bio-Technology-Type-Vermifiltration-Systems-Developed-at-Iit-Bombay?Start=12. https://forum.susana.org/205-vermifilters-or-vermi-digesters-for-wastewater-treatment/23494-soil-bio-technology-type-vermifiltration-systems-developed-at-iit-bombay?start=12

CTN. (1996). *Normas Técnicas de las descargas de aguas residuales a cuerpos receptores y alcantarillado sanitario. Acuerdo N° 058* (Issue December).

Da Silva, F. J. A., De Souza, R. O., De Castro, F. J. F., and Araújo, A. L. C. (2011). Prospectus of waste stabilization ponds in Ceará, Northeast Brazil. *Water Science and Technology*, *63*(6), 1265–1270. https://doi.org/10.2166/wst.2011.106

Dados, N., and Connell, R. (2012). The Global South. *Context*, *11*(February 2012), 12–13. https://doi.org/10.1177/1536504212436479

Dama, P., Bell, J., Foxon, K. M., Brouckaert, C. J., Huang, T., Buckley, C. A., Naidoo, V., and Stuckey, D. (2002). Pilot-scale study of an anaerobic baffled reactor for the treatment of domestic wastewater. *Water Science and Technology*, *46*(9), 263–270. https://doi.org/10.2166/wst.2002.0255

Decezaro, S. T., Wolff, D. B., Araújo, R. K., Faccenda, H. B., Perondi, T., and Sezerino, P. H. (2018). Vertical flow constructed wetland planted with Heliconia psittacorum used as decentralized post-treatment of anaerobic effluent in Southern Brazil. *Journal of Environmental Science and Health – Part A Toxic/Hazardous Substances and Environmental Engineering*, *53*(13), 1131–1138. https://doi.org/10.1080/10934529.2018.1530106

Denny, P. (1997). Implementation of constructed wetlands in developing countries. *Water Science & Technology*, *35*(5), 27–34.

Department of Water Affairs-South Africa. (2013). *Government Gazette N° 36820, 6 September 2013*. http://extwprlegs1.fao.org/docs/pdf/saf126916.pdf

Doorn, M. R. J., Towprayoon, S., Manso Vieira, S. M., Irving, W., Palmer, C., Pipatti, R., and Wang, C. (2006). Volume 5: Waste. Wastewater Treatment and Discharge. *2006 IPCC Guidelines for National Greenhouse Gas Inventories*, *5*, 1–56. www.ipcc-nggip.iges.or

Edokpayi, J. N., Odiyo, J. O., and Durowoju, O. S. (2016). Impact of wastewater on surface water quality in Developing Countries: A case Study of South Africa. *Water Quality*, 13. https://www.intechopen.com/books/advanced-biometric-technologies/liveness-detection-in-biometrics

El-Deeb Ghazy, M. M., El-Senousy, W. M., Abdel-Aatty, A. M., and Kamel, M. (2008). Performance evaluation of a waste stabilization pond in a rural area in Egypt. *American Journal of Environmental Sciences, 4*(4), 316–326. https://doi.org/10.3844/ajessp.2008.316.325

FAO. (2012). *AQUASTAT – FAO's Global Information System on Water and Agriculture.* Http://Www.Fao.Org/Aquastat/En/Databases/Maindatabase/Metadata.

Fuchs, V. J., Mihelcic, J. R., and Gierke, J. S. (2011). Life cycle assessment of vertical and horizontal flow constructed wetlands for wastewater treatment considering nitrogen and carbon greenhouse gas emissions. *Water Research, 45*(5), 2073–2081. https://doi.org/10.1016/j.watres.2010.12.021

Gallego-Schmid, A., and Tarpani, R. R. Z. (2019). Life cycle assessment of wastewater treatment in developing countries: A review. *Water Research, 153*, 63–79. https://doi.org/10.1016/j.watres.2019.01.010

Gallegos Valqui, D. G. (2019). Evaluación del lombrifiltro como tratamiento primario del sistema de tratamiento de aguas residuales del proyecto Manchay Verde [Universidad Nacional Tecnológica de Lima Sur]. In *Universidad Nacional Tecnológica de Lima Sur.* http://repositorio.untels.edu.pe/handle/UNTELS/166

Garfí, M., Flores, L., and Ferrer, I. (2017). Life Cycle Assessment of wastewater treatment systems for small communities: Activated sludge, constructed wetlands and high rate algal ponds. *Journal of Cleaner Production, 161*(February 2018), 211–219. https://doi.org/10.1016/j.jclepro.2017.05.116

Gensh, R., Jennings, A., Renggli, S., and Reymond, P. (2018). *Compendium of Sanitation Technologies in Emergencies* (1st Editio). German WASH Network (GWN), Swiss Federal Institute of Aquatic Science and Technology (Eawag), Global WASH Cluster (GWC) and Sustainable Sanitation Alliance (SuSanA).

Global Wetland Technology. (2020). *Global Wetland Technology.* Https://Www.Globalwettech.Com/References/Ref-Municipal.Html. https://www.globalwettech.com/references/ref-municipal.html

Godin, D., Bouchard, C., and Vanrolleghem, P. A. (2012). Net environmental benefit: Introducing a new LCA approach on wastewater treatment systems. *Water Science and Technology, 65*(9), 1624–1631. https://doi.org/10.2166/wst.2012.056

GoI. (2014). *Program Percepatan Pembangunan Sanitasi Permukiman (PPSP).* Http://Www.Ampl.or.Id/Program/Program-Percepatan-Pembangunan-Sanitasi-Permukiman-Ppsp-/1.

Gokalp, Z., Uzun, O., and Calis, Y. (2014). Kayseri'deki doğal aritma (yapay sulak alanlar) sistemlerinde karşilaşilan yaygin başarisizliklar. *Ekoloji, 44*(92), 38–44. https://doi.org/10.5053/ekoloji.2014.924

Gondwe, E., Mwanuzi, F. L., and Mbwette, T. S. A. (1997). Septic tank-soakpit systems in Dar es Salaam, Tanzania. *Journal of Environmental Engineering, 93.*

Gopolang, O. P., and Letshwenyo, M. W. (2018). Performance Evaluation of Waste Stabilisation Ponds. *Journal of Water Resource and Protection, 10*(11), 1129–1147. https://doi.org/10.4236/jwarp.2018.1011067

Gorito, A. M., Ribeiro, A. R., Almeida, C. M. R., and Silva, A. M. T. (2017). A review on the application of constructed wetlands for the removal of priority substances and contaminants of emerging concern listed in recently launched EU legislation. *Environmental Pollution, 227*, 428–443. https://doi.org/10.1016/j.envpol.2017.04.060

Goyal, B., and Mohan, D. (2013). Case study on evaluation of the performance of waste stabilization pond system of Jodhpur, Rajasthan, India. *Water Practice and Technology, 8*(1), 95–104. https://doi.org/10.2166/wpt.2013.011

Gutterer, B., and Reuter, S. (2016). *Key Elements for a New Urban Agenda Integrated Management* (B. Gutterer and S. Reuter (eds.)). BORDA e.V.

Harada, H., Dong, N. T., and Matsui, S. (2008). A measure for provisional-and-urgent sanitary improvement in developing countries: Septic-tank performance improvement. *Water Science and Technology, 58*(6), 1305–1311. https://doi.org/10.2166/wst.2008.715

Harder, D. S., Sajise, A. J. U., and Galing, E. M. (2013). Willingness to pay for sanitation services in Dagupan City, Philippines. *Journal of Water Sanitation and Hygiene for Development, 3*(2), 165–180. https://doi.org/10.2166/washdev.2013.064

He, Y., Peng, L., Hua, Y., Zhao, J., and Xiao, N. (2018). *Treatment for Domestic Wastewater from University Dorms Using a Hybrid Constructed Wetland at Pilot Scale.* 8532–8541.

Hegg S., S. S. (2018). Evaluación de optimizaciones de sistemas individuales de tratamiento de aguas residuales. In *Tecnológico de Costa Rica.* Instituto Tecnológico de Costa Rica.

Henze, M., M. van Loosdrecht, M. C., Ekama, G. A., and Brdjanovic, D. (2008). Biological wastewater treatment: Principles, Modelling and Design. In *IWA Publishing.* IWA Publishing.

Herrera-Cárdenas, J., Navarro, A. E., and Torres, E. (2016). Effects of porous media, macrophyte type and hydraulic retention time on the removal of organic load and micropollutants in constructed wetlands. *Journal of Environmental Science and Health – Part A Toxic/Hazardous Substances and Environmental Engineering, 51*(5), 380–388. https://doi.org/10.1080/10934529.2015.1120512

Ho, L., and Goethals, P. L. M. (2020). Municipal wastewater treatment with pond technology: Historical review and future outlook. *Ecological Engineering, 148*(March). https://doi.org/10.1016/j.ecoleng.2020.105791

Hoffmann, H., Platzer, C., Winker, M., and von Muench, E. (2011). Technology review of constructed wetlands Subsurface flow constructed wetlands for greywater and. *Deutsche Gesellschaft Für Internationale Zusammenarbeit (GIZ) GmbH.*

IAC Vietnam, and SINH THAI CICE. (2015). Environmental and Social Impact Assessment. In *PEOPLE COMMITTEE OF CAN THO CITY ODA*. https://doi.org/10.1201/9781351183666-2

INNOQUA. (2018). *Demonstration Sites: Strategy and Sites*. Https://Innoqua-Project.Eu/Demo-Sites/. https://innoqua-project.eu/demo-sites/

ISO 14040. (2006). *Environmental Management – Life cycle assessment – Principles and framework.*

IUCN. (2016). Nature-based Solutions to address global societal challenges. In G. Walters, E. Cohen-Shacham, S. Maginnis, and P. Lamarque (Eds.), *IUCN*. International Union for Conservation of Nature (IUCN). https://doi.org/10.2305/IUCN.CH.2016.13.en

K'oreje, K. O., Kandie, F. J., Vergeynst, L., Abira, M. A., Van Langenhove, H., Okoth, M., and Demeestere, K. (2018). Occurrence, fate and removal of pharmaceuticals, personal care products and pesticides in wastewater stabilization ponds and receiving rivers in the Nzoia Basin, Kenya. *Science of the Total Environment, 637–638*(May), 336–348. https://doi.org/10.1016/j.scitotenv.2018.04.331

Kerstens, S. M., Legowo, H. B., and Hendra Gupta, I. B. (2012). Evaluation of DEWATS in Java, Indonesia. *Journal of Water Sanitation and Hygiene for Development, 2*(4), 254–265. https://doi.org/10.2166/washdev.2012.065

Kihila, J., Mtei, K. M., and Njau, K. N. (2014). Wastewater treatment for reuse in urban agriculture; the case of Moshi Municipality, Tanzania. *Physics and Chemistry of the Earth*, 72(November), 104–110. https://doi.org/10.1016/j.pce.2014.10.004

Kimwaga, R., Mwegoha, W., Mahnge, A., Nyomora, A., and Lugali, L. (2013). Factors for Success and Failures of Constructed Wet? lands in the Sanitation Service Chains. *VLIR UOS South Initiatives*, 14,39.

Klinger, M., Ulrich, L., Wolf, T. A., Reynaud, N., Philip, L., and Lüthi, C. (2020). Technology, Implementation and Operation of Small-Scale Sanitation in India – Performance Analysis and Policy Recommendations. *4S Project Report, Vol. I* (July).

Kooy, M., and Bakker, K. (2008). Technologies of government: Constituting subjectivities, spaces, and infrastructures in colonial and contemporary Jakarta. *International Journal of Urban and Regional Research, 32*(2), 375–391. https://doi.org/10.1111/j.1468-2427.2008.00791.x

Kramer, A., Post, J., and Reseach, A. (2003). Guidelines and Standards for Wastewater Reuse. In Technical University Hamburg-Harburg (Ed.), *EMWater: Efficient Management of Wastewater, its Treatment and Reuse in the Mediterranean Countries*. European Union – EU. https://cgi.tu-harburg.de/~awwweb/wbt/emwater/

Kumar, C., and Ghosh, A. K. (2019). Fabrication of a vermifiltration unit for wastewater recycling and performance of vermifiltered water (vermiaqua) on onion (Allium cepa). *International Journal of Recycling of Organic Waste in Agriculture*, *8*(4), 405–415. https://doi.org/10.1007/s40093-019-0247-9

Kumar, D., and Asolekar, S. R. (2016). Significance of natural treatment systems to enhance reuse of treated effluent: A critical assessment. *Ecological Engineering*, *94*, 225–237. https://doi.org/10.1016/j.ecoleng.2016.05.067

Kumar, S., and Dutta, V. (2019). Constructed wetland microcosms as sustainable technology for domestic wastewater treatment: an overview. *Environmental Science and Pollution Research, March*. https://doi.org/10.1007/s11356-019-04816-9

Kumar, T., Rajpal, A., Arora, S., Bhargava, R., Hari Prasad, K. S., and Kazmi, A. A. (2016). A comparative study on vermifiltration using epigeic earthworm Eisenia fetida and Eudrilus eugeniae. *Desalination and Water Treatment*, *57*(14), 6347–6354. https://doi.org/10.1080/19443994.2015.1010230

Kyambadde, J., Kansiime, F., Gumaelius, L., and Dalhammar, G. (2004). A comparative study of Cyperus papyrus and Miscanthidium violaceum-based constructed wetlands for wastewater treatment in a tropical climate. *Water Research*, *38*(2), 475–485. https://doi.org/10.1016/j.watres.2003.10.008

Lamech Mogambi Ming'ate, F. (2015). Concepts of the Global South – Voices from around the world. The Global South: What does it mean to Kenya? In *The global South*. https://kups.ub.uni-koeln.de/6399/%0Akups.ub.uni-koeln.de/6399/1/voices012015_concepts_of_the_global_south.pdf

Lao People's Democratic Republic. (1999). *Resolution No. 02-99/NA*.

Larame, J. E. (2015). Pilot for Low Cost Sanitation: 'P-LOCSAN – Per-capita production rates.' *PhD Data Collection*.

Laramee, J., Tilmans, S., and Davis, J. (2018). Costs and benefits of biogas recovery from communal anaerobic digesters treating domestic wastewater: Evidence from peri-urban Zambia. *Journal of Environmental Management*, *210*, 23–35. https://doi.org/10.1016/j.jenvman.2017.12.064

Legifrance. (2009). *Arrêté du 7 septembre 2009 fixant les prescriptions techniques applicables aux installations d'assainissement non collectif recevant une charge brute de pollution organique inférieure ou égale à 1,2 kg/j de DBO5*. Journal Official Du Gouvernement N°0234. https://www.legifrance.gouv.fr/jorf/id/JORFTEXT000021125109/

Li, M., Wu, H., Zhang, J., Ngo, H. H., Guo, W., and Kong, Q. (2017). Nitrogen removal and nitrous oxide emission in surface flow constructed wetlands for treating sewage treatment plant effluent: Effect of C/N ratios. *Bioresource Technology, 240*, 157–164. https://doi.org/10.1016/j.biortech.2017.02.054

Ligy, P. (2018). A Roadmap for Small-Scale STPs in India: Fullfilling their Potential for Healthy and Water-Secure Cities. Performance of systems: Evaluation Results. In *National Workshop on Small-Scale Sanitation Systems*. https://www.eawag.ch/fileadmin/Domain1/Abteilungen/sandec/schwerpunkte/sesp/4S/India/c_philip_performance_evaluation.pdf

Liu, D., Ge, Y., Chang, J., Peng, C., Gu, B., Chan, G. Y. S., and Wu, X. (2009). Constructed wetlands in China: Recent developments and future challenges. *Frontiers in Ecology and the Environment, 7*(5), 261–268. https://doi.org/10.1890/070110

Looker, N. (1998). Municipal Wastewater Management in Latin America and the Caribbean: A discussion paper on trends, challenges and the market. *The Round Table on Municipal Water, Vancouver, Canada, 15–17 March 1998.*, 100.

Lopsik, K. (2013). Life cycle assessment of small-scale constructed wetland and extended aeration activated sludge wastewater treatment system. *International Journal of Environmental Science and Technology, 10*(6), 1295–1308. https://doi.org/10.1007/s13762-012-0159-y

Lorenzo-Toja, Y., Vázquez-Rowe, I., Chenel, S., Marín-Navarro, D., Moreira, M. T., and Feijoo, G. (2015). Eco-efficiency analysis of Spanish WWTPs using the LCA+DEA method. *Water Research, 68*, 651–666. https://doi.org/10.1016/j.watres.2014.10.040

Luth, R. P., Germain, P., Lecomte, M., Landrain, B., Li, Y. S., and Cluzeau, D. (2011). Earthworm effects on gaseous emissions during vermifiltration of pig fresh slurry. *Bioresource Technology 2, 102.*

Lüthi, C., Panesar, A., Schütze, T., Norström, A., Mcconville, J., Parkinson, J., Saywell, D., and Ingle, R. (2011). Sustainable Sanitation in Cities: A Framework for Action. In C. Lüthi. and T. Schütze (Eds.), *Forum American Bar Association* (1st Editio). Papiroz Publishing House.

Macedo, S. L., Araújo, A. L. C., and Pearson, H. W. (2011). Thermo-tolerant coliformbacteria decay rates in a full scale waste stabilization pond system in Northeast Brazil. *Water Science and Technology, 63*(6), 1321–1326. https://doi.org/10.2166/wst.2011.110

Machado, A. P., Urbano, L., Brito, A. G., Janknecht, P., Rodriguez, J. J., and Nogueira, R. (2006). Life Cycle Assessment of Wastewater Treatment Options for Small and Decentralized Communities: Energy-Saving Systems versus Activated Sludge. *10th International Conference on Wetland Systems for Water*

Pollution Control, May 2014, 1203–1213. http://espace.library.uq.edu.au/view/UQ:187150

Machado, A. P., Urbano, L., Brito, A. G., Janknecht, P., Salas, J. J., and Nogueira, R. (2007). Life cycle assessment of wastewater treatment options for small and decentralized communities. *Water Science and Technology, 56*(3), 15–22. https://doi.org/10.2166/wst.2007.497

Mahenge, A. S., and Malabeja, M. D. (2018). Performance Analysis of Anaerobic Baffled Reactor and Constructed Wetland for Community Based Wastewater in Dar Es Salam, Tanzania. *International Journal of Advanced Engineering Research and Science, 5*(7), 231–239. https://doi.org/10.22161/ijaers.5.7.30

Mara, D. (2003). *Domestic Wastewater Treatment in Developing Countries* (Typesetting by MapSet Ltd (ed.)). Cromwell Press.

Mara, D., and Horan, N. (2003). *Handbook of Water and Wastewater Microbiology*. Elsevier.

Martin-Hurtado, R., and Nolasco, D. (2016). *Managing Wastewater as a Resource in Latin America and the Caribbean.*

Mateo-Sagasta, J., Raschid-Sally, L., and Thebo, A. (2015). *"Global wastewater and sludge production, treatment and use", in Wastewater: Economic Asset in an Urbanizing World* (P. Drechsel, M. Qadir, and D. Wichelns (eds.)). Springer.

Mbwele, L., Rubindamayugi, M., Kivaisi, A., and Dalhammar, G. (2004). Performance of a small wastewater stabilisation pond system in tropical climate in Dar es Salaam, Tanzania. *Water Science and Technology, 48*(11–12), 187–191. https://doi.org/10.2166/wst.2004.0836

Mehta, M., Mehta, D., and Yadav, U. (2019). Citywide Inclusive Sanitation Through Scheduled Desludging Services: Emerging Experience From India. *Frontiers in Environmental Science, 7*(November), 1–10. https://doi.org/10.3389/fenvs.2019.00188

Minister of Natural resources and Environment. (2008). *National Technical Regulation on Domestic Wastewater QCVN 14: 2008/BTNMT.*

Ministerio del Ambiente. (2015). Texto Unificado de Legislación Ambiental Secundaria Libro VI Anexo1, Norma De Calidad Ambiental Y De Descarga De Efluentes. In *Registro Oficial No. 387* (Issue 097). http://extwprlegs1.fao.org/docs/pdf/ecu155128.pdf

Ministerio del Ambiente (MINAM). (2008). Límites Máximos Permisibles para los efluentes de Plantas de Tratamiento de Aguas Residuales Domésticas o Municipales. *El Peruano, 202472*, 11.

Ministry of Environment and Forestry of Indonesia. (2016). *NOMO:P.68/Menlhk/Setjen/Kum.1/8/2016 (in Indonesian).* 11. http://neo.kemenperin.go.id/files/hukum/19 Permen LHK th 2016 No. P.63 Baku Mutu Air Limbah Domestik.pdf

Ministry of Water – United Republic of Tanzania. (2018). *Guidelines for the Application of Small-Scale, Decentralised Wastewater Treatment Systems – A Code of Practice for Decision Makers* (Issue December).

Miranzadeh, M. B. (2005). Establishment of Design Criteria at Shoosh Wastewater Treatment Plant in South of Tehran, Iran. *Journal of Biological Sciences, 5*(4), 421–423. https://doi.org/10.3923/jbs.2005.421.423

Mitma J. Y. I. (2017). *Efecto del sistem de lombrifiltro en la depuración del DBO y DQO de las aguas residuales domésticas del distrito de Moche.* Universidad César Vallejo.

Molinos-Senante, M., Gómez, T., Garrido-Baserba, M., Caballero, R., and Sala-Garrido, R. (2014). Assessing the sustainability of small wastewater treatment systems: A composite indicator approach. *Science of the Total Environment, 497–498*, 607–617. https://doi.org/10.1016/j.scitotenv.2014.08.026

Molle, P., Lombard Latune, R., Riegel, C., Lacombe, G., Esser, D., and Mangeot, L. (2015). French vertical-flow constructed wetland design: Adaptations for tropical climates. *Water Science and Technology, 71*(10), 1516–1523. https://doi.org/10.2166/wst.2015.133

Monstadt, J., and Schramm, S. (2017). Toward The Networked City? Translating Technological ideals and Planning Models in Water and Sanitation Systems in Dar es Salaam. *International Journal of Urban and Regional Research, 41*(1), 104–125. https://doi.org/10.1111/1468-2427.12436

Morella, E., Foster, V., and Banerjee, S. G. (2008). *Africa Infrastructure Country Diagnostic; Climbing the ladder: The State of Sanitation in Sub-Saharan Africa* (Issue June).

MoW. (2018). *Guidelines for the Application of Small-Scale, Decentralised Wastewater Treatment Systems A Code of Practice for Decision Makers* (Issue December).

Muga, H. E., and Mihelcic, J. R. (2008). Sustainability of wastewater treatment technologies. *Journal of Environmental Management, 88*(3), 437–447. https://doi.org/10.1016/j.jenvman.2007.03.008

NCWSC. (2019). *Environmental and social impact assessment project report for the proposed Lenana school sewer relief in Nairobi City County of the Nairobi Metropolitan Region* (Issue August).

NEMA. (1999). *The national environment (standards for discharge of effluent into water or on land) regulations.*

NEMA. (2006). *Environmental protection (Standards for effluent discharge) Regulations. LEGAL NOTICE NO. 120.* http://extwprlegs1.fao.org/docs/pdf/ken84962.pdf

NGT. (2019). *Item No. 4 Court. No. 1.*

Nguyen, V., Pham, N. T., Nguyen, T. H., Morel, A., and Tonderski, K. (2007). Improved septic tank, a promising decentralized wastewater treatment alternative in Vietnam. In IWA (Ed.), *Join International Conference of the National Onsite Wastewater Recycling Association (NOWRA) and International Water Association (IWA) Water for all Life.*

Nivala, J., Abdallat, G., Aubron, T., Al-Zreiqat, I., Abbassi, B., Wu, G. M., van Afferden, M., and Müller, R. A. (2019). Vertical flow constructed wetlands for decentralized wastewater treatment in Jordan: Optimization of total nitrogen removal. *Science of the Total Environment, 671*, 495–504. https://doi.org/10.1016/j.scitotenv.2019.03.376

Nolasco, D., Rodriguez, D., Serrano, H., Delgado, A., and Gustavo, S. (2020). *The linear to circular shift for wastewater in Latin America.* Https://Www.Thesourcemagazine.Org/the-Linear-to-Circular-Shift-for-Wastewater-in-Latin-America/. https://www.thesourcemagazine.org/the-linear-to-circular-shift-for-wastewater-in-latin-america/

Noyola, A., Padilla-Rivera, A., Morgan-Sagastume, J. M., Güereca, L. P., and Hernández-Padilla, F. (2012). Typology of Municipal Wastewater Treatment Technologies in Latin America. *Clean – Soil, Air, Water, 40*(9), 926–932. https://doi.org/10.1002/clen.201100707

NRWC. (1995). Environmental Pollution and Legislative Regulations. Law 48-1982 and Decree 8-1983. In *Ministry of Public Works and Water Resources.*

Oakley, S. M., Pocasangre, A., Flores, C., Monge, J., and Estrada, M. (2000). Waste stabilization pond use in Central America: The experiences of El Salvador, Guatemala, Honduras and Nicaragua. *Water Science and Technology, 42*(10–11), 51–58. https://doi.org/10.2166/wst.2000.0607

Oke, I. A., Oladepo, K. T., and Babatola, J. O. (2009). An engineering failure analysis of a biological treatment plant in Nigeria. *Journal of Failure Analysis and Prevention, 9*(3), 197–207. https://doi.org/10.1007/s11668-009-9227-0

Paico Revilla, D. A. (2017). Sistema Tohá, para el tratamiento de aguas residuales de la Universidad Cesar Vallejo. *Universidad César Vallejo*, 1–79.

Pandey, K. M. (2015). Constructed wetlands for wastewater treatment in Nepal. In *Norwegian University of Life Sciences.* Norwegian University of Life Sciences.

Parra-Saldivar, R., Bilal, M., and Iqbal, H. M. N. (2020). Life cycle assessment in wastewater treatment technology. *Current Opinion in Environmental Science and Health, 13*, 80–84. https://doi.org/10.1016/j.coesh.2019.12.003

Pasqualino, J. C., Meneses, M., and Castells, F. (2011). Life Cycle Assessment of Urban Wastewater Reclamation and Reuse Alternatives. *Journal of Industrial Ecology, 15*(1), 49–63. https://doi.org/10.1111/j.1530-9290.2010.00293.x

Patel, J., and Gajera, Y. M. (2018). *Treatment of Sewage by Vermifiltration: A Review.* Indian Water Works Association – IWWA. http://iwwa-ahmedabad.org/images/downloads/papers/2018/1808.pdf

Peña, F. J., Madrid, L., Schneider, J., da Silva, M. M., Caruzzo P., M., Campos S, F., I., and Luiz T., A. (2019). Vermifiltration: The use of earthworms as a new alternative for sewage treatment. *Revista DAE, 67.*

Pescod, M. B. (1992). *Wastewater treatment and use in agriculture* (FAO (ed.)). FAO.

PSI. (2012). *The Palestinian Treated Wastewater Standard (Technical. Specification). TS 34/2012.*

Rahi, M. A., Faisal, A. A. H., Naji, L. A., Almuktar, S. A., Abed, S. N., and Scholz, M. (2020). Biochemical performance modelling of non-vegetated and vegetated vertical subsurface-flow constructed wetlands treating municipal wastewater in hot and dry climate. *Journal of Water Process Engineering, 33*(October 2019), 101003. https://doi.org/10.1016/j.jwpe.2019.101003

Ramesh, G., Goel, M., and Das, A. (2017). A study on comparison of horizontal and vertical flow wetland system in treating domestic wastewater. *International Journal of Civil Engineering and Technology, 8*(5), 1302–1310.

Ramos-Mejía, M., Franco-Garcia, M. L., and Jauregui-Becker, J. M. (2018). Sustainability transitions in the developing world: Challenges of socio-technical transformations unfolding in contexts of poverty. *Environmental Science and Policy, 84*(December 2016), 217–223. https://doi.org/10.1016/j.envsci.2017.03.010

Reyes, J. V. (2016). Determinación de la eficiencia del aserrín y la fibra de coco utilizados como empaques para la remoción de contaminantes en Biofiltros para el tratamiento de aguas residuales. *Enfoque UTV, 7*(3), 41–56.

Reymond, P., Chandragiri, R., and Ulrich, L. (2020). Governance Arrangements for the Scaling Up of Small-Scale Wastewater Treatment and Reuse Systems – Lessons From India. *Frontiers in Environmental Science, 8*(June). https://doi.org/10.3389/fenvs.2020.00072

Reymond, P., Demars, C., Papangelou, A., Tawfik, M. H., Hassan, K., Wahaab, R. A., & Moussa, M. (2014). *Small-Scale Sanitation in the Nile Delta: Baseline Data and Current Practices* (Issue August). https://doi.org/10.13140/RG.2.1.3536.7526

Reynaud, N., and Buckley, C. (2015). Field-data on parameters relevant for design, operation and monitoring of communal decentralized wastewater treatment systems (DEWATS). *Water Practice and Technology, 10*(4), 787–798. https://doi.org/10.2166/wpt.2015.097

Reynaud, N. S. (2014). *Operation of Decentralised Wastewater Treatment Systems (DEWATS) under tropical field conditions.* Technical University, Dresden.

Ronteltap, M., Dodane, P.-H., and Bassan, M. (2014). Overview of Treatment Technologies. In L. Strande, M. Ronteltap, and D. Brdjanovic (Eds.), *Faecal Sludge Management: Systems Approach for Implementation and Operation* (1st Editio, pp. 102–103). IWA Publishing.

Rose, C., Parker, A., Jefferson, B., and Cartmell, E. (2015). The Characterization of Feces and Urine – A Review of the Literature to Inform Advanced Treatment Technology. *Critical Reviews in Environmental Science and Technology*, *45*(17), 1827–1879. https://doi.org/10.1080/10643389.2014.1000761

Rothmund, K., and Becker J. A. M. (2018). Evaluation of the viability and proposal of treatment of domestic effluent in universities through a vermifiltro with eyes of the species Eisenia Andrei. *11° Simpósio Internacional de Qualidade Ambiental*, 1–13.

Royal Government of Cambodia. (2009). *Sub Decree on Water Pollution Control N° 27 ANRK.BK* (Issue October).

Samal, K., Dash, R. R., and Bhunia, P. (2017). Treatment of wastewater by vermifiltration integrated with macrophyte filter: A review. *Journal of Environmental Chemical Engineering*, *5*(3), 2274–2289. https://doi.org/10.1016/j.jece.2017.04.026

Sapkota, N. (2016). *Environmental Performance Evaluation of Decentralized Wastewater Treatment Systems Using Life Cycle Analysis* [Norwegian University of Life Science – Faculty of Environmental Science and Technology]. https://brage.bibsys.no/xmlui/bitstream/handle/11250/2435169/Sapkota_2016.pdf?sequence=6

Sato, N., Okubo, T., Onodera, T., Agrawal, L. K., Ohashi, A., and Harada, H. (2007). Economic evaluation of sewage treatment processes in India. *Journal of Environmental Management*, *84*(4), 447–460. https://doi.org/10.1016/j.jenvman.2006.06.019

Schellenberg, T., Subramanian, V., Ganeshan, G., Tompkins, D., and Pradeep, R. (2020). Wastewater Discharge Standards in the Evolving Context of Urban Sustainability–The Case of India. *Frontiers in Environmental Science*, *8*(April). https://doi.org/10.3389/fenvs.2020.00030

Sehar, S., Naz, I., Khan, S., Naeem, S., Perveen, I., Ali, N., and Ahmed, S. (2016). Performance Evaluation of Integrated Constructed Wetland for Domestic Wastewater Treatment. *Water Environment Research*, *88*(3), 280–287. https://doi.org/10.2175/106143016x14504669767814

SEMARNAT. (1998). NOM-001-SEMARNAT-1996 Límites Máximos Permisibles De Contaminantes En Las Descargas De Aguas Residuales En Aguas Y Bienes Nacionales. In *Norma Oficial Mexicana*.

Singh, A., Sawant, M., Kamble, S. J., Herlekar, M., Starkl, M., Aymerich, E., and Kazmi, A. (2019). Performance evaluation of a decentralized wastewater treatment system in India. *Environmental Science and Pollution Research*, *26*(21), 21172–21188. https://doi.org/10.1007/s11356-019-05444-z

Singh, R., Bhunia, P., and Dash, R. R. (2017). A mechanistic review on vermifiltration of wastewater: Design, operation and performance. *Journal of*

Environmental Management, 197(July), 656–672. https://doi.org/10.1016/j.jenvman.2017.04.042

Singh, R., Samal, K., Dash, R. R., and Bhunia, P. (2019). Vermifiltration as a sustainable natural treatment technology for the treatment and reuse of wastewater: A review. *Journal of Environmental Management, 247*(May), 140–151. https://doi.org/10.1016/j.jenvman.2019.06.075

Singh, S., Haberl, R., Moog, O., Shrestha, R. R., Shrestha, P., and Shrestha, R. (2009). Performance of an anaerobic baffled reactor and hybrid constructed wetland treating high-strength wastewater in Nepal-A model for DEWATS. *Ecological Engineering, 35*(5), 654–660. https://doi.org/10.1016/j.ecoleng.2008.10.019

Sinha, R. K., Bharambe, G., and Chaudhari, U. (2008). Sewage treatment by vermifiltration with synchronous treatment of sludge by earthworms: A low-cost sustainable technology over conventional systems with potential for decentralization. *Environmentalist, 28*(4), 409–420. https://doi.org/10.1007/s10669-008-9162-8

Sinha, R. K., Patel, U., Soni, B. K., and Li, Z. (2014). wastes and wastewaters, remediation of contaminated soils and mitigation of global warming: A review. *Journal of Environment and Waste Management, 1*(1), 11–25.

Sperling, von M. (2007). Wastewater characteristics, Treatment and Disposal. In Aptara Inc. (Ed.), *Biological Wastewater Treatment series* (Vol. 4, pp. 1–28, 183–188). IWA Publishing.

Starkl, M., Brunner, N., Amerasinghe, P., Mahesh, J., Kumar, D., Asolekar, S. R., Sonkamble, S., Ahmed, S., Wajihuddin, M., Pratyusha, A., and Sarah, S. (2015). Stakeholder views, financing and policy implications for reuse of wastewater for irrigation: A case from Hyderabad, India. *Water (Switzerland), 7*(1), 300–328. https://doi.org/10.3390/w7010300

State Minister of Environment. (2003). *NOMOR 112 TAHUN 2003.*

Strande, L., Ronteltap, M., and Brdjanovic, D. (2014). *Faecal Sludge Management: Systems Approach for Implementation and Operation.* IWA Publishing.

Su, X., Chiang, P., Pan, S., Chen, G., Tao, Y., Wu, G., Wang, F., and Cao, W. (2019). Systematic approach to evaluating environmental and ecological technologies for wastewater treatment. *Chemosphere, 218*, 778–792. https://doi.org/10.1016/j.chemosphere.2018.11.108

Suriyachan, C., Nitivattananon, V., and Amin, N. T. M. N. (2012). Potential of decentralized wastewater management for urban development: Case of Bangkok. *Habitat International, 36*(1), 85–92. https://doi.org/10.1016/j.habitatint.2011.06.001

Taherkhani, A., Attar, H. M., Mirzaee, S. A., Ahmadmoazzam, M., Jaafarzadeh, N., Hashemi, F., and Jalali, M. (2018). Performance evaluation of waste stabilization ponds on removal of Listeria spp.: A case study of Isfahan, Iran. *Journal of Water and Health, 16*(4), 614–621. https://doi.org/10.2166/wh.2018.031

Taylor, P., Ayaz, S. Ç., Akta, Ö., and Akça, L. (2015). Full-scale hybrid constructed wetlands incorporated with an initial anaerobic stage for domestic wastewater treatment in a drinking water catchment area. *Desalination and Water Treatment, September*, 0–13. https://doi.org/10.1080/19443994.2015.1024751

TBS. (2005). *TZS 860: 2005 MUNICIPAL AND INDUSTRIAL WASTEWATERS – General Tolerance Limits for Municipal and Industrial Wastewaters.*

Tchobanoglous, G., and Leverenz, H. (2013). The rationale for decentralization of wastewater infrastructure. In T. A. Larse, K. M. Udert, and J. Lienert (Eds.), *Source Separation and Decentralization for Wastewater Management* (pp. 101–115). IWA Publishing.

Tchobanoglous, George, Burton, F. L., and Stensel, H. D. (2003). *Wastewater Engineering: Treatment and Reuse* (Metcalf and Eddy Inc. (ed.); 4th Editio). McGraw-Hill Companies, Inc.

The United Republic of Tanzania: National Audit Office. (2018). *Performance Audit Report on Provision of Sewage Services in Urban Areas.*

The United Republic of Tanzania – Ministry of Water. (2016). *Environmental and Social Management Framework: Second Water Sector Support Projct (WSSP-II) & Water Sector Development Programe II* (Issue October).

Tilley, E., Ulrich, L., Lüthi, C., Reymond, P., Schertenleib, R., and Zurbrügg, C. (2008). *Compendium of Sanitation Systems and Technologies.*

Tomar, P., and Suthar, S. (2011). Urban wastewater treatment using vermi-biofiltration system. *Desalination, 282*, 95–103. https://doi.org/10.1016/j.desal.2011.09.007

Tsuzuki, Y., Koottatep, T., Sinsupan, T., Jiawkok, S., Wongburana, C., Wattanachira, S., and Sarathai, Y. (2013). 'A concept for planning and management of on-site and centralised municipal wastewater treatment systems, a case study in Bangkok, Thailand. I: Pollutant discharge indicators and pollutant removal efficiency functions.' *Water Science & Technology, 67*(9). https://iwaponline.com/wst/article/67/9/1923/17295/A-concept-for-planning-and-management-of-on-site

Tyagi, V. K., Sahoo, B. K., Khursheed, A., Kazmi, A. A., Ahmad, Z., and Chopra, A. K. (2011). Fate of coliforms and pathogenic parasite in four full-scale sewage treatment systems in India. *Environmental Monitoring and Assessment, 181*(1–4), 123–135. https://doi.org/10.1007/s10661-010-1818-4

Uehara, M., and Vidal, W. (1989). *Operacao e manutencao de lagoas anaerobias efacultativas* (CETESB (ed.)). CETESB série manuais.

UN-Water. (2017). *Wastewater: The untapped resource.* UNESCO. http://unesdoc.unesco.org/images/0024/002471/247153e.pdf

UN. (2019). World Urbanization Prospects: The 2018 Revision. In *United Nations, Department of Economic Affairs, Population Division* (Vol. 12). https://doi.org/10.4054/demres.2005.12.9

UNDP. (2004). *Forging a Global South* (Issue December).

UNEP – IETC. (2000). International Sourcebook on Environmentally Sound Technologies for Wastewater and Stormwater Management. In *IETC Technical Publication Series* (Vol. 45, Issue 17). UNEP. https://doi.org/10.1021/es2014264

UNEP, and WHO. (1997). *Water Pollution Control – A Guide to the Use of Water Quality Management Principles* (R. Helmer and I. Hespanhol (eds.); 1st Editio). E & FN Spon. https://doi.org/10.4324/9780203477540

UNICEF, and WHO. (2019). Progress on household drinking water, sanitation and hygiene 2000–2017: Special focus on inequalities. In *Unicef/Who.* United Nations Children's Fund (UNICEF) and World Health Organization (WHO). https://washdata.org/sites/default/files/documents/reports/2019-07/jmp-2019-wash-households.pdf

Vairavamoorthy, K., and Nolasco, N. (2020). *Needs and opportunities for water and sanitation in Latin America and the Caribbean.* Https://Www.Thesourcemagazine.Org/Needs-and-Opportunities-for-Water-and-Sanitation-in-Latin-America-and-the-Caribbean/. https://www.thesourcemagazine.org/needs-and-opportunities-for-water-and-sanitation-in-latin-america-and-the-caribbean/

Vasconcellos, G. R., von Sperling, M., and Ocampos, R. S. (2019). From start-up to heavy clogging: Performance evaluation of horizontal subsurface flow constructed wetlands during 10 years of operation. *Water Science and Technology, 79*(7), 1231–1240. https://doi.org/10.2166/wst.2019.062

VEN. (2015). Mekong Delta's largest waste water facility opens. *Http://Ven.vn/Mekong-Deltas-Largest-Waste-Water-Facility-Opens-16528.Html.* http://ven.vn/mekong-deltas-largest-waste-water-facility-opens-16528.html

Vietnam Investement Review. (2020). Can Tho to double capacity of German-funded wastewater treatment plant. *Https://Www.Vir.Com.vn/Can-Tho-to-Double-Capacity-of-German-Funded-Wastewater-Treatment-Plant-73085.Html.* https://www.vir.com.vn/can-tho-to-double-capacity-of-german-funded-wastewater-treatment-plant-73085.html

Wang, H., Wang, T., Zhang, B., Li, F., Toure, B., Omosa, I. B., Chiramba, T., Abdel-Monem, M., and Pradhan, M. (2014). Water and Wastewater Treatment in Africa – Current Practices and Challenges. *Clean – Soil, Air, Water, 42*(8), 1029–1035. https://doi.org/10.1002/clen.201300208

Wang, X., Liu, J., Ren, N. Q., Yu, H. Q., Lee, D. J., and Guo, X. (2012). Assessment of multiple sustainability demands for wastewater treatment alternatives: A refined evaluation scheme and case study. *Environmental Science and Technology*, *46*(10), 5542–5549. https://doi.org/10.1021/es300761x

WaterAid. (2019). *Functionality of wastewater treatment plants in low- and middle-income countries. Desk review.* https://washmatters.wateraid.org/sites/g/files/jkxoof256/files/functionality-of-wastewater-treatment-plants-in-low--and-middle-income-countries-desk-review_1.pdf

WEPA. (2016). *Tolerance Limits for Wastewater to be Discharged into Inland Surface waters from Combined Wastewater Treatment Plant in Nepal.* Http://Www.Wepa-Db.Net/Policies/Law/Nepal/Mo03.Htm. http://www.wepa-db.net/policies/law/nepal/mo03.htm

WEPA. (1990). *DENR Administrative Order No. 35 Series of 1990.* http://www.wepa-db.net/policies/law/philippines/dao35_1990.htm

WHO. (2006). Guideline for the safe use of wastewater, excreta and greywater. Volume 2: Wastewater use in agriculture. In *WHO Library Cataloguing* (Vol. 1). https://doi.org/10.1007/s13398-014-0173-7.2

WHO, and UNICEF. (2000). Global Water Supply and Sanitation Assessment 2000 Report. In *http://www.who.int/water_sanitation_health/monitoring/jmp2000.pdf*. https://doi.org/http://www.who.int/water_sanitation_health/monitoring/globalassess/en/

WHO, and UNICEF. (2015). *Progress on Sanitation and Drinking Water – 2015 Update and MDG Assessment.*

WHO, and UNICEF. (2017). *Progress on Drinking Water, Sanitation and Hygiene: 2017 Update and SDG Baselines.* https://doi.org/10.1111/tmi.12329

Williams, A. R., and Overbo, A. (2015). *Unsafe Return of Human Excreta to the Environment: A Literature Review* (P. Kolsky and J. Bartram (eds.); Issue June). The Water Institute at UNC.

Wolter, D. (2018). *Communal Wastewater Characteristics in Developing Countries – BORDA's Internal Document.*

World Bank and Agence Francaise de Developpement (afd). (2014). *Environmental, Social, and Cultural Heritage Impact Assessment (ESCHIA) to Support Additional Financing for the Hebron Wastewater Management Project.* http://documents1.worldbank.org/curated/en/839551468327906091/pdf/E42530V10REVIS00Box385338B00PUBLIC0.pdf

World Bank Group. (2019). *Improved Wastewater Management And Equitable Access To Sanitation Services In Guayaquil.* https://www.worldbank.org/en/results/2019/10/15/improved-wastewater-management-and-equitable-access-to-sanitation-services-in-guayaquil

World Bank Group. (2020a). *Danang Sustainable City Development Project (SCDP)*. https://projects.worldbank.org/en/projects-operations/project-detail/P123384

World Bank Group. (2020b). *Nairobi Metropolitan Services Improvement Project*. https://projects.worldbank.org/en/projects-operations/project-detail/P107314

WWAP (UNESCO World Water Assessment Programme). (2019). The United Nations World Water Development Report 2019: Leaving no one behind. In *UNESCO Digital Library*. UNESCO. https://doi.org/.1037//0033-2909.I26.1.78

Xing, M., Li, X., and Yang, J. (2010). Treatment performance of small-scale vermifilter for domestic wastewater and its relationship to earthworm growth, reproduction and enzymatic activity. *African Journal of Biotechnology*, *9*(44), 7513–7520. https://doi.org/10.5897/ajb10.811

Xing, Meiyan, Li, X., and Yang, J. (2010). Treatment performance of small-scale vermifilter for domestic wastewater and its relationship to earthworm growth, reproduction and enzymatic activity. *African Journal of Biotechnology*, *9*(44), 7513–7520. https://doi.org/10.5897/ajb10.811

Zacharia, A., Ahmada, W., Outwater, A. H., Ngasala, B., and Van Deun, R. (2019). Evaluation of Occurrence, Concentration, and Removal of Pathogenic Parasites and Fecal Coliforms in Three Waste Stabilization Pond Systems in Tanzania. *Scientific World Journal*, *2019*(October). https://doi.org/10.1155/2019/3415617

Zhang, D. Q., Jinadasa, K. B. S. N., Gersberg, R. M., Liu, Y., Ng, W. J., and Tan, S. K. (2014). Application of constructed wetlands for wastewater treatment in developing countries – A review of recent developments (2000–2013). *Journal of Environmental Management*, *141*, 116–131. https://doi.org/10.1016/j.jenvman.2014.03.015

Zhang, D. Q., Jinadasa, K. B. S. N., Gersberg, R. M., Liu, Y., Tan, S. K., and Ng, W. J. (2015). Application of constructed wetlands for wastewater treatment in tropical and subtropical regions (2000–2013). *Journal of Environmental Sciences (China)*, *30*, 30–46. https://doi.org/10.1016/j.jes.2014.10.013

Zhejiang RenXin HuanKeYuan Co Ltd. (2019). *Environmental and Social Impact Report for Flood Risk Management Project In Ninghai Central Urban Area* (Issue August).

DOI: 10.1561/9781680837810.ch4

Chapter 4

Primary and Secondary Treatment: Vermifiltration

By David Tompkins

4.1 Introduction

The degree to which wastewater must be treated prior to discharge from the treatment system will depend on several factors – principally the quality requirements dictated by the needs of the receiving environment or the next user of this resource. In centralised wastewater treatments, a combination of primary and secondary processes is commonly applied, the first acting as a physical screen for suspended pollutants and the second utilising microbial processes to consume dissolved pollutants.

Similar step-wise treatments may also be applied where wastewater is processed in decentralised systems, and a number of these have been examined in detail in Chapter 3. Septic tanks and Imhoff tanks are common primary treatment options (Sasse, 1998). Both are essentially sedimentation tanks in which settled sludge is partially stabilised by anaerobic digestion prior to removal and separate treatment/disposal. Wastewater leaves the tank relatively untreated, although the settlement process may remove up to 50% of solids to deliver an overall process removal of up to 40% of BOD (Tilley *et al.*, 2014). Secondary treatment may then be achieved by passing the effluent through a soil filter/drainage bed or a constructed wetland. Other decentralised techniques – such as package plants, anaerobic filters and anaerobic baffled reactors provide a combination of primary and secondary treatment, although all tend to accumulate sludges over time.

Vermifilters (or 'lumbrifilters') are capable of delivering both primary and/or secondary treatment in single units. The addition of earthworms to a filtration process means that sludge accumulation is minimised, while the aerobic conditions serve to minimise odour potential and support active microbial communities capable of reducing BOD by more than 90% (Lourenço and Nunes, 2017). This chapter explores design and operation principles for vermifiltration, how it has been commercialised, and the challenges faced in demonstrating the technology under field conditions within the INNOQUA project.

4.1.1 What is Vermifiltration?

Vermifilters are engineered natural systems, based on the interaction between earthworms and microorganisms, in which earthworms degrade and homogenise organic wastes, increasing their surface area and facilitating subsequent bio-chemical degradation of pollutants by the microbial biofilm established on a filter bed (Arora and Kazmi, 2015). Vermifiltration relies on passive aeration through the filter bed to maintain aerobic conditions – requiring no blowers. Where ground conditions are suitable, filtration can take place entirely under gravity flow.

Vermifilters are used to treat blackwater, greywater, primary (settled) sewage and a range of industrial effluents – the treated effluent then being suitable for discharge, reuse or further treatment. These microbial-earthworm eco-filters (MEEs) have been shown to provide more consistent wastewater treatment performance than conventional biofilters that do not include earthworms. There is some evidence that they can also remove both nutrients and pathogens, and produce little excess sludge (Jiang *et al.*, 2016). Vermifiltration principles have also been applied to the development of 'Tiger Toilets' – a variation on traditional pit latrines in which earthworms consume and stabilise faecal material, dramatically reducing solids' accumulation rates (Furlong *et al.*, 2016).

Vermifiltration systems all follow similar design principles:

1. Filter media are built up in a series of layers – normally of increasing particle size with depth. These layers may be mineral (sand, gravel or man-made equivalents) or organic (compost, bark, sawdust);

2. A distinct uppermost 'bedding' layer is normally included, comprising an organic substrate or an organic-matter rich soil – to suit the requirements of epigeic[1] earthworm species such as *Eisenia fetida*, *Eisenia andrei*, *Perionyx sansibaricus* or *Lumbricus rubellus*. Vegetation is sometimes established in this top layer;
3. Wastewater is introduced to the top of the filter using a distribution system. Wastewater can be introduced as greywater, blackwater or settled sewage – each of which requires different filter media, depth and operating volume to optimise treatment;
4. Wastewater percolates through the filter bed, where treatment takes place in an established biofilm as it does in aerobic trickling filter systems. Earthworms graze on the microbial biomass and solids introduced in the wastewater – moderating the microbial community and helping to maintain aerobic conditions. The treated effluent collects within the lowest gravel layer or a separate sump, from where it may be collected for discharge, further treatment or reuse;
5. The filter bed requires little maintenance, since a healthy earthworm population will maintain a network of channels throughout the medium. The surface may eventually become clogged with earthworm casts that can be harvested for reuse in agriculture or horticulture;
6. Vermifiltration systems require no external power, although pumps are commonly used to introduce wastewater and/or remove treated effluent. They may be operated as single units, or as multiple units in series – depending on site and wastewater-specific circumstances.

The key components of a basic vermifilter are shown in Figure 4.1, while the key mechanisms are shown in Figure 4.2 and the key trophic pathways in Figure 4.3.

4.2 Design and Operating Principles

Vermifilters combine three key elements:

1. The ability of specific types of earthworm to consume organic solids;
2. The ability of bacteria in attached-growth biofilms to break down dissolved organic pollutants in wastewater; and
3. Aerobic conditions, facilitated by selection of appropriate filter media.

1. Found predominantly on the soil surface in leaf-litter, compost or manure. These species are widely used in vermi-composting.

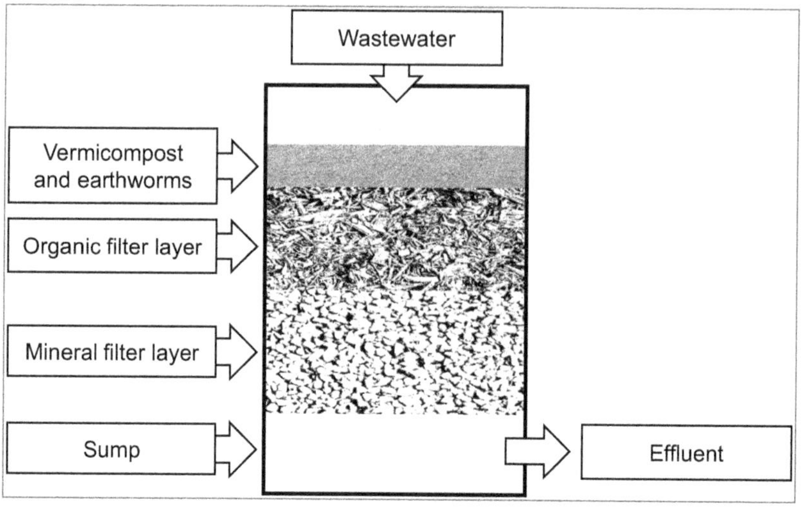

Figure 4.1. Schematic for a vermifilter, adapted from Singh *et al.* (2017).

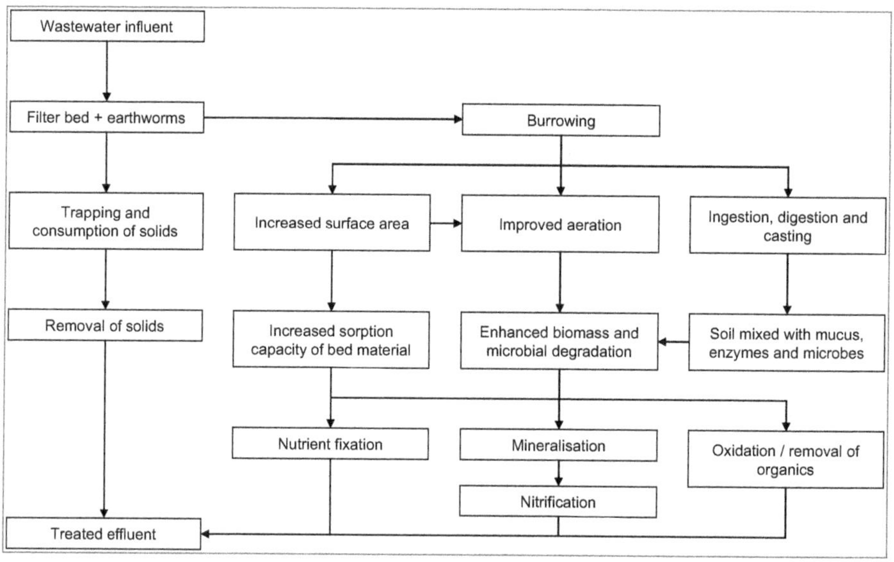

Figure 4.2. Mechanisms of vermifiltration, as set out by Singh *et al.* (2019).

Despite their apparent simplicity, interactions between earthworm species, bedding material type, filter media type, type of wastewater, temperature, pH, hydraulic loading and other factors will affect the performance of any individual system. Variations in filter bed depth and wastewater dosing regime are also important, with intermittent dosing more likely to produce stable aerobic conditions, increasing nitrification activity and reducing the potential for ammonia toxicity to earthworms – which will in turn allow them to exploit the filter media

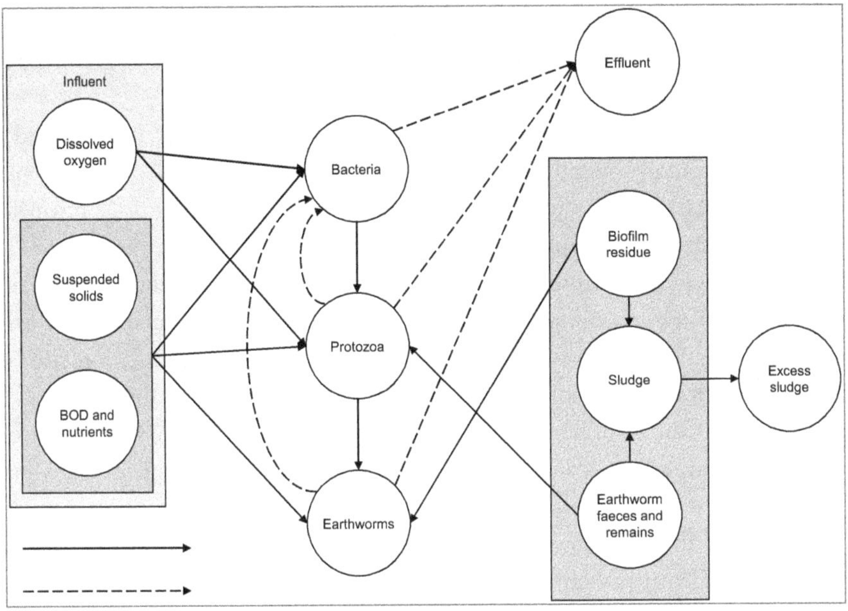

Figure 4.3. Basic trophic pathways in vermifiltration, as set out by Jiang *et al.* (2016).

more thoroughly. Earthworm feeding and casting behaviours also lead to changes in microbial community structure, which in turn lead to changes in wastewater treatment performance Jiang *et al.* (2016), Singh *et al.* (2017), and Liu *et al.* (2012). Detailed exploration and understanding of these interactions remains the focus of research activities that are reported in the scientific literature. Key elements – earthworms, bacteria and filter media – are explored in the following sections.

4.2.1 The Role of Earthworms

Earthworms are functionally classified according to their feeding and burrowing habits. Detritivores feed at or near the soil surface on plant litter and animal manures, while geophages feed within the soil and derive their nutrition from soil organic matter and dead roots ingested within large quantities of soil (Curry and Schmidt, 2007). Various subdivisions within these two broad groups are commonly used: epigeic, anecic, and endogeic (Bouché, 1977).

- Endogeic (soil feeders) and Anecic species (burrowers) live in the soil and consume a mixture of soil and organic matter; and
- Epigeic species inhabit organic soil horizons, in or near the surface litter, and feed primarily on coarse particulate organic matter (Domínguez and Edwards, 2011).

Further subdivisions of the endogeic group have been proposed, based on feeding strategies in relation to organic matter: polyhumic, mesohumic and oligohumic etc. (Curry and Schmidt, 2007).

Epigeic species occupy unpredictable and unstable habitats, characterised by highly variable environmental conditions, food availability, and predation pressures. In response to these pressures, they have high rates of consumption and digestion, as well as high reproductive and metabolic rates. When conditions are unfavourable, epigeic earthworms suffer high mortality, population density oscillates widely, and the reproduction rate increases greatly. Under unfavourable environmental conditions, high reproduction rates help to ensure population survival, and the formation of cocoons can enable the worms to resist until conditions become more favourable. Since organic matter acts as both the habitat and food during vermicomposting – with no requirement for soil – only epigeic earthworms are suitable for use in these processes (Domínguez, 2018). As an epigeic species, *Eisenia fetida* is abundant in garden compost heaps and also found in grasslands improved with farmyard manure, as well as sewage filter beds. It is not common in organic horizons in woodland soils (Natural England, 2014).

Earthworms have no specialized respiratory organs; they obtain oxygen by diffusion through the body wall and lose carbon dioxide by diffusion. However, earthworms are sensitive to anaerobic conditions, and their respiration rates decrease by around 55 to 65% when oxygen levels reach 0.25 of their normal partial pressure (Edwards and Lofty, 1977). Individuals of *E. fetida* and other species have been reported to migrate in large numbers from a water-saturated substrate in which the oxygen conditions had been depleted or in which carbon dioxide or hydrogen sulphide had accumulated (Edwards and Lofty, 1977). However, they can live for long periods in adequately aerated water, such as in trickling filters in wastewater treatment plants (Domínguez, 2004).

The habits of epigeic species suit them for engineered processes such as vermicomposting and vermifiltration, which require managed degradation of organic substrates such as food wastes and sewage sludge. Five earthworm species have been used extensively in vermicomposting: *Eisenia andrei*, *Eisenia fetida*, *Dendrobaena veneta*, *Perionyx excavatus*, and *Eudrilus eugeniae* (Domínguez and Edwards, 2011).

The optimum temperature for growth of *Eisenia fetida* and *Eisenia andrei* is 25°C, and although these species can tolerate a wide range of moisture conditions, the optimum moisture content is 85%. In optimum conditions, hatchlings can reach sexual maturity in as little as 21 days, with an ultimate lifespan of up to five years. Feeding is reduced at temperatures below 10°C, while development of young earthworms and cocoon production cease below 4°C. In extremes of temperature, earthworms migrate to deeper layers where they hibernate – although they cannot survive long periods under freezing conditions. The unfavourable effect of high

Table 4.1. Some of the optimal conditions for breeding of *E. fetida* and *E. andrei* in organic wastes, as reported by Domínguez and Edwards (2011).

Parameters	Requirements
Temperature	15°C to 20°C (optimum)
	4°C to 30°C (range)
Moisture	80–90% (optimum)
	60–90% (range)
Oxygen	Aerobic conditions
NH_3-N content	Less than 1 mg g^{-1} (fresh weight basis)
pH	5–9

temperatures (above 30°C) on most species of earthworms is not entirely a direct effect, because these warm temperatures also promote chemical and microbial activities in the substrate; increased microbial activity tends to consume available oxygen, reducing earthworm survival (Domínguez and Edwards, 2011).

Earthworms are very sensitive to ammonia and cannot survive in organic wastes containing NH_3-N at concentrations greater than 1 mg g^{-1} (such as fresh poultry manure). The optimal conditions for breeding *E. fetida* and *E. andrei* are summarised in Table 4.1.

Research into the potential use of earthworms to break down sewage sludge began in the 1970s, when it was demonstrated at laboratory scale that aerobic sewage sludge would be ingested by *E. fetida* and egested as casts – and that in the process the sludge was decomposed and stabilised around three times as fast as non-ingested sludge, apparently because of the increases in rates of microbial decomposition in the casts (Domínguez *et al.*, 2000). Other epigiec species have been identified in the moist, organic-rich environments of septic tank drainage fields – including *Aporrectodea caliginosa*, *Aporrectodea trapezoides*, *Diplocardia conoyeri* and *Diplocardia meansi* (Hawkins *et al.*, 2008).

Kaplan *et al.* (1980) examined the impacts on *E. fetida* of various substrate characteristics in activated and anaerobically digested sludges – identifying optimum temperature, moisture and pH conditions that have been widely reported since (including by Domínguez and Edwards (2011) as presented in Table 4.1). They further reported that concentrations of soluble salts in excess of 0.5% were lethal, and that anaerobically digested sludges were toxic to earthworms, due to their low oxidation-reduction potentials. For vermifiltration systems, these findings indicate that care must be taken with wastewater salinity, NH_3-N concentration and pH, if earthworm productivity is to be maximised. Aerobic conditions must also be maintained – indicating that filter media and reactor configurations must be chosen with care. There are risks that organic media (such as woodchips) will begin to actively compost – creating microbial hotpots that lead to localised anaerobiosis and excessive temperatures, both of which are antagonistic to optimal earthworm performance, but both of which may contribute to aeration of vermifilter beds through chimney effects.

In vermicomposting the earthworms comminute (break down and mix) and aerate the waste. In traditional composting the temperature in the composting mass can reach in excess of 70°C, but vermicomposting is carried out at much lower temperatures, typically below 25°C. These temperatures are themselves inadequate to cause thermal inactivation of any pathogens present in the substrate, and it is sometimes recommended that vermicomposts be subjected to additional pre or post-composting processing to ensure that they are adequately sanitised before use (The Composting Association, 2003). However, earthworms themselves exert significant effects on microbial populations, which result in pathogen kill at mesophilic temperatures. Eastman *et al.* (2001) report data from a field experiment in which two 6 m long windrows of treated sewage sludge were inoculated with faecal coliforms, *Salmonella* spp., enteric viruses and helminth ova. One row was then seeded with *Eisenia fetida* at a ratio of 1:1.5 wet weight earthworm biomass to sewage sludge. All of the pathogen indicators in the vermi-processing row were decreased to below levels in the control row within 144 hours, showing (*inter alia*) a $6.4\log_{10}$ reduction in faecal coliforms, $8.6\log_{10}$ reduction in *Salmonella* spp., and $1.9\log_{10}$ reduction in helminth ova. Swati and Hait (2018) suggest that pathogen reduction through earthworm activity may be due to direct actions such as mechanical disintegration through ingestion and grinding action in the gizzard, inhibition due to intestinal enzymatic action, secretion of coelomic fluids with antibacterial properties, and selective grazing, as well as indirect actions such as stimulation of other microbial species.

It is apparent that in addition to valuable mineralisation and comminution functions (Gomez-Brandon *et al.* (2013) and Singh *et al.* (2017)), earthworms are influencing microbial community structures within their habitats – including biofilms, as normally present in vermifilters (Di *et al.*, 2016). Whether this contributes significantly to the increased treatment efficiencies within vermifilters (as compared with conventional biofilters) is unknown, since the browsing habits of earthworms also contribute to improved aerobic conditions within filters. However, in their fruit and vegetable waste vermicomposting study, Huang *et al.* (2017) showed increased populations of both ammonia oxidising bacteria and archaea in material processed by earthworms – suggesting a combined effect of bacterial community influence and increased aeration in vermifilters, leading to increased functionality and wastewater treatment efficacy. These interactions are difficult to correct for experimentally,

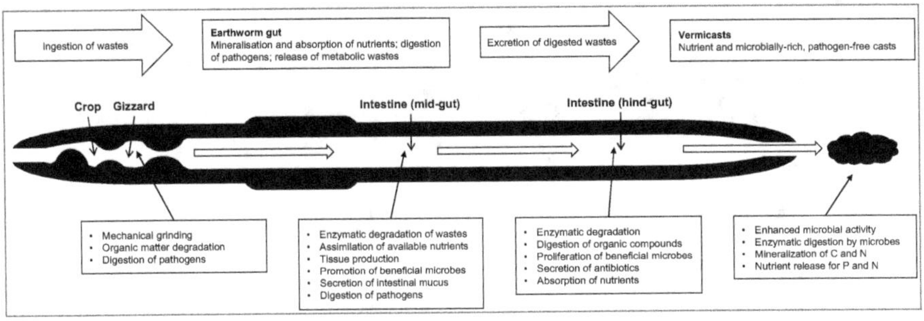

Figure 4.4. Various processes associated with earthworm stabilisation of organic waste materials. Adapted from Swati and Hait (2018).

and research normally compares biofilters with and without earthworms to gain an understanding of overall differences in treatment effects.

Fundamental impacts of earthworm feeding processes on organic waste materials are outlined in Figure 4.4. A proportion of the ingested material is absorbed into the earthworm and the remainder is ejected as casts (or 'vermicasts').

In vermifiltration systems, earthworms also play a vital role in reducing clogging within physical filter media – not through creation of permanent or semi-permanent burrows, but simply through their browsing habits, which create temporary burrows. This is considered in more detail in the next section, but is illustrated by the studies of Wang *et al.* (2010) in which the impacts of *Eisenia fetida* on the hydraulic characteristics of intermittent sand filters (ISFs) were examined. In a simple experiment, two pilot-scale ISFs were constructed, one of which contained earthworms and one of which did not. Although operated at the same organic loading (approximately 300 g COD m^{-2} d^{-1}) the ISF without earthworms clogged after 53 days – and partially recovered after a seven-day rest interval, after which is clogged again in 40 days. By contrast, no standing water was ever observed on the filter with earthworms (Wang *et al.*, 2010).

4.2.2 The Roles of Support Media and Biofilms

Vermifilters share some characteristics with trickling filters, which have a long history of deployment in centralised and semi-centralised wastewater treatment. A trickling filter is an aerobic fixed film reactor in which the filtration media support the growth of a biological film. Media traditionally consist of rocks, slag, or synthetic materials. Rock and slag trickling filters generally have one to two metres of media depth. Plastic media trickling filters are normally constructed much deeper because of their lighter weight and better ventilation capabilities once packed (USEPA, 1991).

Once wastewater starts to flow over the medium, microorganisms in the water gradually attach themselves to the supporting media to form a biological film

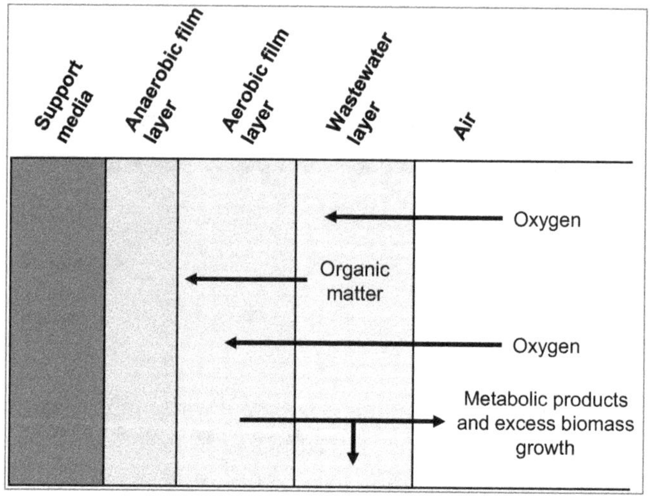

Figure 4.5. Cross-section of a biofilm. Adapted from Mara and Horan (2003).

approximately 0.1 to 0.2 mm thick. This community comprises a mixture of aerobic, anaerobic, and facultative bacteria – as well as fungi, algae and protozoa (USEPA, 2000a). Bacteria attach themselves to the substrate by means of the exopolymers they produce. The initial colonisation of a solid takes place on a number of preferential sites and it is from these sites that the biofilm develops until the total surface area of the substrate is coated. At the same time, new cells are produced and cause the film to thicken. The oxygen and nutrients carried in the wastewater diffuse through the film until it achieves a thickness such that oxygen no longer reaches the deepest cell layers (Figure 4.5). The thickness of the aerobic and anaerobic layers will vary depending on reactor and substrate types (SUEZ, 2020).

As the biological film continues to grow, microorganisms near the surface lose their ability to cling to the media, and a portion of the film falls off the filter. This process is known as sloughing. The sloughed solids are collected in an underdrain system and transported to a clarifier for removal from the wastewater (USEPA, 2000a).

All media filters require primary clarification (removal of suspended solids) of the wastewater to avoid clogging. In a trickling filter, the influent wastewater is distributed on the top surface and trickles downwards. As the water flows downwards, soluble organic matter is removed by aerobic heterotrophic microorganisms in the attached biofilm. Aeration occurs through natural convection of air through ventilation ports connected to the underdrain system at the filter base. The filter medium is unsaturated, that is, after the liquid has trickled down, the porous spaces are occupied by air, thus guaranteeing aerobic conditions (Oakley and Von Sperling, 2017). These principles are illustrated in Figure 4.6.

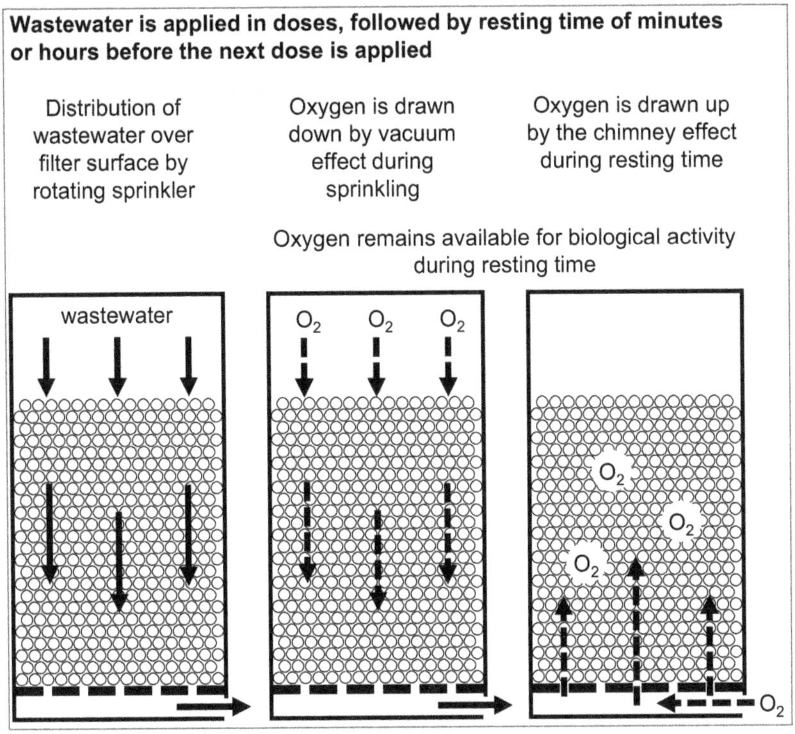

Figure 4.6. Principles of trickling filters. Adapted from Sasse (1998).

Organic loading and flow rates through trickling filters can be varied according to the strength of the wastewater, the intended BOD removal, the nature of the support media and the requirement for periodic flushing (to remove sloughed biomass and minimise clogging). 80% BOD removal is possible, although recirculation may be required to achieve this (depending on hydraulic retention time and wastewater strength). Although incoming wastewater can be applied to the surface of the filter media under gravity, it is more usually applied through pumps – which are also required where recirculation is necessary. Despite their relative simplicity, some authors do not consider trickling filters suitable for decentralised applications (Sasse, 1998). Trickling filters are normally operated with daily flushing cycles, where flow rates are increased up to 15 times of normal operation to manage biomass accumulation and sloughing into the outflow. Such operations can be implemented manually – although automatic operation is more convenient (Daigger and Boltz, 2011).

In addition to removal of BOD, aerobic conditions mean that trickling filters are also capable of nitrification – converting NH_4^+-N to NO_3^--N – providing conditions are correct. Trickling filters may be engineered specifically for nitrification, or for providing combined nitrification and BOD_5 removal (they are also designed

Table 4.2. Classification of different trickling filter types, as set out by Daigger and Boltz (2011). n/a = Not Applicable.

Parameters	Roughing Filter	Carbon Oxidising Filter	Carbon Oxidising and Nitrification Filter	Nitrification Filter
Flow pattern	Vertical flow	Cross flow or vertical flow	Cross flow or vertical flow	Cross flow
Wastewater source	Primary effluent	Primary effluent	Primary effluent	Secondary effluent
Hydraulic loading ($m^3 day^{-1} m^{-2}$)	52.8–178.2	14.7–88.0	14.7–88.0	35.2–88.0
BOD_5 load ($kgm^{-3}day^{-1}$)	1.6–3.52	0.32–0.96	0.06–0.24	n/a
$NH_4{}^+$-N load ($gm^{-2}day^{-1}$)	n/a	n/a	0.2–0.1	0.5–2.4
Removal efficiency (%) or effluent quality (mgL^{-1})	50–75% filtered BOD_5	20–30 mgL^{-1} BOD_5	<10 mgL^{-1} BOD_5 <3 mgL^{-1} $NH_4{}^+$-N	0.5–3 mgL^{-1} $NH_4{}^+$-N
Filter depth (m)	0.9–6.1	≤ 12.2	≤ 12.2	≤ 12.2

to provide preliminary treatment of higher-strength effluents as 'roughing' filters) (Table 4.2).

Nitrification is a two-step process: $NH_4{}^+$-N is first oxidised to $NO_2{}^-$-N by *Nitrosomonas* bacteria, and then to $NO_3{}^-$-N by *Nitrobacter* bacteria. Both types of nitrifying bacteria are autotrophic. Oxidation of soluble BOD by heterotrophic bacteria depletes oxygen, and the nitrifiers are unable to compete with these relatively faster growing heterotrophs until soluble BOD concentrations in the liquid phase are low enough (USEPA, 1991), and for this reason nitrifying trickling filters often operate in recirculation modes, with each pass reducing BOD until conditions are favourable for nitrification. Organic loading, hydraulic loading, temperature, pH, dissolved oxygen concentration and filter media all influence the kinetics of nitrification (USEPA, 2000a).

Trickling filters are designed to remove soluble organic matter, and are not expected to have high pathogen removal rates. Oakley and Von Sperling (2017) report maximum removal rates of 1.0 $\log_{10}$ for Salmonella, 0.5 $\log_{10}$ removal of viruses, 0.8 $\log_{10}$ removal of protozoa cysts/oocysts, and 1.4 $\log_{10}$ removal of *E. coli*. The principal removal mechanisms are retention in the biofilm by adsorption, and sedimentation in the sloughing biofilm.

Advantages

- Simple, reliable process.
- Suitable in areas where large tracts of land are not available for a treatment system.
- Effective in treating high concentrations of organics depending on the type of media used, and flow configuration.
- Appropriate for small- to medium-sized communities.
- High degree of performance reliability at low or stable loadings.
- Ability to handle and recover from shock loads.
- Rapidly reduce soluble BOD_5 in applied wastewater.
- Efficient nitrification.
- Low power requirements.
- Requires only a moderate level of skill and technical expertise to manage and operate the system.

Disadvantages

- Additional treatment may be needed to meet more stringent discharge standards.
- Regular operator attention needed to check mechanical performance and manage clogging.
- Relatively low organic loadings required depending on the media.
- Limited flexibility and control in comparison with activated-sludge processes.
- Potential for fly and odour problems.
- Autotrophic bacteria (nitrifiers) are more sensitive to "shock loads" than other bacteria.
- Predation (i.e. fly larvae, worms, snails) decreases the nitrifying capacity of the system (USEPA, 2000a) and (USEPA, 2000b).

When earthworms are introduced into trickling filters, they have a number of effects – not least reducing or eliminating the sloughing of excess biomass. This is ascribed to the constant browsing of the biofilm by the earthworms, which also helps to ensure that the media do not clog, reducing maintenance and operational costs SUEZ (2020) and Liu *et al.* (2013). Where biofilm does slough from media, its viscosity and 'clogging potential' are reduced through earthworm consumption and casting (Ye *et al.*, 2018).

The balance between biomass growth and earthworm browsing is key to low or zero sludge production in vermifilters, and widely reported (for example in (Sinha *et al.*, 2010)). The central concept behind vermifiltration is that microorganisms perform biochemical degradation of waste material, while earthworms regulate microbial biomass and activity by directly or/and indirectly grazing on microorganisms (Liu *et al.*, 2012). It must therefore be assumed that the lack of sludge in filtered effluent is an artefact of the balance between solids' loading, biomass growth (microbial and earthworm), biofilm browsing and respiration. Over time, wormcasts can be expected to accumulate within the filter bed, which should lead to increased bed depth. However, even this may not be observed where biological degradation of organic filter media (such as sawdust and woodchips) results in corresponding reductions in filtration volume. Earthworms are known to browse the whole filter depth, where conditions (moisture, oxygen, pH etc) are amenable (Luth, 2011).

Data on long-term field performance of vermifilters are extremely scarce and tend to be limited to systems where pre-settled and filtered wastewater is applied to the filter beds. For example, a 500PE lumbrifiltration trickling filter has been operational in Combaillaux (France) since 2005. This system was established with a 120 cm bed of woodchips, irrigated with wastewater pre-filtered to remove >2 mm particles. No biomass sloughing has been reported at the outlet, even after fifteen years of operation. Sludges from filtration of the incoming wastewater are processed separately via vermicomposting (Naigeon, 2005). This installation is based on the Chilean Sistema Tohá® in which Sepúlveda (2004) highlights the absence of sludge in the final effluent, allowing its subsequent disinfection with a simple UV system.

Within classic trickling filters, the role of the filter media is to provide a surface on which the biofilm can establish. The greater the surface area available within a given volume, then (in principle) the greater the area of biofilm and hence treatment efficiency. However, in excessively divided media there is a risk of clogging through biomass growth or even simple physical adherence between adjacent particles. Plastic filter media are widely used in trickling filters – since they contain a higher fraction of voids than mineral media and are less prone to plugging. Their light weight also allows for construction of deeper filter beds, improving treatment efficiency within similar footprints (Lekang and Kleppe, 2000). Vermifiltration has been tested in multiple media – mineral, organic and plastic (see Section 4.3 for examples). There are few reports of negative impacts of filtration media on earthworm mortality, although the surface roughness of media should be considered when designing systems. Enrique Hernández and Furlong (2017) report increased mortality in vermifilters using granular activated charcoal – which they ascribe to the media causing physical damage to *Eisenia fetida* within their filtration units.

The combinations of different filter media characteristics can lead to very different treatment performance. For organic loads of less than 1 kg BODm^{-3}day^{-1}, well-designed and operated rock-media trickling filters are capable of providing performance approaching that of synthetic-media trickling filters. However, as organic load increases, there is likely to be reduced clogging when synthetic media are used (Daigger and Boltz, 2011).

Although trickling filter media are normally designed to ensure even distribution, this may not provide optimal conditions for both earthworm reproduction and growth. Luth (2011) reports on experimental and full-scale vermifilters treating diluted pig slurry. At full scale (10 m × 4 m × 0.8 m), larger earthworms were found in regions with higher accumulations of organic matter, while more earthworm cocoons were found in drier areas. To provide varied conditions, operational changes were made to ensure that part of the filter was irrigated less frequently than the other. Whether the use of such 'nursery' conditions is strictly necessary to optimise vermifilter performance has not yet been demonstrated.

4.2.2.1 Nutrient removal

Simple sand and peat filters can also deliver excellent solids and BOD reductions through a combination of physical and biological action. Lourenço and Nunes (2020) cite up to 96% removal of suspended solids within recirculating sand filters and peat filters, and up to 93% removal in intermittent sand filters. BOD removals of up to 99% are also cited for these technologies. Under laboratory conditions Corley *et al.* (2006) and Rodgers *et al.* (2004) demonstrated total COD removals of 84% and 90% in peat and sand filters (respectively), with almost complete nitrification in both filter types. Denitrification has also been observed by some vermifiltration researchers (Taylor *et al.*, 2003), and this raises interesting prospects for combined nitrification/denitrification vermifilter systems. Dalahmeh *et al.* (2011) summarises prior research into the use of simple down-flow 'mulch' filters to treat greywater. TSS, BOD and COD removals of up to 91%, 99.9% and 98% are cited, respectively – although the authors also note reports of increased COD in the effluent from some filter media. This tendency (from media such as woodchips) is exploited in denitrifying reactors, where the media act both as a biofilm support and source of dissolved organic carbon – which promotes heterotrophic denitrifying bacteria (Lopez-Ponnada *et al.*, 2017).

The basic principles of denitrifying woodchip bioreactors are illustrated in Figures 4.7 and 4.8. A key feature of such reactors is that the woodchips are retained in a saturated zone, to stimulate and maintain anoxic conditions suitable for optimum denitrification. In theory a vermifilter reactor could be extended downwards to include a permanently saturated zone beneath the aerobic filter bed. Dissolved oxygen levels have been shown to increase with depth in experimental vermifiltration systems (Taylor *et al.*, 2003), and this would need to be managed within the denitrification zone. Various authors report good nitrogen removal in periodically-flooded or even unsaturated woodchip filters (for example, Ruane *et al.*, 2012, Carney *et al.*, 2013 and Murnane *et al.*, 2016), suggesting that it might be possible to

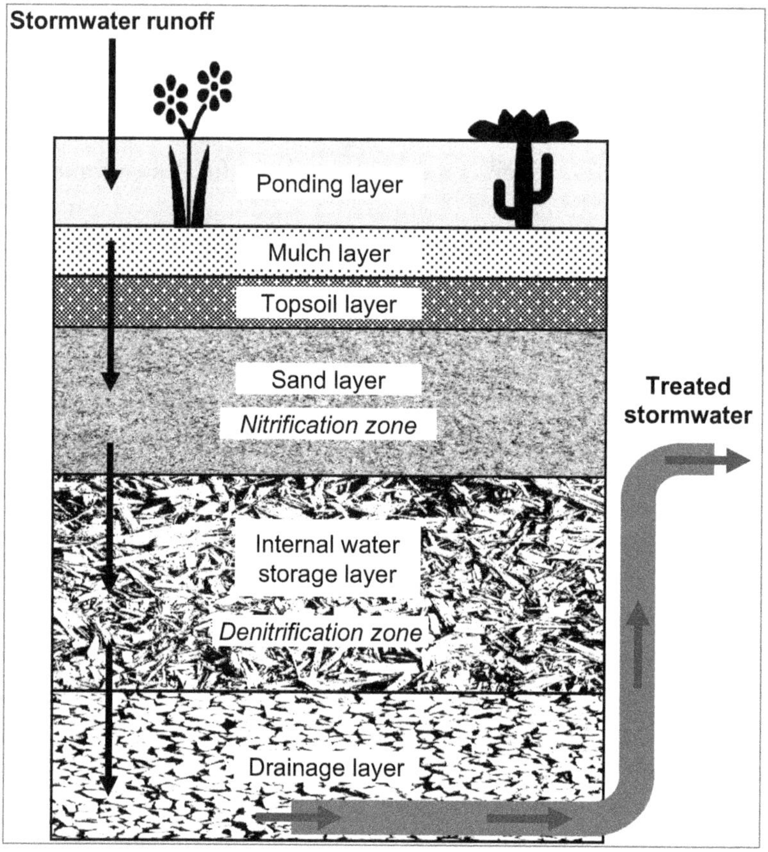

Figure 4.7. Basic design of a storm water treatment swale, in which both nitrification and denitrification take place in different layers. Adapted from Lopez-Ponnada *et al.* (2017).

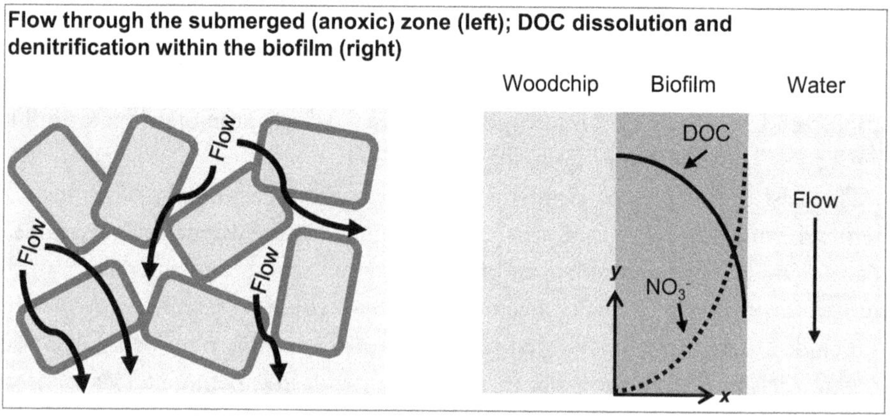

Figure 4.8. Principles of denitrification in a woodchip bioreactor. DOC = Dissolved Organic Carbon. Adapted from Lopez-Ponnada *et al.* (2017).

successfully design and operate a nitrifying/denitrifying vermifilter without compromising other performance aspects. The biochemical mechanisms of denitrification are expected to be similar in unsaturated systems to those maintained under anoxic conditions, although nitrogen removal may not be quite so consistent.

The type and size of the woodchips, hydraulic loading rate, and intervals between wetting have been shown to affect the hydrolysis rate of lignocellulosic substrates, which then influences the amount and bioavailability of dissolved organic carbon required for denitrification. Maintenance of saturated conditions during non-operational periods is also a critical design aspect that controls the overall performance. Higher NO_3^--N removal and longer woodchip media longevity is expected from these designs compared with bioreactors that are designed to be unsaturated except during operation (Lopez-Ponnada *et al.*, 2017). Commercial decentralised denitrification systems using these principles are available.

In addition to its roles as biofilm carrier and potential carbon source for denitrification, filter media can also be used to remove dissolved and particulate phosphorus from wastewater. The phosphorus-adsorption capabilities of many different media have been trialled in a range of applications relevant for decentralised wastewater treatment. Vohla *et al.* (2011) summarise data on different filter media used for phosphorus removal in constructed wetlands, while (Pratt *et al.*, 2012) consider standalone reactive filters.

In most cases, phosphorus is bound to the filter media through adsorption and precipitation reactions with calcium (Ca), aluminium (Al) and iron (Fe), although biological uptake can also be significant in planted filter systems or those where biofilms establish (Vohla *et al.*, 2011). Although very effective adsorbents, pH correction of the treated effluent may be needed – particularly when industrial waste materials such as blast furnace slag are used (Table 4.3). In addition to the media listed in Table 4.3, researchers have examined the performance of many others, including: bauxite (Altundoğan and Tümen, 2002), dolomite (Karaca *et al.*, 2004), peat (Kõiv *et al.*, 2009), zeolite (Sakadevan and Bavor, 1998), iron ore (Grüneberg and Kern, 2001) and lightweight expanded clay aggregate (leca) (Öövel *et al.*, 2007).

Units for phosphorus removal are normally downstream of primary and/or secondary treatment units, to ensure that suspended solids are removed and don't blind the filter media (Hedström, 2006). Hedström (2006) summarises experiences in Norway with single household P removal filter beds. These typically contain 40 m^3 of reactive filter media and are expected to last up to 15 years, before the filter media require replacement (Figure 4.9). P removal rates of 90% can be expected through these simple downflow reactors. Compact upflow reactors are also available, but require more frequent media replacement (as frequently as every year). Establishing

Table 4.3. Phosphorus and phosphate removal performance for a number of natural and artificial media in full-scale constructed wetlands (or trickling filters), as summarised by Vohla *et al.* (2011).

Substrates	Descriptions	P Removal	pH of Outflow
Gravel	Dairy farm wastewater; Mean influent concentration 15 mg P L^{-1}; sampled over 5 years; HLR 21.4–71.7 mm d^{-1}; HRT 1.95–6.54 d	Total P removal 184–296 gm^{-2}	–
Gravel	Vertical sub-surface flow wetland; HLR 3 m^3 d^{-1}	$PO_4{}^{3-}$-P removal efficiency: 4.33%	–
Sand	Horizontal sub-surface flow sand filter; 5 years; HLR 1.0–6.3 m^3 d^{-1}	Total P removal efficiency: 78.4%	–
Fly ash	Eutrophic river water treatment through a three-stage filter (one stage filled with fly ash); HLR 60 m^3d^{-1}; sampled over 14 weeks	Total P removal efficiency: up to 83%	–
Blast furnace slag	Trickling filters; wastewater flow of ~60 m^3d^{-1}; sampled over seven months	Total P removal efficiency: up to 99%	Up to 11.52

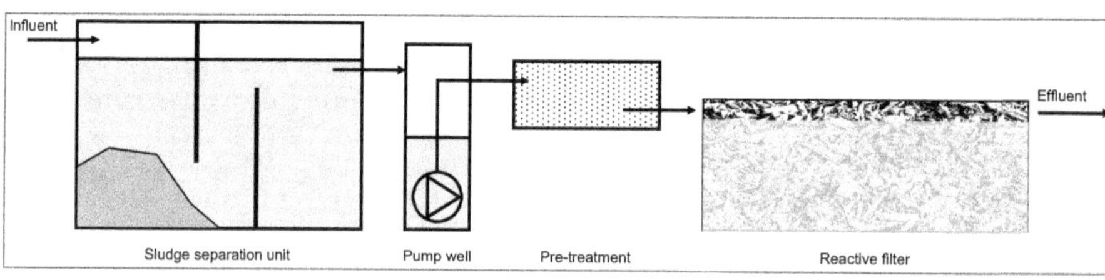

Figure 4.9. Outline of decentralised wastewater treatment system with reactive filter (for P removal). Adapted from Hedström (2006).

vegetation in the top layer of downflow filters is common, to improve their amenity value – and increase phosphorus removal.

A key point in P removal (unlike nitrogen removal), is that the adsorption reactions that remove phosphorus from the liquid phase will eventually exhaust all available adsorption sites – allowing breakthrough of phosphorus. The media therefore need to be periodically replaced.

4.3 Commercial Vermifiltration Systems

Vermifiltration has been successfully demonstrated at experimental scale in many countries, including: Burkina Faso (Adugna *et al.*, 2014); India (Arora and Kazmi, 2015); Jordan (Dalahmeh *et al.*, 2011); China (Liu *et al.*, 2012); Portugal (Lourenço and Nunes, 2017); Zimbabwe (Manyuchi *et al.*, 2013); Australia (Sinha *et al.*, 2008). The technology has also been deployed commercially – particularly in Latin America, India and Australasia – to treat blackwater, greywater and industrial effluents. These commercial systems tend to fall into two broad categories:

1. Primary. Those capable of processing whole wastewater (Figure 4.10). These also generate vermicastings requiring periodic removal and the treated effluent is usually disposed to ground – either via constructed drainage/transpiration fields or (in the case of Tiger Toilets) directly into the soil surrounding the constructed filter pit. These systems are used in Africa, Asia and Australasia; and
2. Secondary. Those that require preliminary filtration or clarification of wastewater before vermifiltration (Figure 4.11). Such systems are commonly

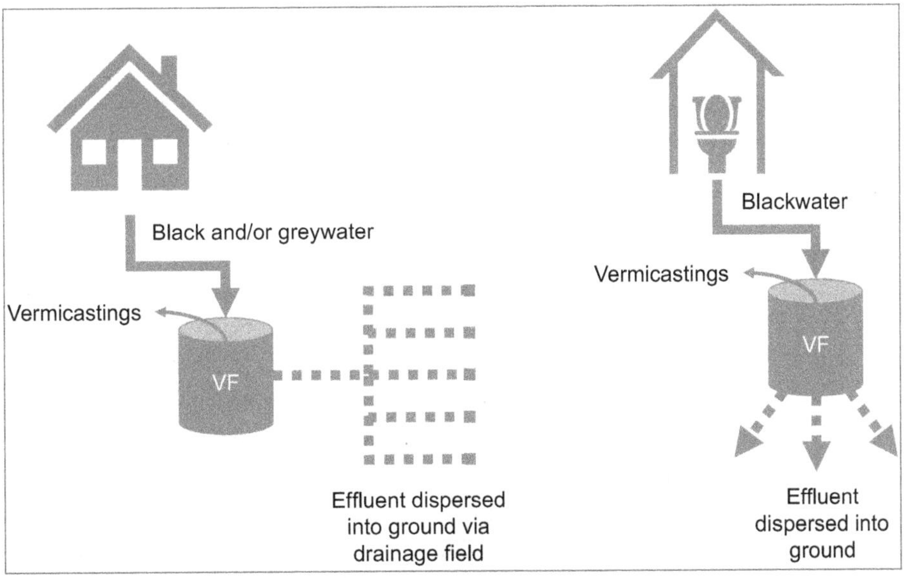

Figure 4.10. Primary vermifilters (denoted by VF). [L] A typical Australasian system; [R] A Tiger Toilet.

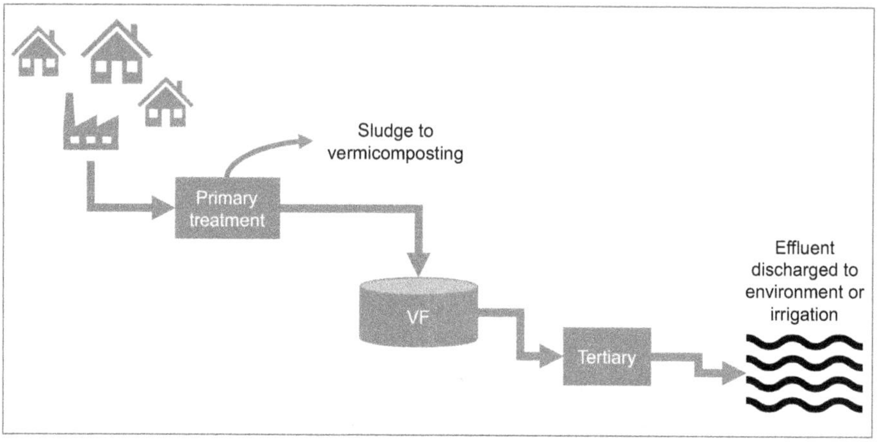

Figure 4.11. Secondary vermifilters. Illustration based on typical Sistema Tohá implementation.

derived from the Chilean Sistema Tohá®. Primary sludges may be processed by vermicomposting, but the vermifilter itself produces little or no sludge. The treated wastewater may be suitable for reuse in irrigation following a final disinfection step. These systems are used in Latin America and Europe.

A number of commercial vermifiltration systems have been identified (Table 4.4), and are described in the following sections.

4.3.1 Primary Vermifilters

4.3.1.1 Biofil

The Biofil digester can be used with households' existing toilet systems or it can be purchased as an integrated toilet-digester system. Wastewater and faeces enter at the top of the unit and are deposited on the vermibed – a layer of organic materials (such as fine woodchips) that provides the earthworm habitat. Liquids percolate down through this layer and through a porous concrete filter, both of which retain solids. The solids are degraded by bacteria and earthworms, while liquids soak away directly into surrounding soil (Figure 4.12). The basic digester (600 mm × 600 mm × 1800 mm) can accommodate 25 users in the micro-flush setting and 10 users when operated as a flush unit (Stichting Waterputten voor Kwasiakan, 2017). This system is very similar to the Tiger Toilet, described in the next section.

In a 2016 study, Amoah *et al.* (2016) sampled effluent from a number of installed units, and tested against the Ghanaian Environmental Protection Agency (EPA) standards. Results are summarised in Table 4.5, and while non-compliant, the authors suggested that the units were overloaded – although still reducing incoming loads significantly (Amoah *et al.*, 2016).

Table 4.4. Commercial vermifiltration systems from various countries.

Companies	Countries	Types
Wastewater Wizard*	UK (Scotland)	Primary
Biofil	Ghana	Primary
Biolytix	New Zealand	Primary
NaturalFlow	New Zealand	Primary
Autoflow	New Zealand	Primary
Simple Waste Water Solutions*	New Zealand	Primary
Wormsmart	Australia, New Zealand and India	Primary
Worm Farm Waste Systems	Australia	Primary
Zenplumb	Australia	Primary
Aqua Clarus*	Australia	Primary
CPlantae	Mexico	Primary
Biofiltro	Chile, Perú, USA and New Zealand	Secondary
Sistema Tohá®	Chile, Bolivia, Ecuador, India and Mexico	Secondary
Transplast	Chile	Secondary
Solsan*	Chile	Secondary
LombriTek association	France	Secondary
Organic Solutions	India	Secondary
TBF Environmental Solutions	India	Primary & Secondary
Transchem Agritech	India	Primary & Secondary

*Due to lack of available design or performance information, these systems are not described in this chapter.

4.3.1.2 Tiger toilet

This system is supplied to the Indian market by TBF Environmental Solutions Pvt Ltd (see Section 4.3.3.1). It combines a flushing WC with a small vermifilter and is based on research previously undertaken to provide sanitation solutions to the humanitarian sector. The vermifilter can either be constructed *in situ* with local materials as a replacement for a conventional pit latrine, or supplied within a plastic reactor unit that can be placed within a pit. Within the reactor, incoming solids accumulate on the surface of the bedding material – where they are subsequently consumed by the earthworms. Liquids drain through gravel layers, where microbiological treatment takes place as described in previous sections. Partially treated effluent is discharged into surrounding soil.

Table 4.5. Performance of biofil digester units ($n = 108$). From Amoah *et al.* (2016).

Parameters	Range (Median)	Mean ± Standard Error	Ghana EPA Standards
BOD (mg L^{-1})	10.0–984 (144)	219 ± 20.0	50
TSS (mg L^{-1})	12.0–3440 (254)	500 ± 55.0	50
NH_4^+-N (mg L^{-1})	2.0–368 (109)	117 ± 5.9	1
PO_4^{3-}-P (mg L^{-1})	0.2–36.1 (5.8)	8.9 ± 0.8	2
E. coli (CFU 100 mL^{-1})	$0.0 - 1.0 \times 10^8$ (1.0×10^7)	$1.7 \times 10^7 \pm 1.8 \times 10^7$	1×10^1

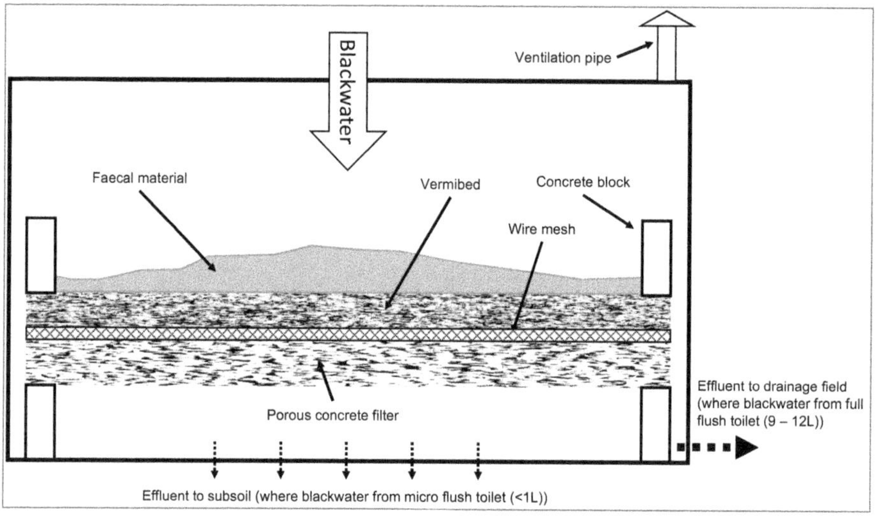

Figure 4.12. Cross-section through Biofil vermifilter unit, adapted from Amoah *et al.* (2016).

The design and performance of different Tiger Toilet configurations are described in the literature, with examples presented in Figures 4.13 and 4.14.

Ten in-field prototypes were constructed in brick, as illustrated in Figure 4.13. The bedding layer consisted of 10 cm of compost while the drainage layer was made of graded aggregate (60 cm) below an uppermost layer of sand. The vermifilters had a diameter of 1.2 m by 1.25 m deep and were set in the ground. This was connected to a separate pour flush squat pan via an inspection chamber. The worm density was 2 kg/m^2 (Furlong *et al.*, 2014). COD reduction in these units (influent vs effluent) was 57% (n = 10) (Furlong *et al.*, 2016). COD removals in 18 smaller experimental containerised units were estimated at 89–94% (Furlong, 2018). The internal diameter of these reactors was 400 mm, with a height of 800 mm. Under the rim of the lids a row of 1 mm holes were drilled at 20 mm intervals to allow for airflow

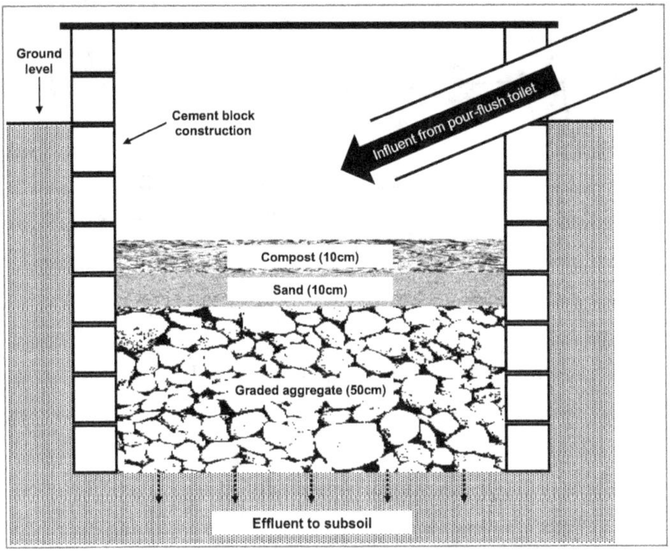

Figure 4.13. Tiger toilet constructed *in situ*. Adapted from Furlong *et al.* (2017).

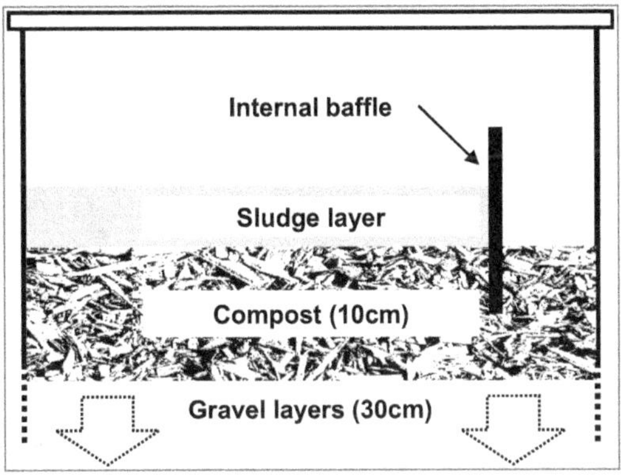

Figure 4.14. Experimental vermifilter, showing internal baffle to prevent bed smothering. Adapted from Furlong (2018).

(Furlong, 2018). In these smaller units it was necessary to incorporate a baffle to prevent smothering of the whole bed surface with faecal material; this would have compromised passive air flow through the filter beds (Figure 4.14).

4.3.1.3 Biolytix

The Biolytix BioPod system treats incoming domestic wastewater from a single household through a layered vermifilter. Plastic filter media are used in all treatment layers, but two layers additionally include 'coco peat', to increase the surface area of the treatment process and to retain moisture. The filtration media together occupy more than 2 m^3 of the reactor volume, with each discrete layer separate

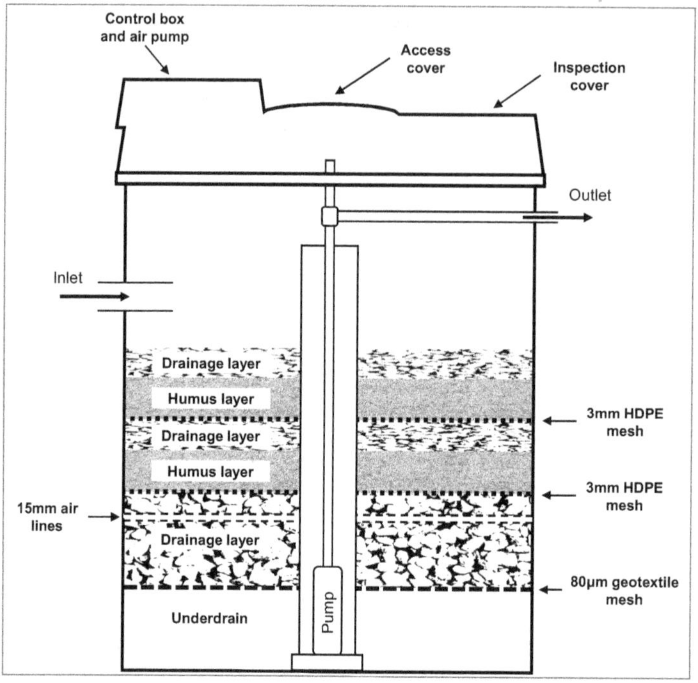

Figure 4.15. Cross-section diagram of Biolytix BF6 system, adapted from Biolytix (2015).

by plastic filter screens with a 3 mm aperture size. The filter bed is separated from the underdrain by a geotextile filter layer with a nominal pore size of 80 μm. The underdrain is filled with plastic media to 400 mm depth (Figure 4.15), giving a total bed depth of 1050 mm. An air pump is used to provide additional air to the bed at the rate of 350 L per hour.

During commissioning, the filter is inoculated with a kilogram of tiger worms (*Eisenia fetida*), which then populate the whole filter bed. Treated effluent is pumped from the sump to a disposal/drainage field.

Indicated maximum loadings of the system are:

- Flow rate: 1600 L per day
- BOD_5 load of 700 g per day
- SS load of 700 g per day

Effluent quality at the point of discharge to the drainage field:

- BOD_5 <20 mgL^{-1}
- Suspended solids <30 mgL^{-1}

Information from Biolytix (2015).

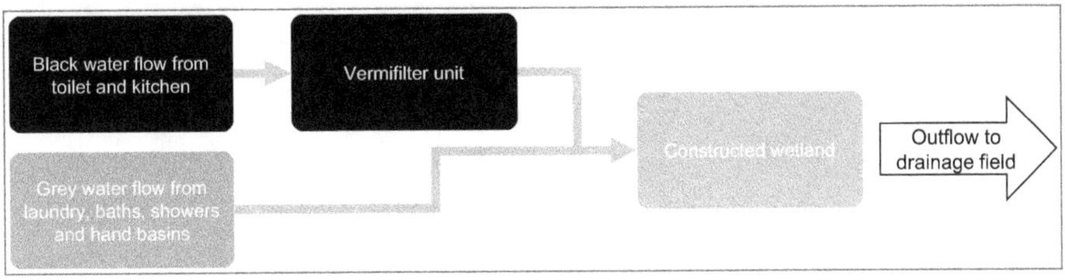

Figure 4.16. NaturalFlow and Autoflow treatment approach. Adapted from Waterflow (2019).

4.3.1.4 Autoflow and NaturalFlow

In both of these systems, separate black and grey water are treated from single households. Only the blackwater (which combines flow from the household kitchen and toilets) is passed into a vermifilter, where Tiger worms consume the solid material trapped within a filter medium. Bark and a mix of soil with straw are used in the NaturalFlow and Autoflow systems, respectively. The partially treated vermifiltrate is then combined with the greywater flow ahead of treatment in a constructed wetland (Figure 4.16). Gradual accumulation of worm casts is expected in both vermifiltration systems, which will require removal at intervals of around five years.

Following treatment in the combined vermifiltration/constructed wetland system, a discharge quality of 20 mgL^{-1} (BOD_5) and 30 mgL^{-1} (TSS) is achieved prior to disposal in a drainage field.

4.3.1.5 Worm Farm and Zenplumb

Worm Farm and Zenplumb are both single chamber solutions, marketed as domestic 'septic' system alternatives. Worm Farm also design bespoke systems for larger applications, such as schools, hotels and other businesses. The domestic units from both companies are designed to accept combined black and greywater flows, and may also incorporate entry points for direct introduction of other organic matter (such as kitchen scraps) (Figure 4.17). Treated effluent flows through the filter and is either gravity fed or pumped to transpiration/drainage fields. No performance data could be identified for either of these systems (Baumgartner, 2013).

4.3.1.6 CPlantae and WormSmart

In the CPlantae system from Mexico, black and/or greywater is treated via a vermifilter and constructed wetland. The system is targeted at both domestic and business users, based on a standard vermifiltration module that has an internal volume of 1.2 m^3, partially filled with 'organic substrate' (0.5 m^3) and expanded polystyrene (0.25 m^3) (Figure 4.18). Aeration is maintained with a solar driven fan. At start-up the unit is inoculated with 1 kg (fresh weight) of earthworms.

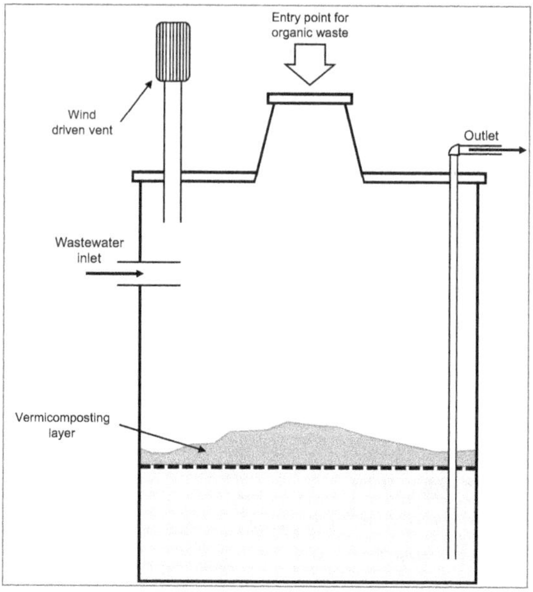

Figure 4.17. Worm Farm system, adapted from A&A Worm Farm Waste Systems (2020).

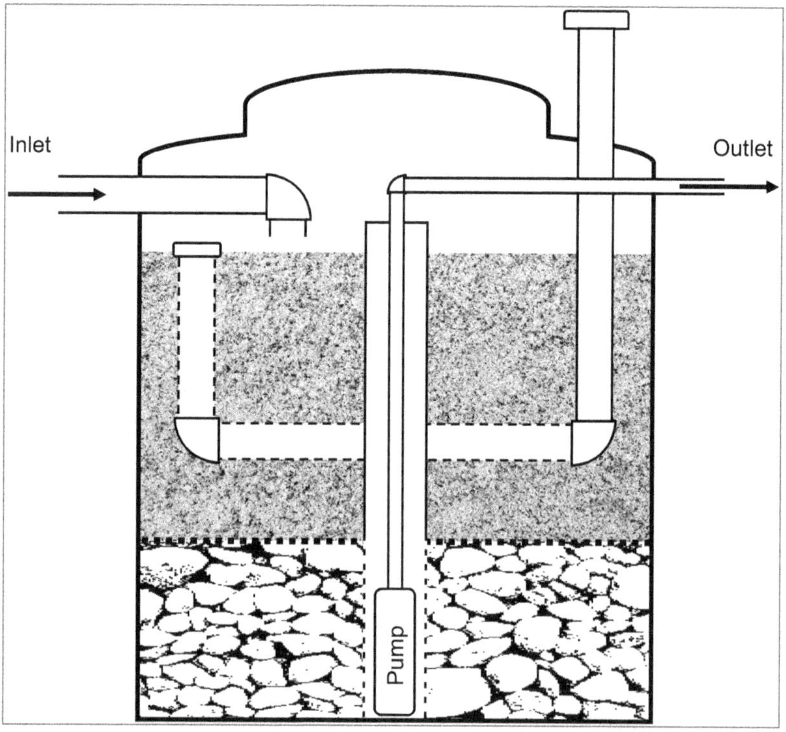

Figure 4.18. CPlantae 'WormPod', adapted from CPlantae (2017).

Retained solids are converted to vermicompost, while effluent passes through the filter bed prior to discharge. Total liquid retention time within the vermifilter is minimal (minutes), and it appears that the majority of treatment is performed within the receiving constructed wetland. Users are required to agitate the composting media once per month during the first six months of operation – after which no further intervention is required, other than removal of accumulated vermicompost at three yearly intervals. It is claimed that treated water complies with local NOM-001-SEMARNAT-2017 requirements after the initial six month acclimation phase (CPlantae, 2017).

The WormSmart solution from New Zealand is similar, although the partially treated effluent passes from the vermifilter and into a second (foam-filled) physical filter unit before discharge to a drainage or irrigation field. Various configurations are supplied without the secondary filtration phase, but under tests with domestic wastewater, final effluent from a combined system achieved a BOD_5 of <20 g/m^3 and TSS of <30g/m^3 (Worm Smart, 2018).

4.3.2 Secondary Vermifilters

4.3.2.1 Sistema Tohá®

Biofiltro, Lombritek and Sistema Tohá® systems have common origins in research undertaken at the University of Chile (Santiago), and all use similar design and operating principles. Sistema Tohá® and Lombritek take a bespoke approach to wastewater treatment, while Biofiltro also offer modular systems. In all cases, wastewater is first filtered and/or settled to remove suspended solids >2 mm. Clarified wastewater is then irrigated onto a bed that may contain combinations of woodchips, wood shavings, sawdust and bark, within which the earthworm population is established. The effluent then trickles down over gravel and/or rock layers where it is further treated within an established microbial biofilm. The systems are passively aerated, and final effluent from Tohá systems is sufficiently treated to be passed through UV prior to discharge, allowing its use in a number of irrigation applications.

4.3.2.2 Biofiltro

The Biodynamic Aerobic (BIDA®) System is sold in a number of containerised configurations to treat different flow and loads in a variety of sectors (domestic wastewater, dairy wastewater, food and beverage industry effluents) (Table 4.6). All systems operate by spraying or trickling settled and filtered wastewater onto a bed of wood shavings inoculated with *Eisenia fetida* at densities of up to 12,000

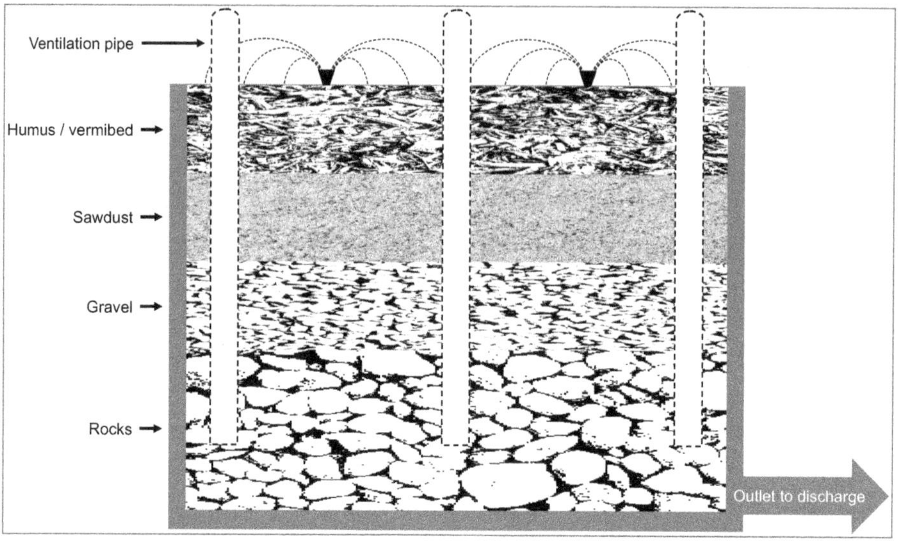

Figure 4.19. Sistema Toha®. Adapted from Sepúlveda (2004).

Table 4.6. Claimed performance characteristics for the BIDA® system in different sectors. From Dore *et al.* (2019).

Sectors	Design Flow ($m^3 day^{-1}$)	BOD Removal (%)	TSS Removal (%)
Dairy	168	52	85
Food	271	87	85
Municipal	13.7	95	93
Slaughterhouses	0.54	91	86
Wine	19.4	90	91

individuals per m^3. The vermifiltration bed is supported on a gravel layer, beneath which filtered effluent is captured in an underdrain. Final effluent can be discharged to irrigation fields or further treated before reuse, depending on local regulatory and customer requirements.

The systems are supplied with monitoring software, allowing remote user access to basic operational and troubleshooting functions. The software also logs water usage and influent and effluent water quality data.

All configurations are shipped in standard 6.06 m (20′) long containers, but their treatment capacity varies according to filter bed size and configuration within the container. For example, the "Wiggle Room" configuration comprises two separate (but stacked) vermifilter layers, which can be run in single pass (the layers operate in parallel) or double-pass (the layers operate in series) (Figure 4.20).

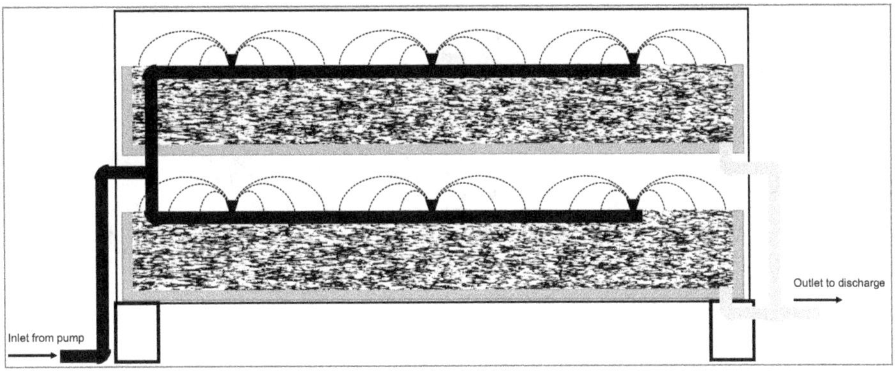

Figure 4.20. Biofiltro Wiggle Room treatment system, showing stacked containerised filters and separate control/pumping unit. Adapted from Biofiltro (2020b).

Influent concentrations can vary from 200 to 35,000 mgL^{-1} (BOD) and 0 to 2,000 mgL^{-1} (TSS), with a maximum daily flow of 5.5 m^3. In single-pass mode >90% BOD removal is claimed within four hours – with >99% BOD removal claimed when the system operates in double-pass mode. Roughly 10 m^3 of 'worm castings' are generated every 18 months, requiring removal for separate use (which could include direct land application, depending on local regulatory requirements).

Filter bed dimensions are constrained within these modular units, and flow rates are dictated by influent strength and treatment requirements. For example, the "Can of Worms" unit is stated to remove up to 99% BOD_5 under daily flows of around 1 m^3 per day when influent BOD_5 is up to 6,000 mgL^{-1}. The same removal can be delivered at a flow of around 5 m^3 per day when influent BOD_5 is up to 1,000 mgL^{-1} (Biofiltro, 2020a). In common with other technology providers, Biofiltro offer bespoke design for applications that are outside the scope of their modular units.

4.3.2.3 Lombritek association

This French enterprise designs and installs bespoke vermifiltration units for use at community scale. The 500PE system at Combaillaux has been operating since 2005. Screened and settled wastewater is irrigated onto the 12 m diameter filter bed which comprise a 10 cm pine bark layer (a mixture of woodchips, pine sawdust and bark – in which *Eisenia andrei* are cultivated) over a 120 cm mixture layer of woodchips. The organic layers sit atop a 10 cm deep gravel drainage bed (20 to 30 mm particle size) (Figure 4.21) (Gautier, 2007). The unit is designed to treated clarified domestic wastewater – the primary sludges being partially de-watered and removed for vermi-composting off site.

BOD_5 removal rates of between 94 and 97% were achieved during initial trial and demonstration phases, and the mandated final effluent quality in this

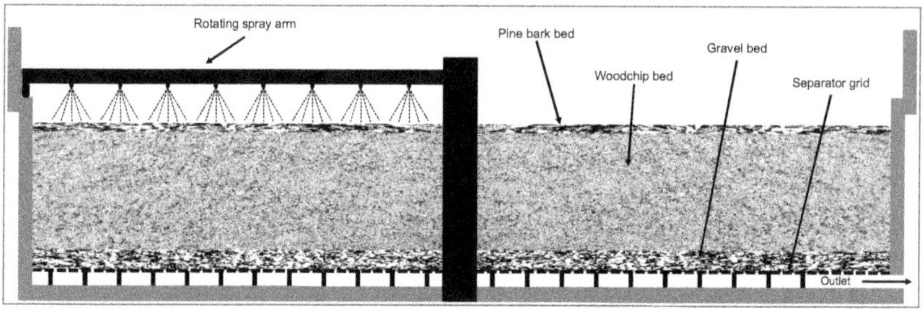

Figure 4.21. Cross-sectional representation of the vermifilter at Combaillaux, adapted from Naigeon (2005).

application was consistently achieved, allowing discharge into local surface waters (Naigeon, 2005):

- 25 mgL^{-1} (BOD_5);
- 125 mgL^{-1} or 75% removal (COD); and
- 35 mgL^{-1} or 90% removal (TSS)

4.3.2.4 Transplast

This company supplies modular vermifiltration units that are sized for loads of between 6 and 24PE. The volume of these units varies from 1,500 to 6,000 L accordingly. The filtration principles within the modules are those described for Sistema Tohá®, above – and the units can be installed above or (partially) below ground. BOD and TSS removals of 95% are claimed. This system uses sodium hypochlorite dosing modules to disinfect the effluent, allowing its use in irrigation (Infraplast, 2011).

4.3.2.5 Organic solutions

Supplying the market in India, this enterprise designs and installs modular and bespoke vermifiltration units, treating flows of municipal wastewater and industrial effluents of between 10 and 2,000 m^3 per day. Filter bed design and irrigation approaches are similar to those noted in other secondary vermifiltration systems, treating wastewater that has first been settled to remove 'large' solids. In addition to earthworms, the filter bed is inoculated at start-up with enzymes and microorganisms. Final effluent quality of BOD_5 <10 mgL^{-1}, TSS <20 mgL^{-1}, COD <50 mgL^{-1} and Faecal Coliforms <100 MPN per 100 mL are claimed (Organic Solutions, 2019).

4.3.3 Combined Primary and Secondary Vermifilters

Few commercial suppliers offer combined primary and secondary vermifiltration systems capable of accepting screened wastewater (from which larger physical contaminants and grit have been removed) and delivering sufficient treatment for the

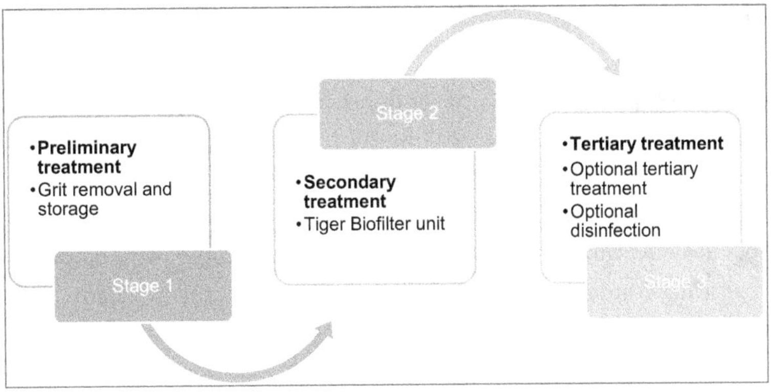

Figure 4.22. Basic overview of TBF Environmental system. Adapted from TBF Environmental Solutions (2018).

final effluent to be discharged to surface water or re-used for irrigation purposes. A single-pass reactor of this kind would offer an alternative to small package plants as well as combined septic tank/drainage field configurations – and it is this niche that the INNOQUA Lumbrifilter units have (in part) been developed to exploit. Of the suppliers that have been identified, two offering bespoke vermifiltration solutions to customers in India are examined here.

4.3.3.1 TBF Environmental Solutions Pvt Ltd

The Tiger Biofilter sewage treatment plant (STP) is modular but scalable, and can be installed for wastewater volumes ranging from 1,500 to 500,000 L per day. Sewage is first settled to remove grit, before being distributed to vermifiltration beds where solids are retained and converted into vermicompost. Drained effluent may then pass on to secondary vermifiltration or tertiary treatment and disinfection, depending on wastewater flows, loads and intended end uses (Figure 4.22).

4.3.3.2 Transchem Agritech

No design information is provided by this company, which has implemented solutions for a number of industrial clients in India, and which states that no solids' removal is required prior to treatment. Claimed performances are listed in Table 4.7.

Table 4.7. Claimed performance for some municipal wastewater characteristics. From Transchem Agritech (2018).

Parameters	Influents	Effluents
COD (mgL^{-1})	200–400	<20
BOD (mgL^{-1})	100–200	<5
Turbidity (NTU)	40–100	Nil

4.4 INNOQUA Lumbrifilter Implementation

The INNOQUA project aimed to demonstrate the utility of a suite of nature-based wastewater treatment solutions suitable for application in a number of decentralised scenarios. In all cases the development of these solutions began with concepts that were tested at laboratory scale before the installation of pilot facilities at operational (municipal) wastewater treatment works in Ireland and Spain. Experience from laboratory and pilot scale experiments informed the subsequent installation of ten demonstration facilities around the world. Introduced in Chapter 1, these were designed to demonstrate system performance under real environmental and operational conditions over a period of months. Operation of the pilot facilities continued through the demonstration phase, and while results from these units are not reported here they are available in published project reports.

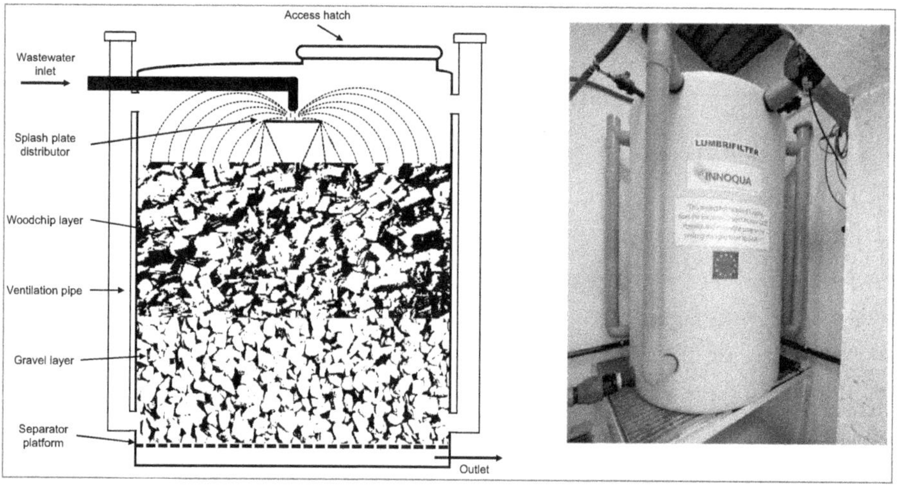

Figure 4.23. Cross-section of simplified INNOQUA lumbrifilter, illustrating the main components (left) and lumbrifilter as installed in a domestic basement in Italy (right). Photo courtesy of Pietro De Cinque.

Vermifilters (referred to as 'Lumbrifilters' within INNOQUA) were installed at all ten of the demonstration sites, and although site-specific requirements were accommodated through the use of different filter media, different earthworm species, different reactor sizes and different dosing regimes – all follow the same basic design, as set out in Figure 4.23. The demonstration sites are shown in Figures 4.25 and 4.26.

Since wastewater can change in concentration from site to site and regulations vary between geographies, sizing of the installed demonstration units was based on one or more of several parameters that included: hydraulic load, organic carbon load (COD or BOD_5), suspended solids load or ammonium load. For example, it may not be possible to achieve the desired hydraulic load as doing so may overload the system and compromise organic carbon or ammonium removal. Some of the key considerations include:

(i) Solids load, which should be matched to the capacity of the earthworms within the system to process solid matter and thus avoid clogging;
(ii) Hydraulic load, which should ensure sufficient residence time for biological removal of soluble contaminants such as organic carbon;
(iii) Where discharge limits require low concentrations of ammonium (e.g. <1 mg NH_4^+-N/L) ammonium may be the limiting factor in designing the system; and
(iv) The system should have capacity to adapt to higher loads for at least short periods of time.

In the design of the systems at the INNOQUA demonstration sites all of the above considerations formed part of the decision-making process. Key design loading ranges for the Lumbrifilter are shown in Table 4.8. Exact loads depended on the regulatory requirements, expected earthworm density, temperature and the relative concentrations of various parameters (e.g. carbon and nitrogen) at each site. At some of the demonstration sites the ranges were purposefully or accidentally exceeded for short periods – for stress testing the system or due to unexpected influent flows, or where on-site conditions indicated that additional loading would be acceptable.

The wastewater distribution system was modified during site testing (Figure 4.24). The original perforated pipe system proved susceptible to fouling and

Table 4.8. Typical flow and loading rates to the Lumbrifilter. The depth of the organic layer was typically 1.0 m.

Flow (L/m^3 active layer per day)	300–600
BOD_5 (g BOD_5/m^3 active layer per day)	120–240
COD (g COD/m^3 active layer per day)	240–480
TSS (g TSS/m^3 active layer per day)	140–280
TKN (g N/m^3 active layer per day)	30–60
TP (g P/m^3 active layer per day)	8–16

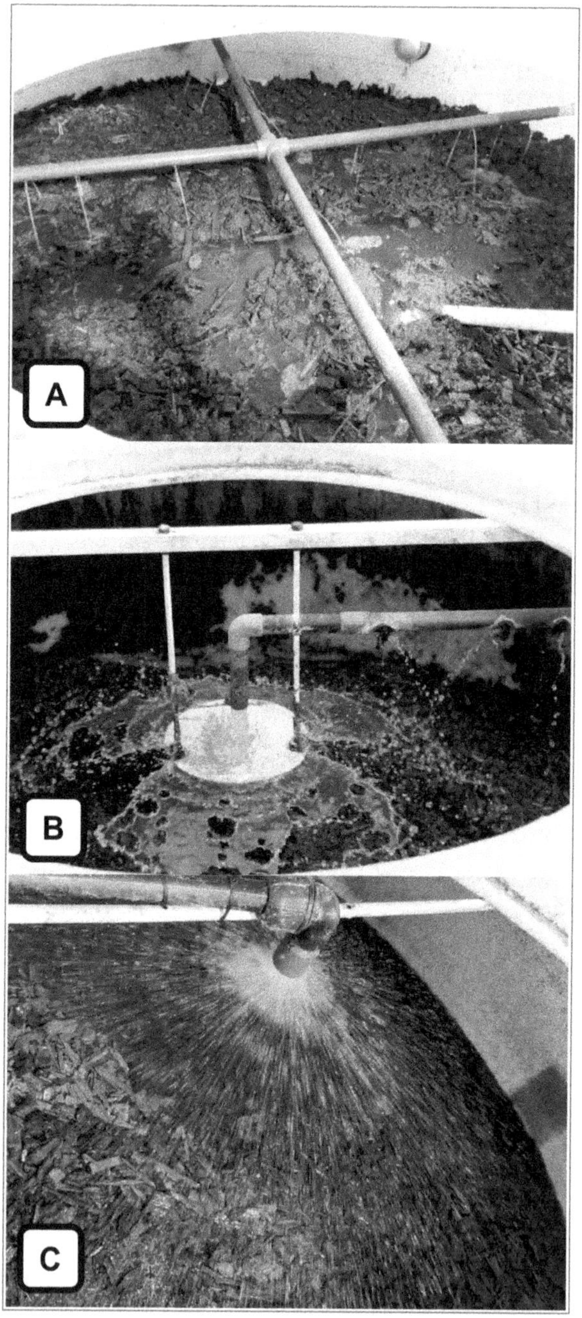

Figure 4.24. Internal views of INNOQUA lumbrifilter – showing evolution of the wastewater dosing mechanism through the project. [A] Perforated pipe; [B] Splash plate and [C] Full cone spray nozzle. Images courtesy of Costel Bumbac (A and C) and EKODENGE (B).

was replaced with a splash plate distribution system. In all cases an 'active' woodchip layer of 1 m depth was used – with the exception of the facility in Ecuador, where site constraints led to the use of a 0.75 cm deep active layer. Selected design parameters and key performance indicators for the ten demonstration facilities are presented in Table 4.9. Section 4.4.1 provides site-specific feedback on the Lumbrifilter, while key 'lessons learned' are set out in Section 4.4.2.

4.4.1 INNOQUA Lumbrifilter Performance

Selected design and performance characteristics for all ten demonstration Lumbrifilter units are presented in Table 4.9. The following sections provide brief site-specific observations on system performance.

4.4.1.1 Italy

Excellent removals of BOD and TSS were consistently observed, with values averaging around 95%. An increase of the COD removal rate was observed during the first four months of operation, stabilising at around 95%. Although the system is not specifically designed to remove TN and TP, removals averaged 74% and 27%, respectively.

4.4.1.2 Ireland

Good removals for BOD (61.9%), COD (59.4%) and TSS (74.9%) were observed. A key driver for trialling the technology on this site was to explore whether the Lumbrifilter could deliver denitrification to reduce return nitrogen loadings to land on a dairy farm, and this was indeed observed (with average removal of 62.3% for TN). Although the system was not designed for phosphorus removal, 33.4% removal was observed. Some variability was noted in system performance over time but this was ascribed to variations in influent quality due to the farmer using varying amounts of water to wash down the milking parlour.

4.4.1.3 France

Very good average removal rates for BOD (89.9%), COD (69.4%) and TSS (76.5%) were observed – and consistent over the first six months of operation. At this point a septic tank was installed upstream of the pumping chamber (from which the Lumbrifilter was dosed) to explore the impacts of primary settlement on system performance. This reduced the loads to the Lumbrifilter, with treated effluent quality remaining consistent throughout.

4.4.1.4 Scotland

Variable removal rates were observed for TSS, BOD, COD and TN, possibly related to numerous operational issues encountered during the demonstration

Table 4.9. Summary of lumbrifilter, influent and performance characteristics at the 10 INNOQUA demonstration sites. Note that these data include start-up, acclimation and operational phases. SD = Standard Deviation.

Countries		Italy	Ireland	France	Scotland	Turkey	Romania	Ecuador	Tanzania	India	Peru
Wastewater types		Domestic wastewater	Milking parlour washings	Mainly black water	Domestic wastewater	Domestic wastewater	Domestic wastewater	Domestic wastewater	Domestic wastewater	Domestic wastewater	Domestic wastewater
Pre-treatment		Septic tank	Settlement tank	–	–	Septic tank	Settlement tank	Settlement tank	Septic tank	–	Settlement tank
Organic filter medium		Woodchips	Woodchips (soft wood)	Woodchips	Woodchips	Woodchips	Woodchips (hard wood)	Coconut husk chips	Coconut husk chips	Coconut husk chips	Wood shavings
Ambient temperature range	°C	+8.0 to +22.0	+3.0 to +18.0	−5.0 to +27.0	+2.9 to +14.8	+6.0 to +22.0	−25.0 to +25.0	+13.6 to +14.1	+23.8 to +27.9	+20.7 to +27.1	+13.2 to +15.3
Filter surface area	m^2	1.1	1.1	1.7	3.5	3.5	3.5	1.1	3.5	3.5	3.5
Design flow	$m^3\ day^{-1}$	2.0	0.1	0.6	1.5	3.0	2.4	2.0	1.5	1.5	1.0
Measured flow	$m^3\ day^{-1}$	0.4	0.1	0.7	1.2	1.5	1.0	0.36	2.5	1.0	0.7
TSS influent	mgL^{-1} (±SD)	314 ± 208	4025 ± 7562	489 ± 283	67 ± 17	73 ± 128	344 ± 255	267 ± 137	361 ± 267	1998±1029	260 ± 674
TSS removal	% (±SD)	90 ± 10	75 ± 15	78 ± 22	65 ± 23	40 ± 33	64 ± 21	64 ± 29	25 ± 20	81 ± 21	72 ± 33
BOD_5 influent	mgL^{-1} (±SD)	388 ± 149	4628 ± 3089	439 ± 249	130 ± 45	51 ± 59	414 ± 162	505 ± 307	–	1090 ± 456	195 ± 160
BOD_5 removal	% (±SD)	96 ± 3	62 ± 32	90 ± 9	77 ± 19	38 ± 27	77 ± 15	54 ± 29	–	89 ± 10	78 ± 19
COD influent	mgL^{-1} (±SD)	993 ± 468	8590 ± 6140	1094 ± 521	304 ± 85	150 ± 173	912 ± 276	754 ± 360	459 ± 272	2072 ± 903	606 ± 1067
COD removal	% (±SD)	84 ± 14	59 ± 33	71 ± 21	47 ± 28	34 ± 29	71 ± 14	61 ± 20	64 ± 19	79 ± 11	42 ± 22
N-total influent	mgL^{-1} (±SD)	119 ± 29.3	247 ± 46.3	–	33.5 ± 10.1	–	105 ± 38.8	–	10.7 ± 7.13	–	148 ± 379
N-total removal	% (±SD)	49.2 ± 28.7	62.3 ± 14.4	–	18.5 ± 12.4	–	55.4 ± 20.4	–	36.5 ± 23.8	–	44.3 ± 22.7
P-total influent	mgL^{-1} (±SD)	12.2 ± 6.40	49.6 ± 14.3	19.1 ± 5.46	–	4.75 ± 8.37	9.20 ± 3.29	17.3 ± 3.82	49.7 ± 49.6	–	8.68 ± 2.13
P-total removal	% (±SD)	9.2 ± 45.4	33.4 ± 31.2	36.8 ± 18.7	–	39.1 ± 22.4	35.8 ± 15.9	0.86 ± 13.0	35.7 ± 13.8	–	20.6 ± 13.4

Figure 4.25. INNOQUA demonstration sites in (clockwise from A): Ecuador, France, Ireland, India and Ecuador. Images courtesy of Nicolas Salmon (A and E); Jean-Baptiste Dussaussois (B); NUIG (C) and Tatjana Schellenberg (D).

phase. Influent concentrations and ambient temperatures were low during the initial stages, affecting the removal rates. The dosing pump fouled several times due to sanitary products and grease in the settlement chamber, while sediment also accumulated in the sampling chamber installed at the Lumbrifilter outlet, impacting on sample quality.

Figure 4.26. INNOQUA demonstration sites in (clockwise from A): Peru, Tanzania, Scotland, Turkey and Romania. Images courtesy of Nicolas Salmon (A); Evelyn Herrera and Gervaz Lushaju (B); Scottish Water (C); EKODENGE (D) and Costel Bumbac (E).

4.4.1.5 Turkey

System performance at this site was severely impacted by under-loading, due to much lower than expected occupancy of the dwellings providing wastewater for the facility. Provision was made to tanker wastewater from a centralised treatment works to the site during the latter few weeks of the demonstration phase, following which performance improved for some characteristics.

4.4.1.6 Romania

TSS and BOD removal rates were relatively stable over six months of operation, and then started to drop following a change in influent quality, reflecting a change in use in the building providing the wastewater – and occasional unauthorized discharges of paint into the system. However, in terms of mass removal, performance doubled from an average of 406 g BOD_5 removed/day to an average of 811 g of BOD_5 removed/day over time. Removal rates were variable for COD, TP and TN – and as the system was operated at loading rates close to or above the recommended maximum loading rates, any change in the influent quality could be expected to affect removal efficiencies.

4.4.1.7 Ecuador

The Lumbrifilter at this site was designed with a shallower filter layer than other units, as an adaptation to physical restrictions on site. Removal of TSS, BOD and COD was extremely variable from day to day, possibly due to variations in influent quality. Performance improved following the implementation of regular surface raking and inoculation with additional earthworms. To better accommodate site flow and load, the Daphniafilter at this site was converted for use as a secondary Lumbrifilter prior to the conclusion of the INNOQUA project.

4.4.1.8 Tanzania

The Lumbrifilter delivered good average removal for COD (67.7%) despite very significant operational challenges at this site. Hydraulic loadings were initially three times higher than the maximum design capacity, due to inaccurate manual dosing. Outflow was also restricted due to the use of an inline flow meter (subsequently removed), which combined with over-dosing to cause the unit to flood and kill the earthworms. The flooding resulted in raw wastewater discharge into the surrounding compound, creating significant local opposition to the continuation of the study. These issues were addressed by moving the Lumbrifilter within the compound and engaging in intensive liaison with local community representatives. Relocation provided useful opportunities to replace the coconut husk filter media with imported mango woodchips (delivering a final effluent colour that was acceptable to the local community) and to re-inoculate the system with local earthworms (*Eudrilus eugeniae*).

4.4.1.9 India

Removal rates for TSS, BOD and COD averaged 87.7, 75.5 and 76.6% respectively, over the demonstration phase. This level of performance was consistently delivered despite numerous operational challenges and deliberate interventions at this site. High ambient temperatures combined with an excessively free-draining filter medium (coconut husk chips), causing the initial earthworm population to fail within the first two months of operation. To increase retention time, a coconut fibre layer was added at 20 cm depth and a proportion of small coconut chips mixed into the media above this. Furthermore, while the system was originally intended to treat settled wastewater, it was operated with whole wastewater from the second month onwards to understand whether production of primary (faecal) sludge could be avoided. In order to prevent fouling under this arrangement, the dosing system was swapped for a simple splash plate. These changes led to an increase in loading rates for all tested parameters and also resulted in a more variable influent (for example, TSS ranged from 1170 mg/L to 4015 mg/L). The filter surface was prone to clogging under the higher loads, but this was successfully addressed by decreasing the dosing frequency.

4.4.1.10 Peru

Although an average BOD removal of 68.5% is reported for this site it should be noted that samples were collected from a retention tank located after the Lumbrifilter, where effluent could reside for up to 30 hours before being pumped forward to the next treatment stage. Due to difficulties sourcing woodchips on this site, wood shavings were used instead – and found to be extremely prone to compaction. This was partially mitigated through addition of coconut chips, but poor drainage remained problematic and will undoubtedly have impacted on system performance. An inline filter before the dosing pump was prone to fouling, and this resulted in gradual decreases in the volumes dosed at each interval following cleaning – leading to lower than expected moisture concentrations within the top layers of the filter, which resulted in downward earthworm migration and reduced treatment performance. Earthworm viability was also impacted by an influent ammonium peak of 1700 mg N/L, corresponding to laboratory residues flushed into the influent. The configuration of the dosing pump was subsequently changed, and the unit inoculated with additional earthworms.

4.4.2 Lessons Learned

At each stage of Lumbrifilter implementation, observations were logged by the demonstration site managers. These observations were reflected in system and operational upgrades across the demonstration network and provide a resource for others developing or implementing vermifiltration solutions.

4.4.2.1 Preparatory works

- The Lumbrifilter is (principally) sized by surface filter area according to the anticipated flow and pollutant load, based on a standard active layer depth. It is essential that initial wastewater characterisation is as representative of

operational flows and loads as possible to minimise potential for over or under-loading
- Where wastewater flows are combined (rainwater/surface water with wastewater) checks should be made to determine whether the wastewater flow can be isolated. This will minimise the required filter size
- Where the system is relying on an electrical pump for dosing, then a robust electrical supply will be required. An appropriately rated solar panel may suffice, but theft protection measures may be required
- An acclimation period is required before a vermifilter achieves its maximum treatment efficiency, and it may be necessary to provide for alternative safe disposal of (partially) treated wastewater during this period

4.4.2.2 Filter media and earthworms

- Different woody media may be used, including hardwood and softwood chips, as well as coconut husk chips. Wood shavings and some coconut shell chips are not recommended, as they were found to be prone to clogging
- Where possible, media should be washed before use, to remove surface grit and other debris that might otherwise clog the filter
- Since the woody media can influence the colour of the treated wastewater (for example, through leaching of tannins), this can influence system acceptability to some local communities. Different media could be trialled in such situations, in combination with a local outreach programme to explain the functioning of the system
- *Eisenia fetida, Eisenia andrei, Dendrobaena veneta* and *Eudrilus eugeniae* were all found to be suitable for vermifiltration
- Suitable earthworm species may be available locally from vermicomposting or fishing bait suppliers
- Where it is not possible to source sufficient earthworms locally, then it may be necessary to implement a breeding programme. Starter cultures might have to be imported – potentially incurring veterinary or other checks and taxes or other tariffs
- Where sourced commercially, earthworms may be supplied in carrier media that are incompatible with the vermifilter – and which might cause clogging if introduced into the filter with the earthworms. Earthworms should be separated from fine solids before inoculation of the filter

4.4.2.3 System start-up and acclimation

- Once all filter layers are in place, the designated wastewater dosing regime should be run for one or two weeks prior to the introduction of earthworms.

This is essential for wetting the filter layers and stimulating biofilm development to create a favourable habitat for the earthworms

- Assuming a woody media bed of one metre depth, earthworms should be added at a rate of around 3,000 to 5,000 individuals per square metre of filter surface area. Ideally this population should comprise a mix of size classes as well as cocoons
- Stable operation can be expected after an interval of between 20 and 60 days following inoculation with earthworms

4.4.2.4 System operation and optimisation

- Rags and other solid waste must be screened from the incoming wastewater, or excluded from the collection system
- Dribble bar irrigation is extremely prone to clogging – both from suspended solids entrained in the wastewater flow and as a result of biofilm growth. Screened and settled/clarified water can be distributed via spray nozzles under pressure, but a simple splash-plate distributor was found to be adequate on the majority of demonstration sites
- Where dosed by pumping (as distinct from gravity flow), the sizing and design of the pump chamber, as well as the location of the pump within the chamber must be carefully considered to ensure that settled sludges do not impede pump flow and/or lead to blinding of the filter surface
- Where sludge accumulation is noted in balancing tanks/pumping chambers, then this should be periodically removed to prevent pump fouling or filter blockages

4.4.2.5 Routine maintenance

- At several demonstration sites, the filter surface was agitated at weekly or fortnightly intervals (by manual raking) to ensure good percolation. Whether this approach is required at a specific site must be determined by observation during initial system monitoring
- Periodic topping-up of the woody media layer will be required. The interval for this will vary depending on the type of substrate and local operating conditions, with intervals of between two and five months noted across the INNOQUA demonstration sites
- Marking the internal tank surface with depth indicators allows simple identification of the need to replenish woody biomass during routine inspections

4.4.2.6 Other observations

- Where influent flows are periodic (for example, due to interruptions in electricity supply or occupation patterns in the wastewater catchment) it is important that the filter is kept in a moist condition to maximise earthworm viability

- Where ambient temperatures exceed the optimum for earthworm viability, periodic dosing can perform an essential cooling function. Although the woody biomass layer needs to have good percolation properties, excessively free-draining material can rapidly dry out after dosing, impacting this cooling function. In such cases, a proportion of woody media of <5 mm particle size can be mixed through the bulk media to increase hydraulic retention
- When designing monitoring regimes (particularly for experimental facilities), the availability and suitability of local laboratory services should be considered and alternative provisions made, if necessary
- Influent quality can change over time, irrespective of the thoroughness of preliminary characterisation. Users of vermifiltration systems should (ideally) be engaged in the design and implementation phases to ensure that inappropriate materials are not discarded to the wastewater flow
- Although there were virtually no issues of low social acceptance, odours can be generated during dosing events, and it may be necessary to consider filter and vent locations to minimise risks of odour impact on surrounding receptors
- For convenience, the INNOQUA demonstration Lumbrifilter reactors were all supplied by a single European manufacturer. However, project partners at the University of Cuenca (Ecuador) readily replicated the filter using a water tank and other locally available materials Figure 4.27.

Figure 4.27. INNOQUA lumbrifilter created with locally available materials in Cuenca. Images courtesy of Andrés Alvarado, showing (left to right): an external view; gravel layer; splash plate distribution system.

4.5 Discussion and Identification of Future Research Needs

Vermifiltration is a demonstrably robust method of wastewater treatment that has been successfully commercialised in a number of countries and regions, notably Latin America, Australasia and India – where the systems produce final effluent suitable for discharge into soil and/or surface waters, and in some cases suitable for reuse in irrigation or other applications. The basic operating principles of these vermifilters are all the same, creating favourable environments for specific types of earthworm to consume organic solids in wastewater in parallel with microbiological degradation of dissolved organic pollutants. In some cases, vermifilters are promoted as alternatives to septic tanks – designed to accept whole wastewater, with the majority of treatment taking place downstream – while in other cases vermifilters are promoted as alternatives to secondary treatment systems, accepting settled wastewater and delivering sufficient treatment within the unit to allow final effluent to be discharged to surface water or used in irrigation (following disinfection). Very few commercial examples are known to combine both primary and secondary treatment into a single vermifiltration unit.

The potential for combining primary and secondary treatment into a single vermifilter is particularly attractive in those contexts where faecal sludge treatment infrastructure is absent or inadequate. This is the case for several of the countries hosting INNOQUA Lumbrifilter demonstration sites. Excellent performance was delivered at the two demonstration sites processing whole wastewater (France and India), where earthworm capacity and solids' loading were balanced to minimise clogging. It is not yet possible to infer the potential of these installations as zero sludge solutions since fundamental operational aspects of both sites were changed during the demonstration phase and long term (>1 year) data are not available for these configurations. Performance was also excellent at facilities treating septic tank effluent and settled wastewater – potentially allowing vermifiltration to replace drainage fields or constructed wetlands to intensify treatment where septic tanks or other basic infrastructure already exists. Sequential primary and secondary treatment with separate vermifiltration units is also possible, although not studied within INNOQUA.

In addition to removal of suspended solids, COD and BOD, there was also evidence of nutrient removal by the INNOQUA Lumbrifilters – both for nitrogen and phosphorus. The project did not explore mechanisms of removal, although nitrogen removal can be explained by a combination of biomass uptake, nitrification and denitrification, while phosphorus removal can be explained by a combination of biomass uptake and adsorption. Understanding the durability of these removal patterns in the long term will require further research, since it is well established that adsorption sites for phosphorus will eventually reach saturation – after which break-through will occur unless the filter media is replaced. Denitrification requires anoxic conditions that are normally created by avoiding aeration and ensuring soluble COD or BOD concentrations that are sufficiently high to allow microbial activity to utilise free oxygen. Since vermifilters (nominally) operate under aerobic conditions, the occurrence of denitrification suggests the presence of anoxic micro-sites within the filter media – perhaps facilitated by partial breakdown of the woodchips. These mechanisms are exploited in denitrifying reactors designed for decentralised applications, where the media act both as a biofilm support and source of dissolved organic carbon. In principle the combination of organic and mineral filter layers within a vermifilter can be designed to act as a physical filter, a habitat for earthworms, a scaffold for biofilm establishment, a matrix of phosphorus adsorption sites and a source of dissolved organic carbon to support denitrification. Incorporating all of these elements within a single reactor would present significant maintenance challenges – but the flexibility of vermifiltration systems is nonetheless apparent.

Interactions between earthworms and microorganisms were not explored in INNOQUA, although such interactions may be key to vermifiltration systems noted to deliver more than the sum of their parts. Pathogen reduction may be due to mechanical action within the earthworm gut, or due to enzymic inhibition (whether within the earthworm or its casts). Interactions between earthworms and microbial community structure within their surrounding habitat are also evident from previous research. Whether this contributes significantly to the increased treatment efficiencies within vermifilters (as compared with conventional biofilters) is unknown, since the browsing habits of earthworms also contribute to improved aerobic conditions within filters. However, increased populations of specific types of bacteria have been noted in material processed by earthworms – suggesting a combined effect of bacterial community influence and increased aeration, leading to increased functionality and wastewater treatment efficacy. These interactions are difficult to correct for experimentally but are important to understand if the potential for vermifiltration to attenuate pathogens, known chemical hazards and (particularly) chemicals of emerging concern in wastewater is to be robustly explored and exploited.

Care must also be taken to ensure that the fundamental elements of vermifiltration are considered in any future implementation. While earthworms are susceptible to various wastewater characteristics (such as salinity, dissolved ammonia and pH), they are also susceptible to drowning, and this can be induced in units where outflows are not adequately designed, where filter media clog due to excessive solids' loading in wastewater and if excess biofilm accumulates within the filter bed. Since most commercial systems are passively aerated, chimney effects created by localised hotspots of composting substrate might be useful – or might lead to localised anaerobiosis that reduces overall treatment performance. In contrast, filter media that are too free-draining may cause earthworm desiccation or may reduce cooling effects necessary in climates with high ambient temperatures. Thus, despite their apparent simplicity, optimum functioning within vermifilters results from the interaction of earthworms and multiple other factors, including:

- Bedding/filter material type and physical form
- Filter media depth
- Dosing regime
- Hydraulic retention time
- Wastewater characteristics
- Operating temperatures

Optimum functioning will also be an artefact of the specific location and intended use for the treated wastewater. It is this combination of simplicity and complexity that means many commercial suppliers offer both a standard modular range of vermifilters sized for different flows and loads (often for treating domestic wastewater) – and a bespoke design service to treat industrial wastewater from manufacturing facilities.

As in all wastewater treatment systems, upstream management should be considered and appropriate measures adopted to (for example) exclude non-biodegradable solids such as plastics. Where primary clarification is required, then provision must also be made for management of the resulting sludges. The need for routine maintenance will also be linked to the initial choice between vermifiltration for primary or secondary treatment, since the filter media are less likely to clog in secondary treatment systems and may need more surface agitation in primary treatment systems. In both cases woody media will need to be replenished periodically.

Overall, vermifiltration offers the potential to deliver robust aerobic treatment of domestic and other wastewaters in a range of decentralised scenarios on small footprints. As evidenced by INNOQUA project partners at the University of Cuenca, vermifilters can be constructed and filled with locally-sourced materials, inoculated with locally-sourced earthworms and easily maintained by suitably informed users (Figure 4.27). While the COVID-19 pandemic interrupted the planned monitoring of long term continuous performance, the enforced shutdowns and/or absence of routine maintenance at the majority of INNOQUA demonstration sites during the pandemic served to further highlight the resilience of vermifiltration systems. Providing the filter media had been maintained in a moist condition, treatment performance quickly rebounded following re-start. Epigeic earthworms have evolved to exploit conditions of rapidly changing food availability under a broad range of environmental conditions – and providing certain boundaries are respected, will provide extremely resilient service in vermifiltration (and vermicomposting) applications.

Finally, although vermifilters are commercially available in a range of configurations, some aspects remain relatively unexplored and may prove fruitful for future research:

1. The potential to use biochar in place of a mineral filter layer. This could provide a very high surface area for biofilm establishment and/or adsorption of phosphorus and other chemicals of interest;
2. Interactions between earthworms and microbial communities within filter media, and whether these can be exploited to further intensify treatment functions or to target treatment of specific chemicals of interest;
3. Long term treatment performance and potential to remove nutrients, heavy metals and other chemicals of interest;
4. Long term maintenance requirements and costs, when compared with other decentralised wastewater treatment approaches; and
5. Potential for creation of treatment cascades, with an initial 'roughing' vermifilter for primary treatment, followed by a secondary vermifilter, followed by a denitrifying woodchip reactor and finally a P-sorption reactor. This could deliver full decentralised wastewater treatment on a modest footprint and could be combined with final disinfection to allow reuse in a wide range of applications.

Acknowledgements

Editorial feedback from Eoghan Clifford (NUIG, Ireland), Costel Bumbac (ECOIND, Romania) and Alexandre Schaal (NOBATEK/INEF4, France) is gratefully acknowledged.

Section 4.4 (on INNOQUA Lumbrifilter implementation) draws from three project deliverables: D5.1, D5.2 and D5.3. Their lead authors are gratefully acknowledged: Alexandre Schaal (NOBATEK/INEF4, France); Edgar Rubión (Eurecat, Spain); Louise Hannon and Eoghan Clifford (NUIG, Ireland).

The following INNOQUA project partners kindly provided photographs: Andrés Alvarado (University of Cuenca, Ecuador); Costel Bumbac (ECOIND, Romania); EKODENGE (Turkey); Evelyn Herrera and Gervaz Lushaju (BORDA, Tanzania); Jean-Baptiste Dussaussois (NOBATEK/INEF4, France); Nicolas

Salmon (YES Innovation, Ecuador); NUIG (Ireland); Pietro De Cinque (DE5, Italy); Scottish Water (UK); Tatjana Schellenberg (BORDA, Tanzania).

References

A&A Worm Farm Waste Systems (2020). *Septic Tank System.* Worm Farm Waste Systems. Available at: https://www.wormfarm.com.au/septic-tank-system/ (Accessed: 30 August 2020).

Adugna, A. T. *et al.* (2014). 'Greywater treatment by vermifiltration for sub-Saharan urban poor', *Journal of Water, Sanitation and Hygiene for Development.* IWA Publishing, 4(4), pp. 625–632. doi: 10.2166/washdev.2014.021.

Altundoğan, H. S. and Tümen, F. (2002). 'Removal of phosphates from aqueous solutions by using bauxite. I: Effect of pH on the adsorption of various phosphates', *Journal of Chemical Technology & Biotechnology.* John Wiley & Sons, Ltd, 77(1), pp. 77–85. doi: 10.1002/jctb.525.

Amoah, P, Nartey, E. G. and Schrecongost, A. (2016). 'Effect of different income housing zones on effluent quality of Biofil® toilet waste digesters in Accra, Ghana', in *39th WEDC International Conference.* Available at: https://wedc-knowledge.lboro.ac.uk/resources/conference/39/Amoah-2498.pdf (Accessed: 29 August 2020).

Amoah, Philip, Nartey, E. G. and Schrecongost, A. (2016) 'Performance evaluation of biofil toilet waste digester technologies in Ghana: the efficacy of effluent treatment options', *Environmental Technology.* Taylor and Francis Ltd., 37(23), pp. 3002–3013. doi: 10.1080/09593330.2016.1173116.

Arora, S. and Kazmi, A. A. (2015). 'The effect of seasonal temperature on pathogen removal efficacy of vermifilter for wastewater treatment', *Water Research*, 74, pp. 88–99. Available at: https://www.sciencedirect.com/science/article/abs/pii/S0043135415000743?via%3Dihub (Accessed: 5 June 2019).

Baumgartner, R. (2013). *Developing a Testing Protocol for Vermifiltration-based Onsite Wastewater Treatment Systems (VOWTS).* Murdoch University. Available at: https://core.ac.uk/download/pdf/16464393.pdf (Accessed: 30 July 2020).

Biofiltro (2020a). *Can of Worms.* Available at: http://biofiltro.com/wp-content/uploads/2020/06/Self-Contained-Spec-Sheet-May-2020.pdf (Accessed: 29 August 2020).

Biofiltro (2020b). *Wiggle Room.* Available at: http://biofiltro.com/wp-content/uploads/2020/04/Mobile-Unit-2020.pdf (Accessed: 29 August 2020).

Biolytix (2015). *Biolytix BioPod (BF6) Wastewater Treatment System Specifications.* Available at: https://www.biolytix.com/wp-content/uploads/BF6-Specifications-2015.pdf (Accessed: 29 August 2020).

Bouché, M. B. (1977). 'Strategies lombriciennes', *Ecological Bulletins*. Oikos Editorial Office, pp. 122–132. doi: 10.2307/20112572.

Carney, K. N. *et al.* (2013). 'Treatment of separated piggery anaerobic digestate liquid using woodchip biofilters', *Environmental Technology*. Taylor and Francis Ltd., 34(5), pp. 663–670. doi: 10.1080/09593330.2012.710408.

Corley, M. *et al.* (2006). 'The performance of fibrous peat biofilters in treating domestic strength wastewater', *Journal of Environmental Science and Health – Part A Toxic/Hazardous Substances and Environmental Engineering*. Taylor & Francis Group , 41(5), pp. 811–824. doi: 10.1080/10934520600614421.

CPlantae (2017). *CPlantae – Tratamiento de aguas residuales*. Available at: https://www.cplantae.com/ (Accessed: 30 August 2020).

Curry, J. P. and Schmidt, O. (2007). 'The feeding ecology of earthworms – A review', *Pedobiologia*. Elsevier GmbH, pp. 463–477. doi: 10.1016/j.pedobi.2006.09.001.

Daigger, G. T. and Boltz, J. P. (2011). 'Trickling Filter and Trickling Filter-Suspended Growth Process Design and Operation: A State-of-the-Art Review', *Water Environment Research*. Wiley, 83(5), pp. 388–404. doi: 10.2175/106143010x12681059117210.

Dalahmeh, S. S. *et al.* (2011). 'Potential of organic filter materials for treating greywater to achieve irrigation quality: a review.', *Water science and technology: a journal of the International Association on Water Pollution Research*, 63(9), pp. 1832–40. Available at: http://www.ncbi.nlm.nih.gov/pubmed/21902020 (Accessed: 5 June 2019).

Di, W., Xing, M. and Yang, J. (2016). 'Investigation on the difference between biofilm morphologies of the vermifilter and conventional biofilter with the flow cytometer', *Bioresource Technology*. Elsevier Ltd, 216, pp. 308–316. doi: 10.1016/j.biortech.2016.05.033.

Domínguez, J. (2004). 'State of the art and new perspectives on vermicomposting research', in Edwards, C. A. (ed.) *Earthworm Ecology*. Boca Raton, Florida, pp. 401–424.

Domínguez, J. (2018) 'Earthworms and Vermicomposting', in *Earthworms – The Ecological Engineers of Soil*. InTech. doi: 10.5772/intechopen.76088.

Domínguez, J. and Edwards, C. A. (2011). 'Biology and ecology of earthworm species used for vermicomposting', in Edwards, C. A., Arancon, N. Q., and Sherman, R. L. (eds.) *Vermiculture Technology: Earthworms, Organic Waste and Environmental Management*. Boca Raton, Florida, pp. 25–37.

Domínguez, J., Edwards, C. A. and Webster, M. (2000). 'Vermicomposting of sewage sludge: Effect of bulking materials on the growth and reproduction of the earthworm Eisenia andrei', *Pedobiologia*, 44, pp. 24–32. Available at:

https://pdfs.semanticscholar.org/2945/c8883b5f4f18ce4842c1de33d78024 8d0dce.pdf (Accessed: 27 July 2020).

Dore, S. *et al.* (2019). *THE BIOFILTRO BIDA (R) WASTEWATER TREATMENT SYSTEM*. Available at: http://biofiltro.com/wp-content/uploads/2020/04/BioFiltro-Hydrofocus-BIDA-White-Paper.pdf (Accessed: 28 July 2020).

Eastman, B. R. *et al.* (2001). 'The effectiveness of vermiculture in human pathogen reduction for USEPA biosolids stabilization', *Compost Science and Utilization*. Taylor & Francis, 9(1), pp. 38–49. doi: 10.1080/1065657X.2001.10702015.

Edwards, C. A. and Lofty, J. R. (1977). *Biology of Earthworms, Biology of Earthworms*. Springer US. doi: 10.1007/978-1-4613-3382-1.

Enrique Hernández, C. and Furlong, C. (2017) 'Optimisation of faecal sludge processing via vermifiltration', in *LOCAL ACTION WITH INTERNATIONAL COOPERATION TO IMPROVE AND SUSTAIN WATER, SANITATION AND HYGIENE SERVICES*. Loughborough. Available at: https://wedc-knowledge.lboro.ac.uk/resources/conference/40/Enrique-Hernandez-2833.pdf (Accessed: 30 July 2020).

Furlong, C. *et al.* (2016). 'Technical and user evaluation of a novel worm-based, on-site sanitation system in rural India', *Waterlines*, 35(2), pp. 148–162. doi: 10.3362/1756-3488.2016.013.

Furlong, C. (2018). *Final Reports: Assessment the potential of using vermifilters to treat faecal sludge*. Available at: https://emergencysanitationproject.files.wordpress.com/2018/08/vermicomposting-for-sludge-treatment.pdf (Accessed: 30 July 2020).

Furlong, C., Lamb, J. and Bastable, A. (2017). 'Learning from Oxfam's Tiger Worm Toilets projects', in. Loughborough. Available at: https://wedc-knowledge.lboro.ac.uk/resources/conference/40/Furlong-2835.pdf.

Furlong, C., Templeton, M. R. and Gibson, W. T. (2014). 'Processing of human faeces by wet vermifiltration for improved on-site sanitation', *Journal of Water Sanitation and Hygiene for Development*. IWA Publishing, 4(2), pp. 231–239. doi: 10.2166/washdev.2014.107.

Gautier, N. (2007). *Layman's report*. Available at: https://ec.europa.eu/environment/life/project/Projects/files/laymanReport/LIFE03_ENV_F_000257_LAYMAN.pdf.

Gomez-Brandon, M. *et al.* (2013) 'Animal Manures: Recycling and Management Technologies', in *Biomass Now – Cultivation and Utilization*. InTech. doi: 10.5772/53454.

Grüneberg, B. and Kern, J. (2001). 'Phosphorus retention capacity of iron-ore and blast furnace slag in subsurface flow constructed wetlands', in *Water Science and Technology*. IWA Publishing, pp. 69–75. doi: 10.2166/wst.2001.0811.

Hawkins, C. L. *et al.* (2008). 'Earthworm populations in septic system filter fields and potential effects on wastewater renovation', *Applied Soil Ecology*. Elsevier, 40(1), pp. 195–200. doi: 10.1016/j.apsoil.2008.03.001.

Hedström, A. (2006). 'Reactive filter systems for small scale wastewater treatment', *VATTEN*, pp. 253–263.

Huang, K. *et al.* (2017). 'Effects of earthworms on nitrification and ammonia oxidizers in vermicomposting systems for recycling of fruit and vegetable wastes', *Science of the Total Environment*. Elsevier B.V., 578, pp. 337–345. doi: 10.1016/j.scitotenv.2016.10.172.

Infraplast (2011). *Manual de Instalación y Operación Sistema Tohá Lombrisán.*

Jiang, L. *et al.* (2016). *The use of microbial-earthworm ecofilters for wastewater treatment with special attention to influencing factors in performance: A review, Bioresource Technology*. Available at: https://www.sciencedirect.com/science/article/pii/S0960852415015369?via%3Dihub (Accessed: 5 June 2019).

Kaplan, D. L. *et al.* (1980). 'Physicochemical requirements in the environment of the earthworm Eisenia foetida', *Soil Biology and Biochemistry*. Pergamon, 12(4), pp. 347–352. doi: 10.1016/0038-0717(80)90008-5.

Karaca, S. *et al.* (2004). 'Kinetic modeling of liquid-phase adsorption of phosphate on dolomite', *Journal of Colloid and Interface Science*. Academic Press, 277(2), pp. 257–263. doi: 10.1016/j.jcis.2004.04.042.

Kõiv, M. *et al.* (2009). 'The performance of peat-filled subsurface flow filters treating landfill leachate and municipal wastewater', *Ecological Engineering*. Elsevier, 35(2), pp. 204–212. doi: 10.1016/j.ecoleng.2008.04.006.

Lekang, O. I. and Kleppe, H. (2000). 'Efficiency of nitrification in trickling filters using different filter media', *Aquacultural Engineering*. Elsevier, 21(3), pp. 181–199. doi: 10.1016/S0144-8609(99)00032-1.

Liu, J. *et al.* (2012). 'Effect of earthworms on the performance and microbial communities of excess sludge treatment process in vermifilter', *Bioresource Technology*. Elsevier, 117, pp. 214–221. Available at: https://www.sciencedirect.com/science/article/pii/S0960852412007201?via%3Dihub (Accessed: 5 June 2019).

Liu, J. *et al.* (2013). 'Phylogenetic characterization of microbial communities in a full-scale vermifilter treating rural domestic sewage', *Ecological Engineering*. Elsevier, 61, pp. 100–109. doi: 10.1016/j.ecoleng.2013.09.015.

Lopez-Ponnada, E. V. *et al.* (2017). 'Application of denitrifying wood chip bioreactors for management of residential non-point sources of nitrogen', *Journal of Biological Engineering*. BioMed Central Ltd., p. 16. doi: 10.1186/s13036-017-0057-4.

Lourenço, N. and Nunes, L. M. (2017). 'Optimization of a vermifiltration process for treating urban wastewater', *Ecological Engineering*, 100, pp. 138–146.

Available at: https://www.sciencedirect.com/science/article/abs/pii/S0925857416307145?via%3Dihub (Accessed: 5 June 2019).

Lourenço, N. and Nunes, L. M. (2020). 'Review of Dry and Wet Decentralized Sanitation Technologies for Rural Areas: Applicability, Challenges and Opportunities', *Environmental Management*. Springer, 65(5), pp. 642–664. doi: 10.1007/s00267-020-01268-7.

Luth, L. (2011). *Effet de la combinaison de lombrifiltration et du lagunage à macrophytes sur le recyclage des effluents sur le site d'élevage*. Université de Rennes. Available at: https://hal.inrae.fr/tel-02810571/document.

Manyuchi, M. M., Kadzungura, L. and Boka, S. (2013). 'Vermifiltration of Sewage Wastewater using Eisenia Fetida Earthworms for Potential use in Irrigation Purposes', in. Available at: https://www.researchgate.net/publication/256452259_Vermifiltration_of_Sewage_Wastewater_using_Eisenia_Fetida_Earthworms_for_Potential_use_in_Irrigation_Purposes (Accessed: 5 June 2019).

Mara, D. and Horan, N. (2003). *Handbook of Water and Wastewater Microbiology*. London: Academic Press. Available at: https://www.researchgate.net/publication/291139679_Handbook_of_Water_and_Wastewater_Microbiology (Accessed: 7 July 2020).

Murnane, J. G. *et al.* (2016). 'Assessment of intermittently loaded woodchip and sand filters to treat dairy soiled water', *Water Research*. Elsevier Ltd, 103, pp. 408–415. doi: 10.1016/j.watres.2016.07.067.

Naigeon, C. (2005). *Lombri-Station*. Available at: http://www.recyclaqua.agropolis.fr/dossier_presse/lib/DossierPresse.pdf (Accessed: 27 August 2020).

Natural England (2014). *Earthworms in England: distribution, abundance and habitats – NECR145*. Available at: http://publications.naturalengland.org.uk/publication/5174957155811328 (Accessed: 20 April 2020).

Oakley, S. and Von Sperling, M. (2017). 'Media Filters: Trickling Filters and Anaerobic Filters', in Mihelcic, J. R. and Verbyla, M. E. (eds.) *Global Water Pathogen Project*. doi: 10.14321/waterpathogens.64.

Öövel, M. *et al.* (2007). 'Schoolhouse wastewater purification in a LWA-filled hybrid constructed wetland in Estonia', *Ecological Engineering*. Elsevier, 29(1), pp. 17–26. doi: 10.1016/j.ecoleng.2006.07.010.

Organic Solutions (2019). *SEWAGE TREATMENT – Organic Solutions*. Available at: http://organicsolutions.in/sewage-treatment/ (Accessed: 2 March 2021).

Pratt, C. *et al.* (2012). 'Biologically and chemically mediated adsorption and precipitation of phosphorus from wastewater', *Current Opinion in Biotechnology*. Elsevier Current Trends, pp. 890–896. doi: 10.1016/j.copbio.2012.07.003.

Rodgers, M. *et al.* (2004). 'Organic carbon and ammonium nitrogen removal in a laboratory sand percolation filter', *Journal of Environmental Science and Health – Part A Toxic/Hazardous Substances and Environmental Engineering*.

Taylor & Francis Group, 39(9), pp. 2355–2368. doi: 10.1081/ESE-2000 26283.

Ruane, E. M. *et al.* (2012). 'Performance of a woodchip filter to treat dairy soiled water', *Journal of Environmental Management*. Academic Press, 95(1), pp. 49–55. doi: 10.1016/j.jenvman.2011.09.007.

Sakadevan, K. and Bavor, H. J. (1998). 'Phosphate adsorption characteristics of soils, slags and zeolite to be used as substrates in constructed wetland systems', *Water Research*. Elsevier Sci Ltd, 32(2), pp. 393–399. doi: 10.1016/S0043-1354(97)00271-6.

Sasse, L. (1998). *DEWATS: Decentralised Wastewater Treatment in Developing Countris*. Available at: https://sswm.info/sites/default/files/reference_attachments /SASSE 1998 DEWATS Decentralised Wastewater Treatment in Developing Countries_0.pdf.

Sepúlveda, M. G. G. (2004). *ESTUDIO DE FACTIBILIDAD DE LA APLICACIÓN DEL SISTEMA TOHA EN LA PLANTA DE TRATAMIENTO DE AGUAS SERVIDAS DE VALDIVIA*. Available at: http://cybertesis.uach.cl/tesis/uach/2004/bmfcig993e/doc/bmfcig993e.pdf (Accessed: 1 September 2020).

Singh, R. *et al.* (2019). 'Vermifiltration as a sustainable natural treatment technology for the treatment and reuse of wastewater: A review', *Journal of Environmental Management*. Academic Press, pp. 140–151. doi: 10.1016/j.jenvman.2019.06.075.

Singh, R., Bhunia, P. and Dash, R. R. (2017). 'A mechanistic review on vermifiltration of wastewater: Design, operation and performance', *Journal of Environmental Management*. Academic Press, 197, pp. 656–672. Available at: https://www.sciencedirect.com/science/article/pii/S0301479717303900?via%3Dihub (Accessed: 5 June 2019).

Sinha, R. K. *et al.* (2010). 'Earthworms – The environmental engineers: Review of vermiculture technologies for environmental management and resource development', *International Journal of Global Environmental Issues*. Inderscience Publishers, 10(3–4), pp. 265–292. doi: 10.1504/IJGENVI.2010.037271.

Sinha, R. K., Bharambe, G. and Chaudhari, U. (2008). 'Sewage treatment by vermifiltration with synchronous treatment of sludge by earthworms: a low-cost sustainable technology over conventional systems with potential for decentralization', 28(4), pp. 409–420. doi: 10.1007/s10669-008-9162-8.

Stichting Waterputten voor Kwasiakan (2017). *The Biofil Toilet System*. Available at: http://www.stichtingwaterputten.nl/ (Accessed: 29 August 2020).

SUEZ (2020). *Attached growth processes*. Available at: https://www.suezwaterhandbook.com/water-and-generalities/fundamental-biological-engineering-proces ses-applicable-to-water-treatment/aerobic-bacterial-cultures/attached-growth-processes (Accessed: 21 April 2020).

Swati, A. and Hait, S. (2018). 'A Comprehensive Review of the Fate of Pathogens during Vermicomposting of Organic Wastes', *Journal of Environmental Quality*. Wiley, 47(1), pp. 16–29. doi: 10.2134/jeq2017.07.0265.

Taylor, M., Clarke, W. P. and Greenfield, P. F. (2003). 'The treatment of domestic wastewater using small-scale vermicompost filter beds', *Ecological Engineering*. Elsevier, 21(2–3), pp. 197–203. doi: 10.1016/j.ecoleng.2003.12.003.

TBF Environmental Solutions (2018). *The Tiger Biofilter Sewage Treatment System*. Available at: http://www.tbfenvironmental.in/the-tiger-biofilter.html (Accessed: 30 August 2020).

The Composting Association (2003). *Information Sheet 23: Vermicomposting*. Available at: http://www.organics-recycling.org.uk/dmdocuments/23_Vermicomposting.pdf (Accessed: 20 April 2020).

Tilley, E. *et al.* (2014). *Compendium of Sanitation Systems and Technologies*. Available at: https://www.wsscc.org/wp-content/uploads/2015/06/compendium_of_sanitation_systems_and_technologies_2nd_edition_-_english.pdf.

Transchem Agritech (2018). *Bio filter*. Available at: https://www.transchem.in/bio-filter.html (Accessed: 4 September 2020).

USEPA (1991). *Assessment of Single-Stage Trickling Filter Nitrification*. Available at: https://nepis.epa.gov/Exe/ZyNET.exe/P1004T3S.TXT?ZyActionD=ZyDocument&Client=EPA&Index=1991+Thru+1994&Docs=&Query=&Time=&EndTime=&SearchMethod=1&TocRestrict=n&Toc=&TocEntry=&QField=&QFieldYear=&QFieldMonth=&QFieldDay=&IntQFieldOp=0&ExtQFieldOp=0&XmlQuery=.

USEPA (2000a). *Wastewater Technology Fact Sheet: Trickling Filter Nitrification*. Available at: https://www3.epa.gov/npdes/pubs/trickling_filt_nitrification.pdf.

USEPA (2000b). *Wastewater Technology Fact Sheet: Trickling Filters*. Available at: https://nepis.epa.gov/Exe/ZyNET.exe/P10099PA.TXT?ZyActionD=ZyDocument&Client=EPA&Index=2000+Thru+2005&Docs=&Query=&Time=&EndTime=&SearchMethod=1&TocRestrict=n&Toc=&TocEntry=&QField=&QFieldYear=&QFieldMonth=&QFieldDay=&IntQFieldOp=0&ExtQFieldOp=0&XmlQuery=.

Vohla, C. *et al.* (2011). 'Filter materials for phosphorus removal from wastewater in treatment wetlands-A review', *Ecological Engineering*. Elsevier, 37(1), pp. 70–89. doi: 10.1016/j.ecoleng.2009.08.003.

Wang, D. B. *et al.* (2010). 'Effects of earthworms on surface clogging characteristics of intermittent sand filters', *Water Science and Technology*. IWA Publishing, 61(11), pp. 2881–2888. doi: 10.2166/wst.2010.180.

Waterflow NZ (2019). *Naturalflow Eco-wastewater & Sewage Systems*. Available at: https://www.naturalflow.co.nz/wp-content/uploads/2019/02/NFS-Information-Booklet-01-19.pdf (Accessed: 29 August 2020).

Worm Smart (2018). *Performance Certificate: Worm Smart Plus. OSET NTP Trial 13, 2017/2018*. Available at: https://www.wormsmart.nz/wp-content/uploads/2021/02/OSET-NTP-Trial-13-Perform-Cert-Worm-Smart-Plus-27-Nov-18.pdf (Accessed: 2 March 2021).

Ye, J. *et al.* (2018). 'Reduction of clog matter in constructed wetlands by metabolism of Eisenia foetida: Process and modeling', *Environmental Pollution*. Elsevier Ltd, 238, pp. 803–811. doi: 10.1016/j.envpol.2018.03.062.

DOI: 10.1561/9781680837810.ch5

Chapter 5

Daphniafilter: A Nature-based Tertiary Treatment

By Victoria Salvadó, Manuela Hidalgo, Narcís Pous, Teresa Serra and Jordi Colomer

5.1 Introduction

The main purpose of tertiary treatment systems is to provide a final remediation stage to improve effluent quality sufficiently that it meets legislative requirements for reuse or discharge to a receiving environmental body. During tertiary treatment, different types of pollutants such as organic matter, suspended solids, nutrients, pathogens, and micro-pollutants (including priority and emerging contaminants and heavy metals), which may not be adequately removed during secondary treatment are targeted.

Different technologies are available to ensure that treated effluents achieve high levels of quality, potentially rendering them suitable for irrigation of crops, gardens and golf courses, industrial processes and even for human consumption. These technologies vary from green filters (constructed wetlands) to physicochemical treatments (coagulation, flocculation, sedimentation, filtration, membrane treatments, disinfection, etc.). In some cases, these treatments can imply the use of significant energy resources and/or chemicals (physicochemical treatments), therefore entailing economic and environmental costs that limit their general application. Innovative, low-maintenance and low-energy input solutions to treat wastewater should preferably be based on natural depuration systems. Although the implementation of lagoons and ponds (Young *et al.*, 2017) and constructed wetlands (Lutterbeck *et al.*, 2018) has grown over time, they still have the disadvantage of requiring large areas of land and problems in meeting the criteria for discharge over the whole year (Massoud *et al.*, 2009). Sludge accumulation (requiring routine maintenance) can also be problematic. Taken together, these barriers indicate that it is necessary to improve the efficiency of natural-based treatments. In order to achieve this objective, the role of the different living organisms involved in decontamination processes in natural ecosystems should be investigated, as well as the factors affecting these processes. In this context *Daphnia* filtration is a technology which leverages living organisms from natural ecosystems.

This chapter outlines the background to *Daphnia* filtration and its application to water and wastewater treatment and details key results of research carried out at both laboratory and pilot scale.

5.2 Daphnia Filtration – What We Know So Far

Frederick and Egan (1994) defined the concept of environmental biotechnology as the use of living organisms to reduce human impact on the environment. In this respect, it is known that zooplankton, specifically, daphnids, are of fundamental importance in improving the efficiency of water treatment in ponds (Kampf *et al.*, 2007). *Daphnia* is a large planktonic crustacean belonging to the order Cladocera whose diet is based on ingesting algae and other organic detritus, including protists and bacteria (Ebert, 2005). *Daphnia magna* is a common species that has a broad natural distribution (Fig. 5.1), having been identified in countries as far apart as Russia, India and South Africa. Other Cladocera species occupy similar ecological niches in countries as diverse as Australia, Mexico and Sri Lanka (Serra *et al.*, 2019a). *D. magna* thrive in water temperatures of around 20°C, although its distribution proves that it can adapt to a broad temperature range (Elenbass, 2013). However, water temperatures above 6°C are required for *D. magna* reproduction, whilst in mesocosm experiments, populations experienced seasonal fluctuations, with maximum population densities achieved when water

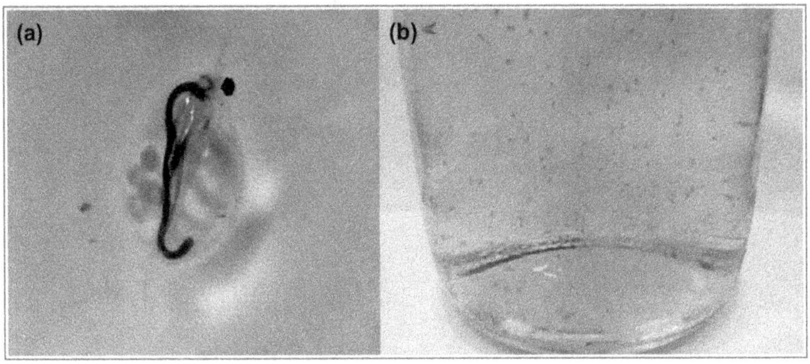

Figure 5.1. (a) *Daphnia magna* and (b) *D. magna* individuals in a vessel.

temperatures were between 15°C and 25°C (Serra and Colomer, 2016). Due to the ease with which *Daphnia* are cultivated (Mittmann *et al.*, 2014) and their capacity as filter feeders they can be considered as promising organisms to treat wastewater.

Laboratory experiments have shown that *D. magna* ingest particles with diameters below 30 μm (Fig. 5.2a) (Pau *et al.*, 2013). Therefore, biological filtration by *D. magna* can reduce the concentration of small particles, which are the most difficult to remove from suspension during clarification. Although sedimentation was shown to be responsible for 60% of the reduction in suspended solids, ingestion by *D. magna* resulted in an additional 10–29% reduction in particles with diameters lower than 30 μm (Pau *et al.*, 2013). Higher reductions were associated with higher population densities of *D. magna* (Fig. 5.2b). From mesocosm experiments, the reduction in the particle concentrations attributed to *Daphnia* filtration ranged between 2.5% and 39%, corresponding to *Daphnia* population densities of between 5 and 100 individuals L^{-1} (hereafter ind L^{-1}). Therefore, the maximum reductions due to both the filtration capacity of *Daphnia* and sedimentation were 99% (in the mesocosm) and 92% (in the laboratory).

In addition to *Daphnia* population densities, mesocosm experiments have demonstrated that hydraulic residence times are also an important factor, with particulate removal efficiencies greatest where residence times exceed 3 h (Serra and Colomer, 2016). This has important implications for system design (sizing) and operation. Conventional tertiary wastewater treatments often rely on chemical coagulants to aggregate particles, facilitating their removal. In contrast, biological filtration by *Daphnia* does not require such chemicals, reducing the cost of treatment and ecological impact when treated wastewater is reintroduced to the environment. A comparison with conventional treatments (lamellar decanters, sand filters and disc filters) has shown that *Daphnia* filtration reduced the concentration of small particles in a similar amount to that found for disc filters and without the need for any backwash (Serra *et al.*, 2014).

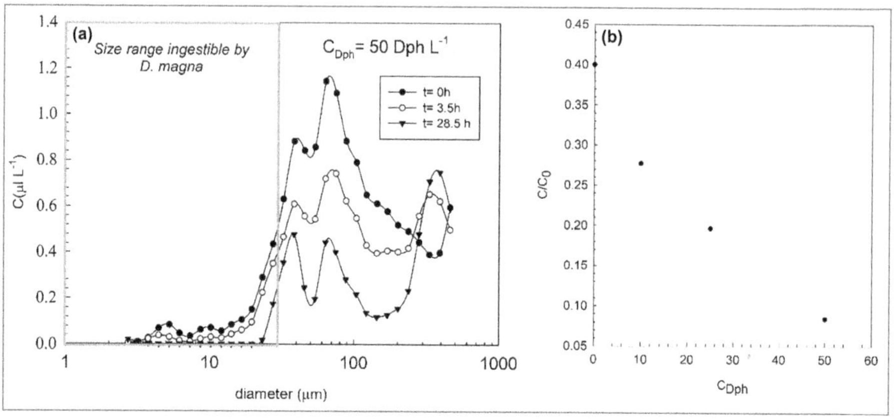

Figure 5.2. (a) Particle volume concentration (C, in μl L^{-1}) versus particle diameter (d, in μm) for different experimental durations (t). The green rectangle represents the ingestible particle range for *Daphnia* with d <30 μm. Particles with d > 30 μm are expected to settle over time under gravity. (b) Evolution of the ratio between the concentration of particles of d <30 μm after 4 h of *Daphnia* filtration (C) and the initial concentration of particles (C_0) versus the concentration of *Daphnia* (C_{Dph}).

Alongside removal of suspended inorganic and organic particulates, the feeding habits of *Daphnia* can reduce concentrations of bacteria such as *E. coli* (Serra *et al.*, 2014) and coliforms (Shiny *et al.*, 2005). These findings suggest that *Daphnia* could be used for clarification (reducing the suspended solid content) and for pathogen removal (reducing the bacterial load). Optimization of these processes would make *Daphnia* biofiltration suitable as a tertiary wastewater treatment (Matamoros *et al.*, 2012; Müller *et al.*, 2018). As illustrated in Table 5.1, which summarises the bacterial inactivation rates of tertiary treatments in operation at different WWTPs of the Costa Brava (Girona, Spain), *Daphnia* biofiltration can enable reclaimed water to be obtained with a microbiological quality that is similar to that produced by conventional treatments, with potential for lower energy and maintenance costs. Estimates of the relative energy consumption (kWh m^{-3}) for the reduction of *E. coli* from the Empuriabrava WWTP indicate a cost of €0.60 in a conventional tertiary systems (*E. coli* reduction between 2.1 and $4.5\log_{10}$) and €0.04 using daphnia cultures (*E. coli* reduction of $2.7\log_{10}$) (Table 5.1).

Furthermore, *Daphnia* have been shown to contribute to reducing BOD content through their consumption of particulate organic matter (Shiny *et al.*, 2005), while in another zooplankton-based reactor, average removal efficiencies of 80% were obtained for emerging and priority organic pollutants (i.e., pharmaceuticals, personal care products, pesticides, antiseptics, fire retardants and plasticisers) (Matamoros *et al.*, 2012). These characteristics make this biotechnology more efficient than some conventional tertiary treatments, such as

Table 5.1. Comparison of the inactivation rates achieved at the mesocosm for *E. coli* and sulphite reducing Clostridia (a type of Clostridia that are characteristic of faecal waters), with those obtained at different tertiary treatment of wastewater treatment plants (Girona, Spain). Data from the Costa Brava water agency (http://www.ccbgi.org/reutilitzacio.php, March 2013).

WWTP Location and Tertiary Treatment		Inactivation Rates (In Logarithmic Units)	
		E. coli	**Sulphite Reducing Clostridia**
El Port de la Selva	Coagulation, flocculation, multilayer pressure filtration UV + Chlorine	3.5	2.9
Empuriabrava	Constructed Wetland	2.2	1.1
Torroella de Montgrí	UV + Chlorine	2.1	1.1
Pals	Chlorine	3.3	1.0
Castell-Platja d'Aro	Sand filtration, UV + Chlorine	4.5	1.3
Tossa de Mar	Coagulation, flocculation, sedimentation, sand filtration, UV + Chlorine	4.5	2.8
Empuriabrava mesocosm	*Daphnia* filtration, 4-day hydraulic retention time	2.7	1.9

coagulation-flocculation-lamellar sedimentation and UV light-chlorination (Matamoros and Salvadó, 2013), but slightly less efficient than others, such as ozonation, Fenton oxidation, and membrane-based systems (Klavarioti *et al.*, 2009) and Dolar *et al.* (2012). The combination of a membrane bioreactor (MBR, composed of an activated sludge bioreactor followed by microfiltration) and a reverse osmosis (RO) system showed high removal rates (99%) for some emerging contaminants, although in some cases, the removal efficiency of the MBR alone was lower than 80% (Dolar *et al.*, 2012). The high removal efficiency of the *Daphnia* reactor, consisting of zooplankton combined with microbial and microalgae biofilms, is explained by the simultaneous occurrence of biodegradation (e.g., ibuprofen, naproxen and furosemide), photodegradation (e.g., diclofenac and ketoprofen), adsorption dynamics, algae and zooplankton uptake (Garcia-Rodríguez *et al.*, 2013).

The results and knowledge gained from the experiments performed in a mesocosm system support the idea that an innovative tertiary treatment based on biological filtration (from here on referred to as the Daphniafilter) is technically feasible and can be competitive in terms of costs and efficiency with respect to current physico-chemical tertiary treatments (Ortiz *et al.*, 2011; Matamoros *et al.*, 2012). In order to design and optimise the performance of a Daphniafilter, it was first essential to understand the sensitivity of *Daphnia* to common wastewater components such as free ammonia (NH_3) (Lyu *et al.*, 2013), ammonium (NH_4^+) and nitrite (NO_2^-) (Serra *et al.*, 2019b), organic matter and metals (Okamoto *et al.*, 2015).

5.3 Laboratory Tests to Optimise the Filtration Capacity of *Daphnia*: Effect of Light, Water Temperature, Organic Matter and Nutrients

The response of *Daphnia* individuals to environmental changes has been studied in terms of their ingestion (through filtration capacity), swimming velocity, heart beat and survival, which are altered by the toxicity (Bownik, 2017), turbulence (Serra *et al.*, 2018), temperature (Müller *et al.*, 2018) and contaminant load (Nørgaard and Roslev, 2016; Serra *et al.*, 2019b) of the water where they thrive. Such contaminants can include microplastics – identified by many authors in water bodies (including municipal wastewaters (Sun *et al.*, 2019)), some of whom have studied the effect of microplastics on freshwater organisms (Rist *et al.*, 2017). Moreover, microplastic ingestion may have physical consequences for zooplankton such as *Daphnia*, disrupting their feeding and digestion, and thus affecting their fitness and performance (Colomer *et al.*, 2019). Light can also be considered crucial for the growth of *Daphnia* as daily and seasonal vertical *Daphnia* migrations through the water column have been observed (Simoncelli *et al.*, 2019). Hence, the photoperiod, which has daily and seasonal time-scale variations that are also latitude-dependent, has also been shown to be a key factor in determining the reproduction and growth rates of zooplankton. Moreover, both individual stressors and combinations of stressors can significantly reduce the filtration capacity of *Daphnia* (Serra *et al.*, 2020).

Extensive laboratory studies have been performed into the behaviour of *Daphnia* in order to assess their usefulness in water treatment; key aspects of these studies are summarised in Fig. 5.3 and outlined in the following sections.

5.3.1 Effect of Photoperiod and the Light Intensity

For continuous light, *D. magna* filtration rates have been found to be highest with photoperiods of 16 h L (Light)/8 h D (Darkness), 12L/8D and 8L/16D (Table 5.2).

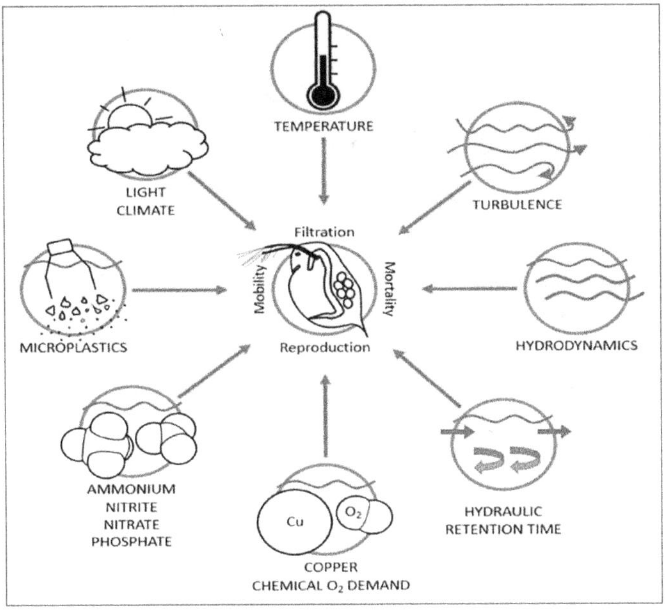

Figure 5.3. Overview of the laboratory tests carried out to study the effect of the different conditions on *D. magna* filtration capacity, mobility and survival.

For shorter periods of light exposure (4L/20D), filtration was lower and only superior to periods of complete darkness (0L/24D). With the maximum light intensity tested (100%), filtration was 2.5 times the rate of filtration in complete darkness ($I = 0$ W m^{-2}) (Table 5.2). For the highest light intensity tested, the *Daphnia* filtration was 2.42 times than that obtained in darkness (Serra *et al.*, 2019a).

5.3.2 The Effect of Water Temperature

The optimal water temperature for *D. magna* filtration has been found to be 20°C (Fig. 5.4a). Both above and below this temperature, the filtration capacity decreases significantly, and this is particularly accentuated as water temperatures become more extreme in both directions. However, experiments have shown that *Daphnia* individuals are able to acclimate to a wide range of temperatures. Acclimation can take several days, during which the filtration capacity of individual *Daphnia* increases significantly (Fig. 5.4b). In the case of 25°C, the filtration capacity after acclimation increased 91.7% compared to shortly after the temperature change, while in the case of 11°C the increase was 136.4%, compared with capacity immediately following the applied temperature change. However, acclimation at temperatures of 29°C or greater is not possible due to high rates of mortality. The final filtration capacities of *Daphnia* individuals that have to acclimate to changes in water temperature are at their highest at 15°C and their lowest at 11°C, but in

Table 5.2. *Daphnia magna* filtration at seventh day of acclimation: (a) at the different photoperiods for 3,940 lux of light intensity and (b) at different light intensities for the 12 L/12 D photoperiod.

Photoperiod Light (L)/Darkness (D) (Hours) 3940 lux During Light Periods	Filtration Capacity (mL ind^{-1} h^{-1})
24 L/0 D	2.51
0 L/ 24 D	1.18
12 L/12 D	2.90
16 L/8 D	2.70
8 L/16 D	3.00
4 L/20 D	1.84
Fixed Photoperiod 12 L / 12 D Varying Light Intensity (lux)	**Filtration Capacity (mL ind^{-1} h^{-1})**
0	1.21
394	1.44
985	1.69
1,970	2.13
2,955	2.59
3,940	2.90

no case do they reach the maximum levels seen at a starting temperature of 20°C (Müller *et al.*, 2018).

5.3.3 The Effect of COD

The effect of COD on *Daphnia* was tested in triplicate in 1L-reactors containing 20 ind L^{-1} in secondary wastewater at a controlled temperature of 20 ± 1°C and photoperiodic illumination (16L/8D) for 24 h. The COD concentration was varied from 0 (control) to 250 mg COD L^{-1}, aligning with the expected value of a low strength urban wastewater (Metcalf and Eddy, 1991), by adding different quantities of CH_3COONa:CH_3CH_2OH, dehydrated meat extract and milk to mineral water. There was 100% mortality in two of the three replicates when they were exposed to the highest COD concentration (250 mg COD L^{-1}). However, for COD concentrations below 160 mg COD L^{-1}, the filtration capacity of *Daphnia* was not inhibited. There are a number of reasons why *Daphnia* could be inhibited by the presence of COD concentrations higher than 160 mg COD L^{-1} in this case: (i) vessels were not mechanically aerated but were open to the air

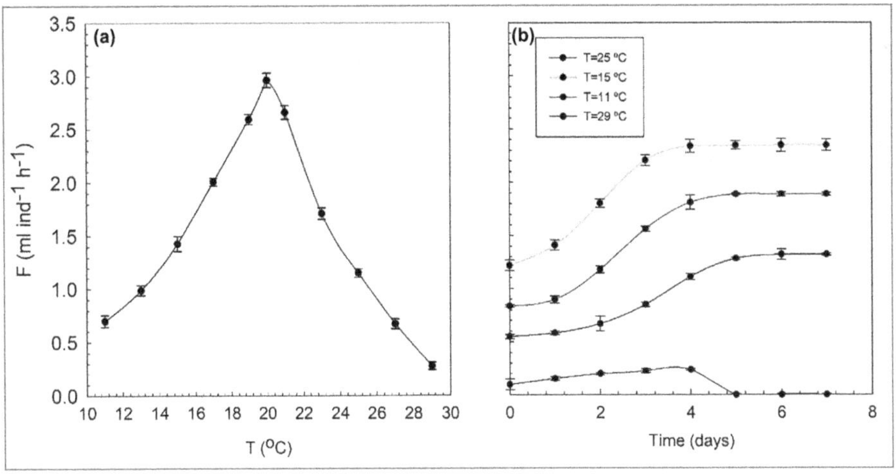

Figure 5.4. (a) *D. magna* filtration capacity, F (ml ind^{-1} h^{-1}) versus water temperature and (b) *D. magna* filtration capacity F (ml ind^{-1} h^{-1}) during acclimation versus the time (in days) exposed at each water temperature.

and oxygen was supplied through passive aeration from the surface. Since microorganisms can grow on the upper surface layer of the wastewater, an oxygen barrier may have been created that harmed the survival of the *Daphnia* individuals and (ii) excessive growth of heterotrophic bacteria could have been a source of toxins that reduced *Daphnia* survival (Rodriguez da Silva *et al.*, 2004). Overall, these findings suggested that a reactor containing *Daphnia* should be externally aerated when the wastewater contains >160 mg COD L^{-1} (Pous *et al.*, 2020). Any such aeration would have to be carefully/gently applied, since background turbulence has a negative impact on mobility and the filtration capacity of *Daphnia* individuals (Serra *et al.*, 2019c).

5.3.4 The Effect of Ammonium, Nitrite, Nitrate and Phosphate

The effect of the presence of nutrient species on the filtration capacity of *Daphnia* can be evaluated by monitoring the inactivation (INACT) factor, defined by Eq. (5.1):

$$INACT(t) = log\left(\frac{F(0,t)}{F(c_x,t)}\right) \tag{5.1}$$

where F(0) is the *Daphnia* filtration rate of suspended solids at $C_x = 0$ (where C is the concentration of a given compound x) and $F(c_x)$ is the *Daphnia* filtration rate for $Cx \geq 0$. For example, in the case of NO_3^--N and PO_4^{3-}-P, and for

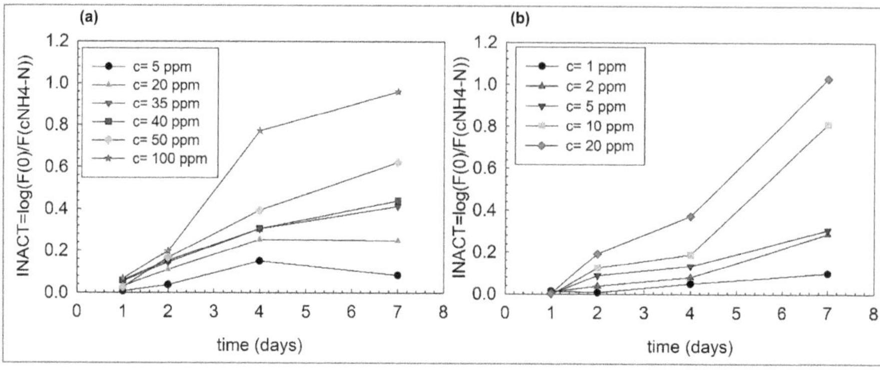

Figure 5.5. Inactivation factor (INACT) versus exposure time (in days) for the different chemical dosages c (in ppm = mg L^{-1}) for (a) NH_4^+-N and (b) NO_2^--N.

the range of concentrations studied, inactivation has been found to be zero. However, the inactivation factor increases with increasing concentrations of NH_4^+-N and, in general for any given concentration, INACT increases with increased exposure time (Fig. 5.5a,b). INACT for *Daphnia* is sensitive to increasing NO_2^--N concentrations but particularly sensitive to exposure time (e.g., for exposure times of more than 1 day and also exposure times of 4 days at concentrations of 2 mg L^{-1} and above (Fig. 5.5b) (Serra *et al.*, 2019b).

These findings indicate that *Daphnia* filtration may not produce satisfactory results for effluents rich in ammonium or nitrites (>40 mg NH_4^+-N L^{-1} and >5 mg NO_2^--N L^{-1}, respectively) when *Daphnia* are exposed to such contaminants for periods longer than 1 day.

5.4 Assessment of the Daphniafilter at Laboratory Scale

The application of any technology to on-site conditions can cause challenges with issues such as flow fluctuation and influent wastewater variations. Thus, a key design factor for on-site systems is the hydraulic residence time (HRT) – a factor that can be analysed in laboratory-based studies. As part of the initial development of the zooplankton system for the INNOQUA project a 2 L laboratory scale Daphniafilter (Fig. 5.6) was assembled to evaluate nutrient removal at different HRTs (fluxes ranging from 0.54 to 3.5 L d^{-1}) from synthetic wastewater (133 ± 64 mg COD L^{-1}, 14 ± 5 mg NH_4^+-N L^{-1} and 7 ± 1 mg PO_4^{3-}-P L^{-1}). The reactor was inoculated with a *D. magna* population of 50 ind L^{-1} and operated under batch conditions for 7 days. The reactor was then switched to continuous flow mode for 11 days at an HRT of 3.7 days (steady-state conditions).

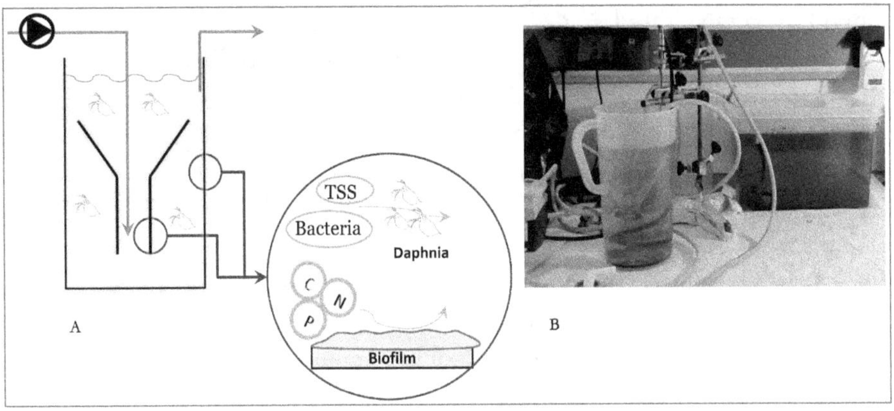

Figure 5.6. (a) Scheme of the Daphniafilter reactor used in the lab-scale experiments and (b) experimental set-up at lab scale. Adapted from Pous *et al.* (2020).

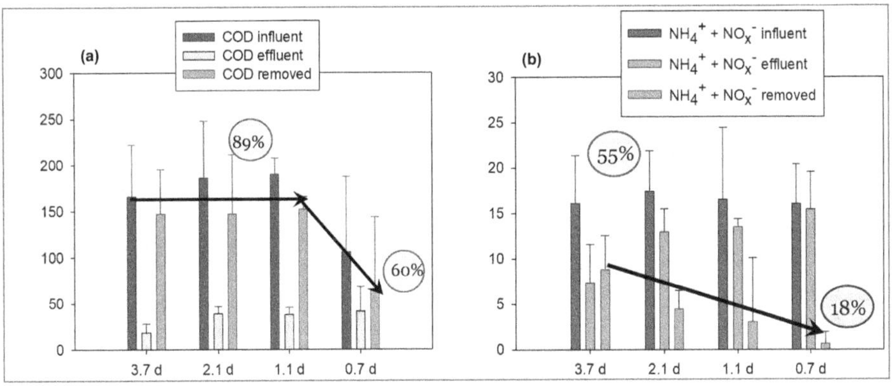

Figure 5.7. Lab-scale Daphniafilter performance at different HRTs: (a) organic matter (COD) removal and (b) nitrogen (NH_4^+ + NO_x^- (NO_2^- + NO_3^-)) removal. Error bars show standard deviation. Adapted from Pous *et al.* (2020).

During this period, the *Daphnia* population evolved and a bacterial/algal biofilm grew on the reactor surface.

Following this, tests for evaluating the effect of HRT on nutrient removal were started. Each HRT (3.7, 2.1, 1.1 and 0.7 days) was tested over 14 days. The complex synthetic wastewater used for these tests implied certain variability on influent nitrogen and COD content. The nutrient removal observed in the zooplankton reactor under continuous flow mode is shown in Fig. 5.7.

The removal of organic matter (COD) was mostly constant over the different HRTs tested and the highest COD removal, 89 ± 2% (equivalent to 39 ± 13 mg COD L^{-1} d^{-1}) was observed at an HRT of 3.7 days (Fig. 5.7a). When the HRT was decreased to 0.7 days, COD removal decreased slightly (74 ± 3%). Nevertheless, this meant that from 3.7 to 0.7 days HRT, the COD removal rate had increased from 39 ± 13 mg COD L^{-1} d^{-1} to 261 ± 98 mg COD L^{-1} d^{-1}. The nitrogen removal performance was more affected by changes to the HRTs (Fig. 5.7b). A total nitrogen removal of 45 ± 13% (8.8 ± 3.8 mg N L^{-1}) was observed at an HRT of 3.7 days, giving a total nitrogen removal rate of 2.3 ± 1.0 mg N L^{-1} d^{-1}. However, in percentage terms, nitrogen removal decreased at higher water-flow rates (7 ± 4% at 0.7 days HRT) although the total nitrogen removal rate at 0.7 and 3.7 days was similar (1.9 ± 0.5 and 2.3 ± 1.0 mg N L^{-1} d^{-1}, respectively). Ammonium removal rate was mostly constant at the HRT range tested (around 3 mg NH_4^+-N L^{-1} d^{-1}). Moreover, an accumulation of both nitrates and nitrites was observed at high HRTs given that only 70% of ammonium removed was denitrified or assimilated (79% at the lowest HRT). These results can be explained by the lack of an electron donor (organic matter) at high HRT, and in the case of lower HRTs, poor nitrification performance leads to low substrate availability (nitrates and nitrites). With regards to phosphate, it can be observed that the removal performance was low at around 12% with an HRT of 3.7 days, but no P removal was observed at lower HRTs (data not shown) (Pous *et al.*, 2020).

In summary, the Daphniafilter presented a significant removal of COD (89%) and total nitrogen (45%) at an HRT of 3.7 days, with low orthophosphate removal (12%). Decreasing the HRT to 1.1 days resulted in a slight decrease in COD removal (74%), while total nitrogen removal decreased to 20%. At lower HRTs, total nitrogen removal rates progressively decreased. Thus, based on this data (albeit with synthetic wastewater) a Daphniafilter operating with an HRT of between 2 and 4 days is adequate for polishing carbon and nitrogen from secondary-treated wastewaters (Pous *et al.*, 2020).

5.5 Basic Operating and Design Principles of a Pilot-Scale Daphniafilter Reactor

Data from the laboratory scale experiments (Sections 5.2 and 5.3) provided the basis for designing a pilot-scale reactor utilising the different wastewater treatment capabilities of zooplankton and bacterial/microalgal biofilms. In order to design a reactor intended to treat a volume of wastewater of 10 population equivalent (PE), a computational fluid dynamic analysis was carried out to optimise the critical parameters and maximise the efficiency of the biological system. Simulations were carried out assuming a range of boundary conditions such as water inlet temperature (8–25°C), water viscosity as a function of the temperature, flow particle characteristics (including volumetric concentration by size) and flow rate conditions. Water velocity is a critical parameter for *Daphnia* (Serra *et al.*, 2018). For this reason, the design specifications took into account not only the influent velocity, but also the different velocities generated inside the reactor to ensure that the maximum tolerance value was not reached at any point. The design also aimed for the control of water temperatures, the promotion of a high concentration of particles below 35 μm in sunlit areas, the promotion of a region with fairly static flow in the lower part of the tank to promote sedimentation of larger particles, as well as the avoidance of preferential flow paths that could drag *Daphnia* out of the tank. The system was designed to ensure aerobic conditions by including a passive aeration system at the inlet.

Light plays an important role in the *Daphnia* life cycle, with alternate periods of light and darkness required, and so water surface exposure to daylight must be ensured. At the same time, however, excessive sunlight must be prevented to avoid algae overgrowth and system overwarming. In this case a perforated cover was sufficient to deliver the necessary dappled shade. The Daphniafilter is discharged by overflow, so no pump is required if the influent is fed gravitationally from a primary/secondary treatment system. At peak flows, a reduced HRT of just 6 h is the lowest that can be tolerated before it becomes necessary to bypass the reactor – or risk complete washout of the *Daphnia* (and wider mesocosm) population. Where peak flow situations exceeding the capacity of the Daphniafilter are a possibility, a buffer tank should be installed to regulate the inflow to the unit. A schematic of the 10 PE Daphniafilter reactor (Salvadó *et al.*, 2019), is presented in Fig. 5.8a,b.

The reactor for the initial pilot-scale system in this project utilised a conical base to minimise the velocity and turbulence at the inlet (located centrally halfway down the reactor depth), and had a total volume of 1,500 L. The reactor contained

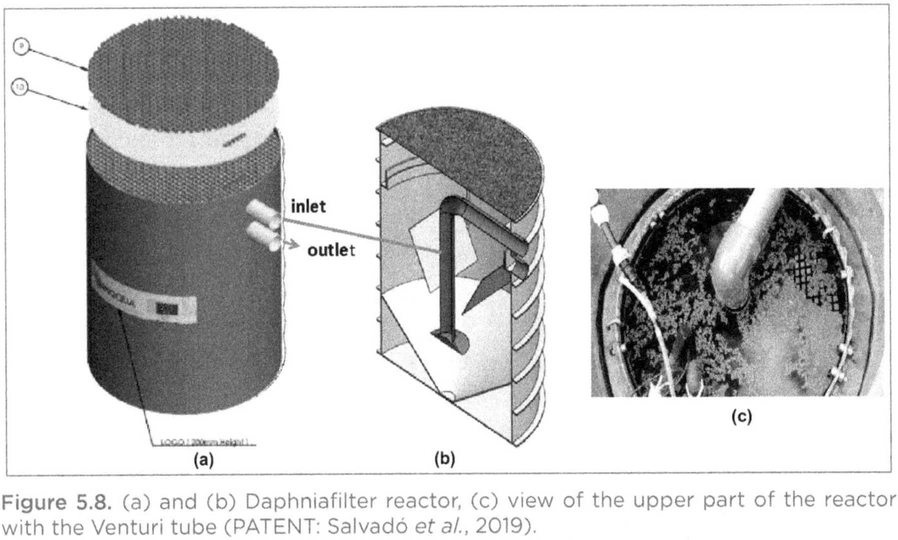

Figure 5.8. (a) and (b) Daphniafilter reactor, (c) view of the upper part of the reactor with the Venturi tube (PATENT: Salvadó *et al.*, 2019).

two flat rectangular plates to increase the internal surface area for bacterial and algal biofilm growth to enhance nutrient removal and encourage sludge settlement. This configuration was designed to operate with velocities lower than 3.5 mm s^{-1} (Serra *et al.*, 2018). The outlet comprised a weir overflow that occupied the entire internal circumference of the reactor to ensure a gentle outflow and to retain *Daphnia* within the reactor.

For ease of transport, handling and installation, the reactor tank was made from high-density polyethylene. The internal support structure was made with alloy steel (AISI 403) and the inlet tube was modified to include a Venturi device to aerate the influent and prevent low oxygen conditions developing (Fig. 5.8c).

The tank was installed on a perfectly horizontal surface to ensure flow conditions as per the design. The inlet and outlet polyethylene pipes provided smooth surfaces that help to prevent the accumulation of biofilm and sludge attachment (that could, ultimately, reduce effluent quality and lead to clogging).

After installation, secondary treated wastewater needs to be circulated inside the reactor for two weeks to promote the growth of a biofilm on the inner walls of the reactor and the lamella surfaces. After the biofilm has been established, *Daphnia* individuals can be introduced into the tank at a minimum population of 0.2 ind L^{-1}. It is recommended that the temperature and quality of the influent are monitored to ensure compatibility between the Daphniafilter and prior primary/secondary treatment.

5.6 Evaluation of the Performance of the Daphniafilter Reactor as a Tertiary System on a Pilot Scale

In order to carry out the prototype validation and demonstration, the 10 PE Daphniafilter prototype was first installed (March 2018) at the municipal wastewater treatment plant of Quart (Girona, Spain), which performs primary and secondary treatment. The system was tested for a total of 412 days, from April 2018 to June 2019. In order to allow bacterial/algal biofilm growth, the reactor was fed for 19 days with secondary treated wastewater with the following characteristics: pH 7.3 $\pm$ 0.3, 1387 $\pm$ 228 μS cm^{-1}, 68 $\pm$ 59 mg COD L^{-1}, 29.5 $\pm$ 14.6 mg $NH_4{}^+$-N L^{-1}, 0.5 $\pm$ 0.7 mg $NO_3{}^-$-N L^{-1}, 0.5 $\pm$ 1.0 mg $NO_2{}^-$-N L^{-1} (total nitrogen = 30.4 $\pm$ 14.3 mg TN-N L^{-1}), 4.4 $\pm$ 7.3 mg $PO_4{}^{3-}$-P L^{-1}, 64 $\pm$ 170 mg TSS L^{-1} and 105 $\pm$ 260 NTU. About 1,000 *Daphnia* individuals (about 0.6 ind L^{-1}) were then added. This reactor was designed to operate normally at 1,500 L d^{-1}, giving a nominal HRT of 1 day, and was tested under three additional conditions: (a) Overflow: the system was operated at 3,000 L d^{-1} (0.5 d HRT) between days 72 and 145 (July–August 2018) and days 233 and 240 (December 2018); (b) Underflow: the system was tested at 750 L d^{-1} (2 d HRT) between days 217 and 222 (December 2018); (c) No flow: the influent pump was stopped between days 146 and 161 (September 2018) and 240 and 264 (December 2018–January 2019). Monthly average temperature varied between 10.4 ± 1.8 and 29.3 ± 1.6°C.

The aim of the Daphniafilter reactor was to produce an effluent suitable for reuse as defined by Spanish water reuse legislation (RD 1620/2007), which was used as a reference. This legislation sets different standards for *E. coli*, TSS, turbidity, and nitrogen depending on the end use. Initially, the capacity of *Daphnia* to reduce the amount of solids present in secondary effluents was assessed. The effect of the wastewater inflow rate on effluent TSS concentrations and turbidity values is presented in Fig. 5.9. The system was resilient enough to keep effluent TSS within Spanish regulatory parameters in autumn and winter at all flow rates tested (Fig. 5.9a). However, the TSS concentrations in the effluent were higher than in the influent during spring and summer at the different flow rates tested. The presence of larger particles such as floccular bacterial aggregates and *Lemna* spp. (duckweed) in the effluent resulted in TSS concentrations that exceeded the regulatory limits at various stages throughout the year.

Turbidity values were acceptable throughout the year at all flow rates and were especially low in winter (Fig. 5.9b). In the Daphniafilter, *Daphnia* was responsible for the removal of small particles (<30 μm) which contributed to the turbidity of

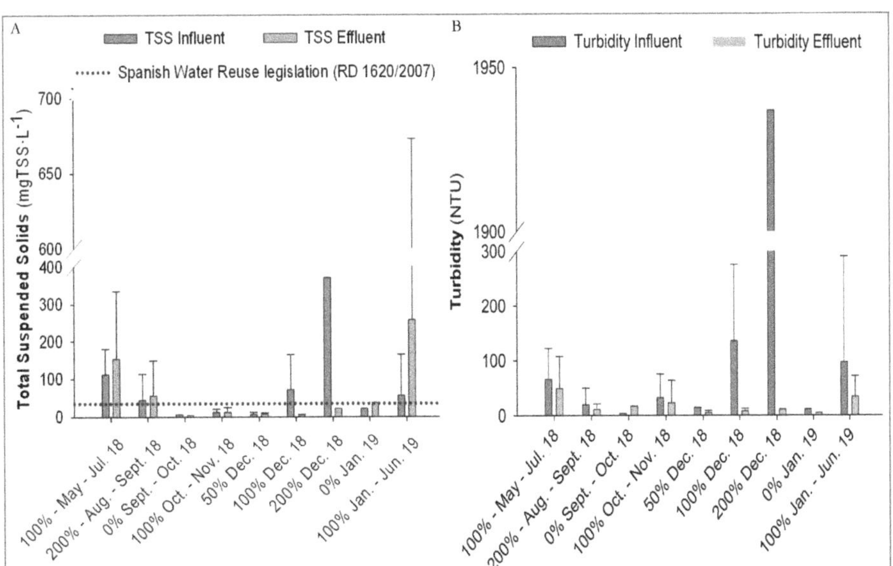

Figure 5.9. Monthly mean data of: (a) Influent and effluent total suspended solid content (TSS), the dotted line represents the TSS standard for water reuse in the Spanish legislation and (b) Turbidity at the influent and the effluent. Error bars show standard deviations. Flow: nominal load 1,500 L d^{-1} (100%), 750 L d^{-1} (50%) and 3,000 L d^{-1} (200%). Adapted from Pous *et al.* (2021).

the influent. The slight variation in effluent turbidity indicates that the filtration activity of *Daphnia* was relatively stable over the whole year, despite potential stressors such as temperature and ammonium concentrations.

In order to remove nitrogen from wastewaters a sequence of biological processes involving nitrification and denitrification are required. More than 95% of nitrogen in the influent to the Daphniafilter was in the form of ammonium (NH_4^+-N), whose concentration (average 30 ± 14 mg NH_4^+-N L^{-1}) varied significantly over time. A gradient of oxygen is expected to occur across the reactor's biofilm, allowing the development of nitrifiers on the outer side of the biofilm (more aerobic) and the growth of denitrifiers in the inner side of the biofilm (more anoxic). However, the lack of mechanical aeration limits nitrification performance, while the low availability of organic matter inside the Daphniafilter makes it difficult to achieve high denitrification efficiency. More detailed study of the nitrification (NH_4^+-N removal) and denitrification (TN removal) performance in the Daphniafilter over time was undertaken and comparison was made with the flow rate of the system (Fig. 5.10). The results obtained demonstrated

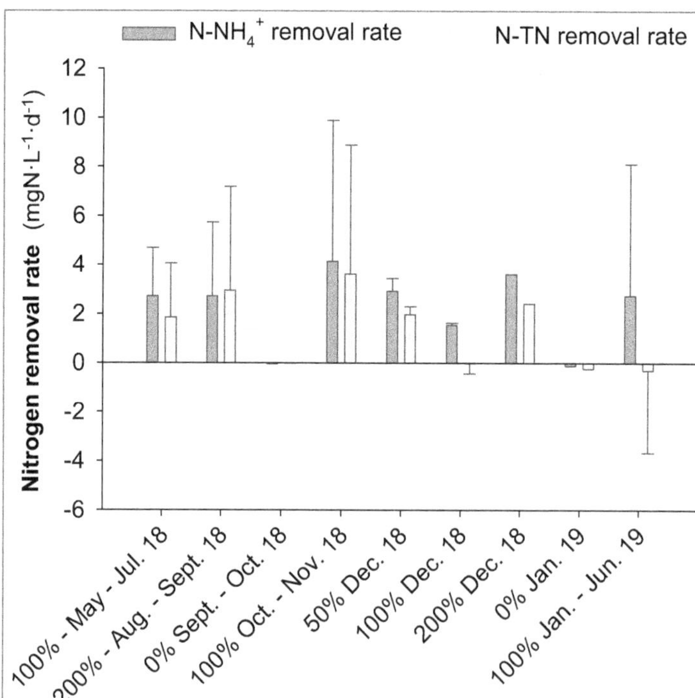

Figure 5.10. Average data from water flow-rate tests on ammonium (NH_4^+-N) and total nitrogen (TN-N) removal. Error bars show the standard deviation. Flow: nominal load 1,500 L d^{-1} (100%), 750 L d^{-1} (50%) and 3,000 L d^{-1} (200%). Adapted from Pous *et al.* (2021).

that denitrification was slower than nitrification, resulting in a small nitrate accumulation within the effluent of between 0.6 ± 0.5 and 4.2 ± 2.0 mg N L^{-1}, as was found in previous laboratory-scale experiments (Section 5.3). The absence of significant differences in nitrogen removal rates at the different water-flow rates tested (750, 1,500 and 3,000 L d^{-1}) support this finding (Pous *et al.*, 2021). Therefore, to obtain higher nitrification rates, the passive aeration system could be improved – and where effluent COD concentrations are too low to support further heterotrophic denitrification – strategies such as the addition of microalgae or aquatic plants (i.e., *Lemna*) could be used to increase nitrogen uptake. The phosphate content in the influent and the effluent was also monitored over the full year of operation but no clear trend could be discerned (data not shown).

Microbial and suspended particles (a different measure than TSS) removals were also monitored at the Daphniafilter prototype at three different flows (750, 1,500 and 3,000L d^{-1}) (Fig. 5.11). At each flow, two wastewater samples of 1 L each were collected at the inlet and the outlet of the Daphniafilter for microbiological tests (total *Coliforms, E. coli* and *Enterococcus*). The *Daphnia* population inside the reactor was determined at each flow rate by collecting a single 4 L sample of water from the system which indicated, levels of 295 ind L^{-1} at 750 L d^{-1}, 209 ind L^{-1} at 1,500 L d^{-1} and 8 ind L^{-1} at 3,000 L d^{-1}. The reduction in the number of *Daphnia* for the higher flow might be due to the unfavourable conditions for *Daphnia* at this rate, with excessive turbulence and water inertia disrupting stable conditions for swimming, and causing the death of individuals. This in turn may have impacted the efficiency (I) of the Daphniafilter in reducing the measured microbiological parameters (Fig. 5.11a).

The efficiency of the system in removing particles was 99% at flows of 750 and 1,500 L d^{-1}, when the *Daphnia* concentrations were high (between 300 and 200 ind L^{-1}, respectively). However, for the highest flow (3,000 L d^{-1}), which had the lowest *Daphnia* concentration (8 ind L^{-1}), the suspended solid removal decreased to 69.2% (Fig. 5.11b). As can be seen in this figure, the efficiency in the reactor decreased lineally as flow increased when it did not contain daphnia.

In order to assess the suitability of the Daphniafilter to deliver tertiary treated wastewater suitable for water reuse, a quality ratio (QR) was calculated for the different scenarios described in the Spanish water reuse legislation following:

$$QR = \frac{C\ P1}{Std\ P1} + \frac{C\ P2}{Std\ P2} + \cdots + \frac{C\ Pn}{Std\ Pn} \tag{5.2}$$

where *QR* is the quality ratio, *C P* is the concentration of each parameter in the effluent from the reactor, *n* is the number of parameters and *Std P* is the standard

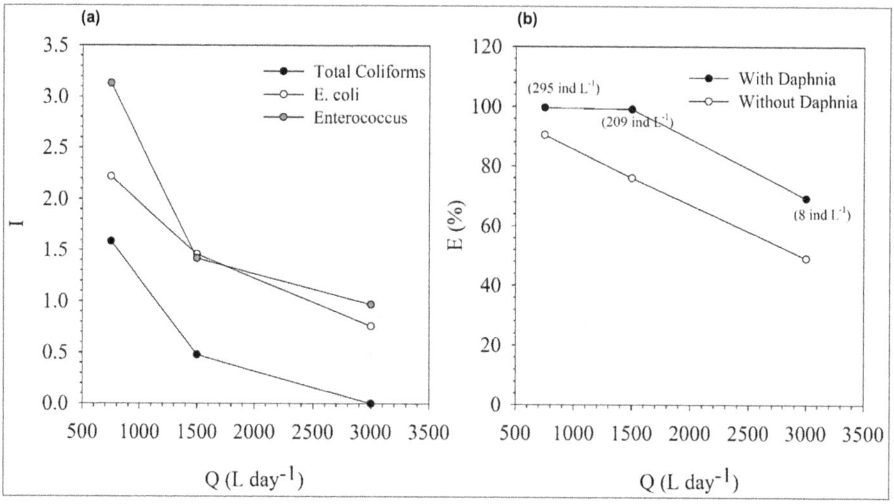

Figure 5.11. (a) Total *Coliforms* (black dots), *E. coli* (white dots) and *Enterococcus* (grey dots) inactivation versus the different flows studied. (b) Daphniafilter efficiency for removing suspended solid particles with *Daphnia* (black dots) and without *Daphnia* (white dots) for the different flows studied.

concentration of each parameter according to the water reuse legislation. When QR < 1, all parameters fulfilled the legal requirements, while at QR > 1, these parameters were not reached by one or more parameter.

The results achieved by operating the system at 1,500 L d^{-1} indicate that the water quality standard is acceptable for the irrigation of forest, crops not aimed at human consumption, and for recreational lakes (QR = 0.4). Higher water-flow rates at 3,000 L d^{-1} were found to have a significant effect on the quality of effluent water when *E. coli* removal decreased as a result of a reduction in the number of *Daphnia* and thus the only application of reuse permitted would be forest irrigation (QR = 0.6). However, the best results were obtained when the system was run at 750 L d^{-1}, where effluent *E. coli* was 400 CFU 100 mL^{-1}. With these results it would be possible to add agricultural irrigation of food products where water is not in direct contact with the edible produce as a reuse application. Taken together, the results suggest that the operation of the reactor should be set at the lowest water-flow rates tested when water reuse is pursued when the Daphniafilter reactor is connected to municipal secondary wastewater treatment (e.g., activated sludge in this case). The year-round performance of the reactor, and particularly its performance in the summer months, can be improved by better temperature and oxygen control and optimisation of the role of *Lemna*, since this aquatic plant accumulates nitrogen and phosphorus (Ennabili *et al.*, 2019) and fixes CO_2 derived from COD oxidation (Mohedano *et al.*, 2019), which could provide additional pH control.

5.7 Evaluation of the INNOQUA Integrated System: Daphniafilter Reactor Connected to Lumbrifilter Reactors on a Pilot Scale

For piloting purposes an integrated system comprising two 5 PE Lumbrifilter systems connected to a 10 PE Daphniafilter reactor (Fig. 5.12), was designed to treat 1,500 L d^{-1} of raw municipal wastewater at the Quart treatment plant (Girona, Spain). The Daphniafilter provided tertiary treatment for secondary treated wastewater from the Lumbrifilter. The system was evaluated over a period of 9 months, the evaluation process comprising tests at three different hydraulic loads, namely 750 L d^{-1}, 1,500 L d^{-1} and 3,000 L d^{-1}. The main objectives of this evaluation were to explore the organic matter, nutrient and pathogen removal capabilities of an entirely nature-based modular wastewater treatment process. The intention was that treated water would be suitable for discharge to surface water and some reuse applications.

Figure 5.13 shows the average removals for COD, TSS, NH_4^+, BOD_5 and turbidity obtained from each treatment step (Lumbrifilter and Daphniafilter) and the average removals from the full treatment plant at the different loads tested. As can be seen, the combination of both technologies to treat raw municipal wastewater resulted in high levels of removal (>85%) of COD, TSS, $NH_4{}^+$ and BOD_5. Lumbrifilter treatment alone delivered average removals of 84 ± 6%, 75 ± 10%, 93 ± 8% and 73 ± 17% for COD, TSS, NH_4^+ and BOD_5, respectively. However, the Daphniafilter provided an important polishing stage, achieving average removal rates of 54.3 ± 19.4% for TSS, 34.5 ± 20% for BOD_5, 31.0 ± 11.7%

Figure 5.12. INNOQUA integrated system at the UDG pilot-site (Quart WWT, Girona, Spain). The system consisted of two 5 PE lumbrifilters (L1 and L2), a mixing tank (MT) and a 10 PE Daphniafilter (D1).

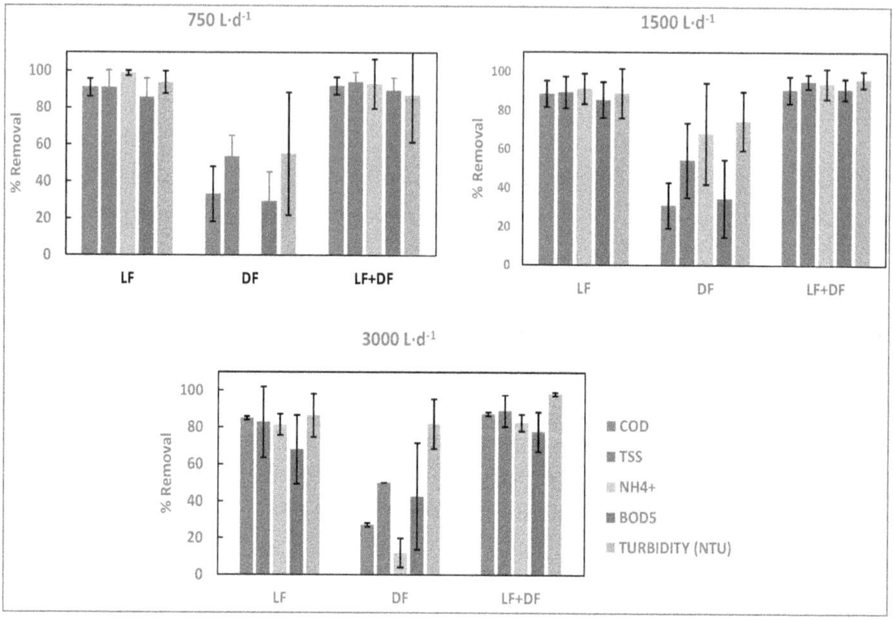

Figure 5.13. Average removal obtained from the 2 × 5 PE lumbrifilter (LF), 10 PE Daphniafilter (DF) and the integrated INNOQUA system (LF+DF) for COD, TSS, NH_4^+, turbidity and BOD_5 under the three different operational loads.

for COD, 68.0 ± 26.2% for NH_4^+-N and 34.5 ± 20% for TP from the Lumbrifilter effluent. The Daphniafilter showed the greatest contribution in terms of turbidity removal (74.5 ± 15.3%) since its main function is to filter fine particles (<30 μm). The average results for NH_4^+-N removal (68.0%) were high although the variability was also high with a standard deviation of ±26.2%.

The capacity of the 10 PE integrated system to treat raw wastewater was slightly affected by changes in the hydraulic operating conditions when three hydraulic loads were tested. Nevertheless, two significant trends can be observed in Fig. 5.13: (i) some changes in performance were observed when the load was doubled from 750 L d^{-1} to 1,500 L d^{-1}. This result is explained by the fact that the overall system was designed to treat a nominal load of 1,500 L d^{-1} and the efficiency at this load was not significantly different to the results obtained when the flow was at 50% of this and (ii) at an influent hydraulic load of 3,000 L d^{-1} (200% of the design value), the efficiency of the system in terms of COD, TSS and NH_4^+ removals was impaired although the overall treatment results remained satisfactory.

5.7.1 Removal of Pathogens and Suspended Particles

Inoculation of *Daphnia* was carried out with a low initial *Daphnia* concentration of 0.1 ind L^{-1} in August 2019, reaching a population of 173 ind L^{-1} after 15 days in

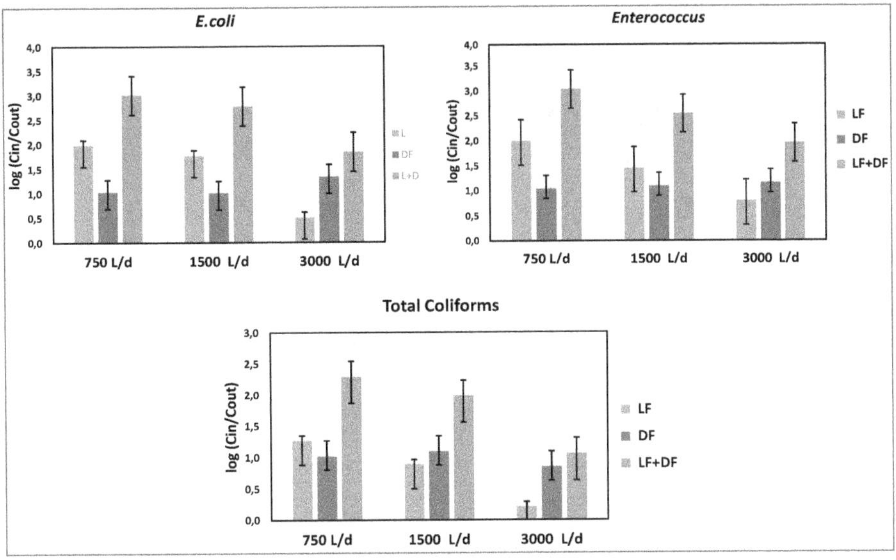

Figure 5.14. Pathogen removal (*E. coli*, Total Coliforms and *Enterococcus*) at the Lumbrifilter (LF) prototype, at the Daphniafilter (DF) prototype and at the integrated system (LF+DF), versus the nominal loads: 750, 1,500 and 3,000 L d^{-1}.

wastewater at 28.1°C. The population continued to increase at the hydraulic rate of 1,500 L d^{-1}, ranging between 200 and 500 ind L^{-1} (depending on the loading rate and day-to-day variation in measurements), until a hydraulic load of 3,000 L d^{-1} (200% design load) was applied, thereafter dropping to 33 ind L^{1}. Turbidity variation with the flow rate can be explained by the number of *Daphnia* individuals thriving in the Daphniafilter during operation at the lowest nominal load (750 L d^{-1}), which varied between 457 and 491 ind L^{-1}, while for a hydraulic load of 3,000 L d^{-1} this number was reduced to 33 ind L^{-1}. Higher hydraulic flows implied an increase of turbulence that could affect both *Daphnia* population and solids' settlement.

The highest bacterial removal was obtained when the LF+DF system was operated at the lowest hydraulic load (Fig. 5.14) with *E. coli*, *Enterococcus* and Coliform populations reduced by up to 3 $\log_{10}$. The efficiency of pathogen removal in the integrated system varied depending on the different flows applied, as can be seen in Fig. 5.14. In the case of the Daphniafilter, pathogen removal efficiency resulted in reductions of 81%, 83% and 85% of *E. coli*, *Enterococcus* and Coliforms, respectively. Removals in the Lumbrifilter were most visibly reduced at a load of 3,000 L d^{-1}, highlighting the benefit of subsequent DF treatment in providing additional treatment where design loads may be exceeded for the secondary treatment system. Overall, the higher the pathogen population in the influent to the DF, the greater

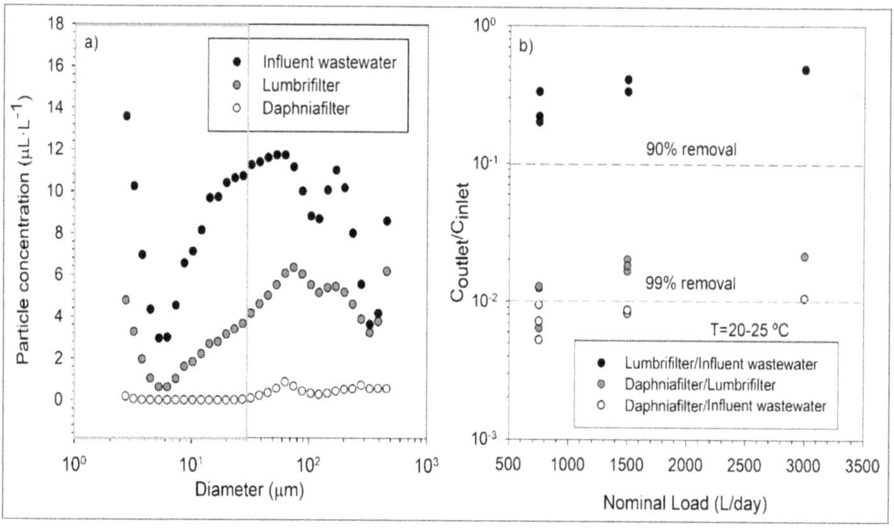

Figure 5.15. (a) Concentration of solid particles in the LF influent (primary effluent), LF effluent and the DF effluent as a function of their size. The green square represents the range of particles ingestible by *Daphnia*. (b) removal of solid particles at various hydraulic loads.

the contribution of the DF unit in eliminating it when using the integrated system, as can be seen in Fig. 5.14 at 3,000 L d^{-1}.

Daphniafilter technology based on the interactions between zooplankton and bacterial/algal biofilm has been demonstrated to be efficient in treating secondary wastewater coming from either a vermifiltration reactor (the lumbrifilter) or a conventional activated sludge secondary treatment process. The results suggest that, in the particular temperature range of this site, seasonality was not a particularly significant factor with regards to nutrient polishing (organic matter and nitrogen) or the removal of solids (TSS and turbidity). Figure 5.15 shows that the concentration of particles in the water is greatly reduced as it passes through the Daphniafilter, which implies a reduction in both pathogens (adhering to fine particulates) and in turbidity. Turbidity reduction at the outlet of the Daphniafilter contributes to improving the efficiency of subsequent UV light treatment when it is required.

The contribution of the Daphniafilter to the reduction of the particle concentration as a function of the particle size can be observed in Fig. 5.15a. This shows a reduction of more than 99% at the design load of 1,500 L d^{-1} (Fig. 5.15b), which represents an additional reduction of around 10% on top of the reduction achieved in the LF. However, as the flow increases there are negative effects of turbulent flows on the mobility and filtration capacity of the *Daphnia* (Serra *et al.*, 2018, 2019c), resulting in a reduction in the removal of solid particles at the highest flow (3,000 L d^{-1}).

The best performance, measured as a decrease in organic carbon, nitrogen, *E. coli* load and suspended solids, was reached when a 10 PE Daphniafilter reactor was fed with secondary wastewater from the Lumbrifilter at a load of 750 L d^{-1}. The quality of the final effluent was suitable for reuse in six categories of Spanish legislation for water reuse such as agricultural irrigation, including products whose edible parts are industrially treated before consumption; irrigation of pastures for the consumption of animals producing milk and meat; industrial uses, including meeting the requirements for the food industry; and for ornamental circulating hydraulic ponds and streams. In the case of the Daphniafilter reactor fed with secondary wastewater from a conventional activated sludge process at 750 L d^{-1}, the treated water can also be reused but for fewer applications than those of the integrated system and it had difficulty in meeting *E. coli* standards, especially in summer months. When the integrated system (LF+DF) was operated at 1,500 L d^{-1}, the quality parameters for water reuse were met, except for *E. coli*. To address the performance shortfall for *E. coli*, the LF+DF effluent can be additionally treated with a UV lamp.

5.7.2 Removal of Metals and Emerging Contaminants

The ability of the integrated system (LF+DF) to remove contaminants such as metal ions and emerging contaminants has also been evaluated. Potentially toxic elements such as Cr, Ni, Cu, Zn, As, Cd, Sn and Pb were analysed at the influent to the integrated system and at the effluent of the Lumbrifilter and Daphniafilter. The results obtained at a flow rate of 1,500 L d^{-1} are summarised in Table 5.3.

Table 5.3. Metal concentrations (μg L^{-1}) at the different stages of the INNOQUA integrated system: influent (INF), Lumbrifilter effluent (LF) and Daphniafilter (DF).

		Cr	Ni	Cu	Zn	As	Cd	Sn	Pb
INF	max	11.9	6.7	44.7	392.2	1.5	0.3	1.1	6.6
	min	5.7	2.7	11.5	41.8	0.7	0.1	0.5	0.8
	median	7.3	5.0	26.1	70.0	1.1	0.1	0.8	2.6
LF	max	8.1	2.8	17.8	65.4	1.6	0.2	0.3	1.0
	min	4.4	0.9	4.9	21.8	0.7	0.1	0.1	<LOQ
	median	5.7	1.7	7.8	34.7	1.0	0.1	0.2	<LOQ
	removal	**22%**	**66%**	**70%**	**50%**	**1%**	–	**75%**	**100%**
DF	max	5.8	3.0	10.9	63.4	1.4	0.1	0.6	0.2
	min	4.6	1.0	4.3	30.8	0.7	0.1	0.1	<LOQ
	median	5.4	1.6	6.1	40.4	1.0	0.1	0.2	<LOQ
	removal	**5%**	**6%**	**22%**	–	**1%**	–	–	–

<LOQ: below limit of quantification.

Table 5.4. Emerging contaminant concentrations (ng L^{-1}) at the different stages of the INNOQUA integrated system: influent (INF), Lumbrifilter effluent (LF) and Daphniafilter (DF). CFB = Clofibric Acid; NPX = Naproxen; IBU = Ibuprofen; DCF = Diclofenac; GMB = Gemfibrozil; TCS = Triclosan.

		CFB	NPX	IBU	DCF	GMB	TCS
INF	max	40.3	9552.8	22043.0	445.0	3461.2	121.6
	min	<LOQ	3907.20	7098.3	208.3	323.0	20.8
	median	28.5	6054.2	9665.8	289.2	1540.1	71.4
LF	max	62.9	3024.0	3310.4	403.7	914.2	38.0
	min	25.2	217.0	112.2	95.3	56.1	2.2
	median	32.5	1027.0	1213.8	197.0	429, 0	9.2
	removal	–	**83%**	**87%**	**32%**	**72%**	**87%**
DF	max	41.4	2419.0	2364.0	240, 0	685.1	20
	min	24.8	146.0	67.0	78.0	60.2	<LOQ
	median	30.3	942.0	737.0	154.1	374.0	10.0
	removal	**7%**	**8%**	**39%**	**22%**	**13%**	–

The integrated system (LF+DF) resulted in the removal of 100% Pb, 77% Cu, 75% Sn, 68% Ni, 57% Cd, 42% Zn and 26% Cr, whilst Cd and As, present at very low concentrations, were not removed.

Several pharmaceutical and personal care products, including anti-inflammatories (diclofenac, naproxen, ibuprofen), antidepressants (paroxetine, fluoxetine, citalopram), antibiotics (sulfametoxazol) and lipid regulators (clofibric acid, gemfibrozil and triclosan), were selected as examples of emerging contaminants and analysed in the influent, LF effluent and LF+DF effluent. The results obtained are collected in Table 5.4.

Values for sulfametazol, carbamazepine, paroxetine, citalopram and fluoxetine are not depicted as they were in most cases below the limit of quantification. As can be seen removal efficiencies for naproxen, ibuprofen, diclofenac, gemfibrozil and triclosan were generally higher in the Lumbrifilter ranging from 72% to 87%. Ibuprofen (39%) and diclofenac (22%) were also significantly removed in the Daphniafilter.

5.8 Performance of the 10 PE Daphniafilter Reactor at the Indian Demo-site

Based on performance data from bench and pilot Daphniafilter experiments, as well as the integrated LF and DF system, further integrated systems were installed at a number of INNOQUA demonstration sites to test performance under real environmental and operational conditions. Operational difficulties were met with at several of the demonstration sites due to the following key issues:

(i) Variable Lumbrifilter effluent sometimes being outside the target range for use as influent to the Daphniafilter, causing repeated crashes of the *Daphnia* populations;
(ii) Difficulties in establishing *Daphnia* populations during the start-up phase and
(iii) Difficulties in establishing and maintaining populations at sites in colder climates (e.g., Ireland and Scotland).

Further work at pilot scale is required to alleviate these issues. Despite these setbacks there were successful demonstrations of the technology outside Spain. At the demonstration site in Bengaluru (India), the Daphniafilter was operated under extreme conditions, exceeding both recommended influent parameters and critical temperature ranges. The effluent from the Lumbrifilter was variable (associated with a high level of variability in the raw wastewater). The average influent temperature was 26°C (with a range from 20.2°C to 38.4°C), considerably exceeding the recommended optimum temperature for *Daphnia* of 20°C (Schellenberg, 2020). Even under these stressful conditions, the Daphniafilter achieved high removals for the main quality parameters once a stable *Daphnia* population was reached (Table 5.5).

Table 5.5. Average values and standard deviation of the quality parameters of the Lumbrifilter (LF) and Daphniafilter (DF) effluents and removal percentage obtained in the Daphniafilter unit of the INNOQUA integrated wastewater treatment systems installed at Bengaluru (India).

		TSS (mg SS L^{-1})	BOD (mg Oxygen L^{-1})	COD (mg Oxygen L^{-1})	NH_4- N (mg NH_4-N L^{-1})
LF	average, STD	271 ± 186	90 ± 76	371 ± 217	15.2 ± 8.5
	max	615	300	803	37.0
	min	36	14	86	2.7
DF	average, STD	16.9 ± 11	10.5 ± 4.7	80.7 ± 16.8	8.9 ± 5.7
	max	48	28	138	24
	min	2	6	54	2.1
Removal	average, STD	80.48 ± 24.14%	60.56 ± 24.4%	76.16 ± 17.61%	50.09 ± 22.64%
	median	93.46	65.41	81.67	44.39

Challenges with establishing and maintaining *Daphnia* populations across the demonstration sites meant that there were fewer than expected opportunities to log long term operational experiences. However at both the Spanish and Indian sites, the growth of aquatic plants such as *Lemna minor* and *Pistia stratiotes* was observed, and as daphnids require a certain amount of sunlight, some removal of these plants had to be carried out to guarantee that the surface coverage did not exceed 75%. Moreover, at the Spanish site accumulated solids were removed from the bottom of the Daphniafilter 11 months after start-up. This operation was undertaken with a simple siphon and neither required that the reactor be emptied nor impacted the *Daphnia* population.

Overall, the Daphniafilter has been shown to require a low initial input of capital, principally related to the reactor itself, *Daphnia* (to inoculate the unit), and a small dosing pump. The use of living organisms (i.e., zooplankton, *Lemna*, and bacteria/microalgae biofilm) and the development of a self-sustained ecosystem reduce the need for maintenance and technical assistance while requiring little operational expenditure beyond electricity to operate the dosing pump. In fact, no maintenance was required during the entire experimental period at the pilot site in Girona, beyond siphoning of some accumulated sludge. It is important to maintain appropriate temperature and dissolved oxygen levels inside the reactor. The oxygen provided by the passive (Venturi) aeration system was sufficient to avoid anoxic conditions but could be further improved to achieve better effluent quality. Burying or semi-burying the reactor underground, as was done in Bengaluru, might be a possible solution to the excessively high summer water temperatures in warmer countries. The adaptive capacity of *Daphnia* also needs to be considered further, particularly with regard to temperature, as successive generations of the community may adapt to the specific conditions of the zooplankton-based reactor environment (Yampolsky *et al.*, 2013).

Members of the Cladocera order can be found in a wide range of different climates around the world (Serra *et al.*, 2019a) and the selection of appropriate species will be critical for the implementation of zooplankton-based reactors in hot countries as has been done in India where *D. pulex* was used (Schellenberg, 2020). Provided they are not completely eliminated, *Daphnia* populations can re-establish very rapidly under optimum conditions. Where flow and HRT are within design limits and additional turbulence due to shock loads is minimised, high removals of carbon, nitrogen, bacteria and particles can be achieved.

The low energy consumption, minimal maintenance, absence of chemicals and the small sludge production make the concept of the zooplankton-based reactor presented here both eco-sustainable and economical. As this study has shown, this alternative tertiary natural depuration system is able to provide reclaimed water that is of a high chemical and biological quality for agricultural irrigation and other non-potable uses – but further beta testing under real conditions is required before commercial implementation can be considered.

References

Bownik, A., 2017. Daphnia swimming behaviour as a biomarker in toxicity assessment: A review. Sci Total Environ 601–602:1–1868. https://doi.org/10.1016/j.scitotenv.2017.05.199

Dolar, D., Gros, M., Rodriguez-Mozaz, S., Moreno, J., Comas, J., Rodriguez-Roda, I. and Barceló, D., 2012. Removal of emerging contaminants from municipal wastewater with an integrated membrane system, MBR-RO. J Hazard Mate 239–240, 64–69. https://doi:10.1016/j.jhazmat.2012.03.029

Ebert, D., 2005. Chapter 2: Introduction to D. Magna biology in Ecology, epidemiology, and evolution of parasitism in D. Magna. Bethesda, MD: National Center for Biotechnology information. https://www.ncbi.nlm.nih.gov/books/NBK2042/

Ennabili, A., Ezzahri, J. and Radoux, M., 2019. Performance of Lemna gibba bioreactor for nitrogen and phosphorus retention, and biomass production in Mediterranean climate. J. Environ. Manage. 252. https://doi.org/10.1016/j.jenvman.2019.109627

Frederick, R.J. and Egan, M. 1994. Environmentally Compatible Applications of Biotechnology: Using living organisms to minimize harmful human impact on the environment. BioScience 44(8), 529–535. https://doi.org/10.2307/1312280

Garcia-Rodríguez, A., Matamoros, V., Fontàs, C. and Salvadó, V., 2013. The influence of light exposure, water quality and vegetation on the removal of sulfonamides and tetracyclines: A laboratory-scale study. Chemosphere 90, 2297–2302. https://doi.org/10.1016/j.chemosphere.2012.09.092

Kampf, R., van der Geest, H., Sala, Ll., Romani, A., Comas, J., Claassen, T., Gerbens, S., Neef, R. and Menkveld, W., 2007. Biological filtration of treated waste water by Daphnia: an alternative for technical filtration, or an addition?. IWA 6th Conference on Wastewater Reclamation and Reuse for Sustainability, Antwerpen, October 2007.

Klavarioti, M., Mantzavinos, D. and Kassinos, D., 2009. Removal of residual pharmaceuticals from aqueous systems by advanced oxidation processes. Environ. Int. 35(2): 402–417. htpps://doi.org/10.1016/j.envint.2008.07.009

Lutterbeck, C.A., Zerwes, F.V., Radtke, J.F., Köhler, A., Kist, L.T. and Machado, Ê.L., 2018. Integrated system with constructed wetlands for the treatment of domestic wastewaters generated at a rural property – Evaluation of general parameters ecotoxicity and cytogenetics. Ecol. Eng. 115, 1–8. https://doi.org/10.1016/j.ecoleng.2018.01.004

Lyu, K., Cao, H., Chen, R., Wang, Q. and Yang, Z., 2013. Combined effects of hypoxia and ammonia to Daphnia similis estimated with life-history traits. Environ. Sci. Pollut. R. 20(8), 5379–5387. https://doi.org/10.1007/s11356-013-1555-7

Massoud, M.A., Tarhini, A. and Nasr, J.A., 2009. Decentralized approaches to wastewater treatment and management: Applicability in developing countries. J. Environ. Manage. 90, 652–659. https://doi.org/10.1016/j.jenvman.2008.07.001

Matamoros, V., Sala, L. and Salvadó, V., 2012. Evaluation of a biologically-based filtration water reclamation plant for removing emerging contaminants: A pilot plant study. Bioresour. Tech. 104, 243–249. https://doi.org/10.1016/j.biortech.2011.11.036

Matamoros, V. and Salvadó, V. 2013. Evaluation of the seasonal performance of a water reclamation pond-constructed wetland system for removing emerging contaminants. Chemosphere 86, 111–117. https://doi.org/10.1016/j.chemosphere.2011.09.020

Metcalf and Eddy, Inc. revised by Tchobanoglous, G., Burton, F.L., Stensel, H.D., 2002. Wastewater engineering. Treatment and reuse. McGraw-Hill, 4th edition.

Mittmann, B., Ungener, P., Klann, M., Stollewerk, A. and Wolff, C., 2014. Development and staging of the water flea Daphnia magna (Straus, 1820; Cladocera, Daphniidae) based on morphological landmarks. EvoDevo 5:12. htpps://doi.org/10.1186/2041-9139-5-12

Mohedano, R.A., Tonon, G., Costa, R.H.R., Pelissari, C. and Belli Filho, P., 2019. Does duckweed ponds used for wastewater treatment emit or sequester greenhouse gases? Sci. Total Environ. 691, 1043–1050. https://doi.org/10.1016/j.scitotenv.2019.07.169

Müller, M.F., Colomer, J. and Serra, T., 2018. Temperature-driven response reversibility and short-term quasi-acclimatation of *Daphnia magna*. PLoS One 13. https://doi.org/10.1371/journal.pone.0209705

Nørgaard, L.S. and Roslev, P., 2016. Effects of ammonia and density on filtering of commensal and pathogenic escherichia coli by the cladoceran Daphnia magna. Bull. Environ. Contam. Toxicol. 97, 848–854. https://doi.org/10.1007/s00128-016-1963-8

Okamoto, A., Yamamuro, M. and Tatarazako, N., 2015. Acute toxicity of 50 metals to *Daphnia magna*. J. Appl. Toxicol. 35, 824–830. https://doi.org/10.1002/jat.3078

Ortiz, J.D., Merseburger, G., Matamoros, V., Salvadó, V., Colom, J., Sala, L., Comas, J. and Kampf, R., 2011. Regeneración de agua mediante filtración por cultivos de *Daphnia* en el sistema de mesocosmos de la EDAR de Empuriabrava. Tecnología del Agua 330, 57–65.

Pau, C., Serra, T., Colomer, J., Casamitjana, X., Sala, Ll. and Kampf, R. 2013. Filtering capacity of Daphnia magna on sludge particles of treated wastewater. Wat. Res. 47: 181–186. https://doi.org/10.1016/j.watres.2012.09.047

Pous, N., Hidalgo, M., Serra, T., Colomer, J., Colprim, J. and Salvadó, V., 2020. Assessment of zooplankton-based eco-sustainable wastewater treatment at laboratory scale. Chemosphere 238, 124683. https://doi.org/10.1016/j.chemosphere.2019

Pous, N., Barcelona, A., Sbardela, L., Hidalgo, M., Serra, T., Colomer, J. and Salvadó, V., (2021). Zooplankton-based reactors for tertiary wastewater treatment: A pilot-scale case study. J. Env. Man. 278, 11153.

Rist, S., Baun, A. and Hartmann, N.B., 2017. Ingestion of micro- and nanoplastics in Daphnia magna – quantification of body burdens and assessment of feeding rates and reproduction. Environ. Pollut. 228, 398–407. https://doi.org/10.1016/j.envpol.2017

Rodriguez da Silva, S.L., Rodriguez, D.O., Sánchez Castillo, P., Conde-Porcuna, J.M. and Cruz-Pizarro, L., 2004. Alterations in the rate of respiration of Daphnia magna below sublethal concentrations of Anatoxin-a. Limnetica, 23(1–2), 159–166.

Salvadó, V., Serra, T., Colomer, J., Pous, N., Font, M., Pijoan, I. and Scheerer, J., 2019. Reactor de depuración para el tratamiento de aguas residuales. ES1234189.

Schellenberg, T. (2020). Personal communication. Data based on investigations carried out under the INNOQUA Research Project funded by the European Union, Horizon 2020.

Serra, T., Colomer, J., Pau, C., Marín, M. and Sala, L., 2014. Tertiary treatment for wastewater reuse based on the *Daphnia magna* filtration – Comparison with conventional tertiary treatments. Water Sci. Technol. 70, 705–711. https://doi.org/10.2166/wst.2014.284

Serra, T. and Colomer, J., 2016. The hydraulic retention time on the particle removal efficiency by *Daphnia magna* filtration on treated wastewater. Int. J. Environ. Sci. Technol. 13, 1433–1442. https://doi.org/10.1007/s13762-016-0985-4

Serra, T., Barcelona, A., Soler, M. and Colomer, J., 2018. Daphnia magna filtration efficiency and mobility in laminar to turbulent flows. Sci. Total Environ. 621, 626–633. https://doi.org/10.1016/j.scitotenv.2017.11.264

Serra, T., Müller, M.F., Barcelona, A., Salvadó, V., Pous, N. and Colomer, J., 2019a. Optimal light conditions for *Daphnia* filtration. Sci. Total Environ. 686, 151–157. https://doi.org/10.1016/j.scitotenv.2019.05.482

Serra, T., Soler, M., Pous, N. and Colomer, J., 2019b. Daphnia magna filtration, swimming and mortality under ammonium, nitrite, nitrate and phosphate. Sci. Total Environ. 656. https://doi.org/10.1016/j.scitotenv.2018.11.382

Serra, T., Müller, M.F. and Colomer, J., 2019c. Functional responses of *Daphnia magna* to zero-mean flow turbulence. Sci. Rep., 9, 3844. https://doi.org/10.1038/s41598-019-40777-2

Serra, T., Barcelona, A., Pous, N., Salvadó, V. and Colomer, J., 2020. Synergistic effects of water temperature, microplastics and ammonium as second and third order stressors on Daphnia magna. Environ. Pollut., 267, 115439. https://doi.org/10.1016/j.envpol.2020.115439

Shiny, K.J., Remani, K.N., Nirmala, E., Jalaja, T.K. and Sasidharan, V.K., 2005. Biotreatment of wastewater using aquatic invertebrates, Daphnia magna and Paramecium caudatum. Bioresour. Technol. 96, 55–58. https://doi.org/10.1016/j.biortech.2004.01.008

Simoncelli, S., Thackeray, S.J. and Wain, D.J., 2019. Effect of temperature on zooplankton vertical migration velocity. Hydrobiologia 829, 143–146. https://doi.org/10.1007/s10750-018-3827-1

Spanish water reuse legislation (RD 1620/2007). https://www.boe.es/diario_boe/txt.php?id=BOE-A-2007-21092

Sun, J., Dai, X., Wang, Q., van Loosdrecht, M.C.M. and Ni, B.J., 2019. Microplastics in wastewater treatment plants: detection, occurrence and removal. Water Res. 152, 21–37. https://doi.org/10.1016/j.watres.2018.12.050

Yampolsky, L.Y., Schaer, T.M.M. and Ebert, D., 2013. Adaptive phenotypic plasticity and local adaptation for temperature tolerance in freshwater zooplankton. Proc. R. Soc. B Biol. Sci. 281. https://doi.org/10.1098/rspb.2013.2744

Young, P., Taylor, M. and Fallowfield, H.J., 2017. Mini-review: high rate algal ponds, flexible systems for sustainable wastewater treatment. World J. Microbiol. Biotechnol. 33, 1–13. https://doi.org/10.1007/s11274-017-2282-x

DOI: 10.1561/9781680837810.ch6

Chapter 6

Tertiary Treatment: Microalgae-based Wastewater Treatment

By Olga Tiron, Elena Manea and Costel Bumbac

6.1 Introduction

To address risks to environmental quality, nutrient limits in treated wastewater are mandated by many countries and regions, particularly where the final effluent is to be discharged into a water body. Various techniques are deployed to remove nutrients in centralised wastewater treatment systems, including chemical dosing (for phosphorus) and biological nutrient removal (for phosphorus and/or nitrogen). These techniques are generally unsuited to decentralised wastewater treatment, since they rely on process controls or chemical interventions that require specialist training and frequent maintenance.

Tertiary treatment in decentralised systems instead relies upon nutrient uptake by vegetation in constructed wetlands or swales (Al-Muyeed, 2017; Capodaglio, 2017), although phosphorus removal by adsorption in media filters is also possible (Bunce *et al.*, 2018). Microalgae offer the potential to intensify wetland treatment techniques, cultivated in bespoke reactors that maximise biomass productivity and nutrient removal. Excess biomass can be periodically harvested for supply to secondary value chains.

This chapter outlines the development of microalgal technologies – from potential protein sources in the mid-twentieth century, through to their potential roles in tertiary wastewater treatment, and barriers to their implementation in this application (Section 6.2). Section 6.3 goes on to explore the interactions between microalgae and bacteria in attached-growth communities, before Section 6.4 considers important operational aspects when exploiting the potential of these communities. Section 6.5 outlines the development of the INNOQUA Bio-Solar Purification (BSP) module from laboratory to pilot scale, presenting results from demonstration facilities in India, Peru and Spain.

6.2 The Potential of Microalgae Biotechnology

Microalgae biotechnology is a wide field of research with extended theoretical and practical applications in multiple sectors (e.g., agriculture, pharmaceutical, food industry, aquaculture, sanitation, bioenergy) and is gaining particular attention in environmental and circular bioeconomy applications (e.g., wastewater treatment, climate change mitigation) (Li *et al.*, 2019; Haarich *et al.*, 2017; Nagarajan *et al.*, 2020). Following initial market developments based on soil-grown crops and woody biomass, microalgal biomass may be considered a third-generation feedstock for the production of biofuels and bioproducts (Chowdhury and Loganathan, 2019). Microalgal biomass is increasingly seen as a feedstock that avoids many of the economic and environmental disadvantages associated with cultivation and processing of first and second generation feedstocks (Ubando *et al.*, 2020; Maity *et al.*, 2014).

Multiple applications for microalgal biomass have been demonstrated at bench and pilot scale, including biofuels, high-value chemicals (pharmaceuticals, cosmetics, etc.), food supplements, bioplastic, as well as fertilisers (Hayes *et al.*, 2017; Khan *et al.*, 2018; Javed *et al.*, 2019; Patil and Kaliwal, 2019). However, full-scale implementation of microalgae biotechnology has been limited, requiring further development to improve economic feasibility in many applications and in some cases to overcome social barriers regarding the origins of algal derivatives. Key aspects of economic feasibility in all applications include cultivation, harvesting and processing (de Carvalho *et al.*, 2020; Tang *et al.*, 2020). Social barriers relate to acceptability of novel practices such as the supply of wastewater-cultivated food-grade proteins (Matassa *et al.*, 2015) as well as multiple regulatory restrictions (Kehrein *et al.*, 2020).

6.2.1 From Opportunity to Implementation

Although the Venezuelan government developed the cultivation of phytoplankton for industrial purposes and extraction of carotene compounds since the early 1930s (Burlew, 1953; Jorgensen and Convit, 1953), it was during the second World War (1939–1945) that widespread interest in microalgae biotechnology was prompted by the need for new protein sources (Goldman, 1979). *Chlorella* spp. (mainly *C. pyrenoidosa*, *C. vulgaris* and *C. ellipsoidea*) and *Scenedesmus* spp. were among the first microalgae tested to show a high tolerance to varying environmental conditions and thus potential suitability for large-scale use. Despite the high proportion of target compounds in microalgae cells, lack of experience in microalgae cultivation meant that experimental studies were limited. However, by 1951, a pilot-scale cultivation system had been implemented by the Carnegie Institution in the USA. This specifically examined the influence of operational conditions (including those determined by local environments) on microalgae growth rates (Cook, 1951). Similar research investigations were also conducted in Germany, England, Israel and Japan where microalgae species were tested in open pond systems or even closed cultivation tubes (or 'photobioreactors') with increased biomass productivity being the focus of much of the work. This progression is shown in Fig. 6.1.

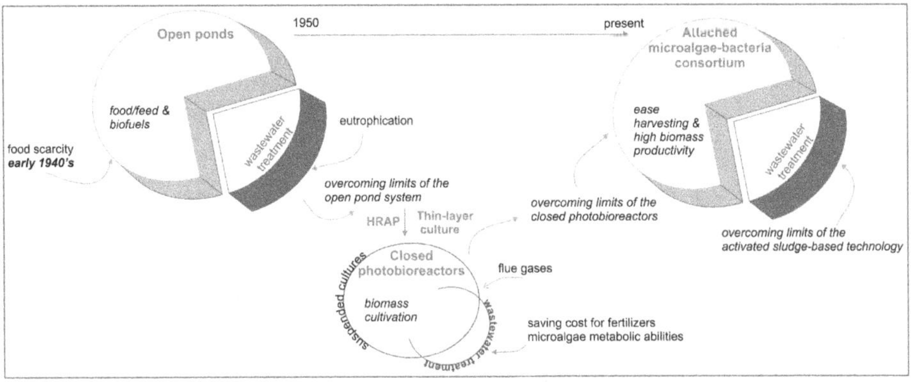

Figure 6.1. The technical evolution of microalgae cultivation systems.

Developments in algal cultivation techniques coincided with growth in the conventional agriculture sector, which caused a decline in interest in the food potential of microalgae. At the same time, their potential to remove nutrients as part of domestic wastewater treatment began to be explored, with cultivation in large-scale algal ponds (Oswald *et al.*, 1955). In addition, it was proposed that excess microalgal biomass from wastewater treatment processes could be used for methane generation via anaerobic digestion (Meier, 1955; Oswald and Golueke, 1960). The potential for positive impacts on both wastewater treatment and links to what is now called the circular economy has prompted ongoing cycles of research and commercial interest in this sector.

Concepts to use microalgae for wastewater treatment generally exploit two aspects:

- the symbiotic relationship between microalgae and bacteria, as primary producers in natural trophic networks; and
- the ability of microalgae to consume inorganic nitrogen and phosphorus from wastewater as an efficient mechanism to prevent or reduce downstream eutrophication.

The core of the symbiotic relationship harnessed in the wastewater treatment process is represented by a nutrient-support exchange between microalgae (which ensure oxygen supply by phototrophy) and aerobic heterotrophic bacteria (which provide macronutrients (mainly nitrogen and phosphorus) and inorganic carbon (CO_2) through degradation of organic matter) (Posadas *et al.*, 2017).

Despite being easy to design and operate, open pond cultivation systems are relatively deep (around 1 m) and host low microalgae concentrations (<500 mg/L) which result in increased costs for biomass harvesting and dewatering, and low organic matter removal efficiency (5–10 g BOD/m^2 pond surface area.day). This leads to relatively long hydraulic retention times (HRT) of between 10 and 40 days, and thus larger systems. In general, wastewater as a nutrient medium can deliver biomass productivity of up to 21 g/m^2.day in an open pond system (Ozkan *et al.*, 2012). The depth of open pond systems can create a downward gradient in O_2 concentrations, reducing algal efficiency and creating a requirement for mixing.

The efficiency of open pond systems was improved through the development of high rate algal ponds (HRAPs) in the 1970s and 1980s. HRAPs are characterised by a shallow depth (<0.5 m) and are equipped with a paddlewheel mixing system which increases photosynthetic activity and organic matter removal efficiency (around 35 g BOD/m^2.d) (Muñoz and Guieysse, 2006). These characteristics allowed HRT to drop below 10 days, while maintaining biomass productivity of between 15 and 25 g/m^2.d (Goldman, 1979). HRAPs also allowed the possibility for selected colony-forming microalgae to be cultured, thus allowing decreased harvesting costs (Mehrabadi *et al.*, 2015). However, despite proving their efficiency for both domestic and agricultural wastewater treatment (Hoffmann, 1998), HRAPs, as with open ponds, require a large land surface (ranging from around 6–10 m^2 per person equivalent) which is about 50 times higher than conventional activated sludge processes (Park *et al.*, 2011; Acién *et al.*, 2016).

Despite their relatively large land area requirements and low biomass productivity, cultivation ponds (open ponds and HRAPs) now account for more than 95% of commercial microalgae biomass cultivation systems, due to their simplicity of design and operation (Acién Fernández *et al.*, 2013).

The algal turf scrubber (ATS) is another configuration exploiting the potential for algae to remove nutrients from water flows. Craggs *et al.* (1996) demonstrated annual ATS removal of phosphorus from agricultural runoff and eutrophic lake water of 0.73 ± 0.28 g P/m^2.day, at an average periphyton productivity (microalgae and bacteria alike) of 35 g/m^2.day. Notably, the authors stressed the influence of the diurnal variation of light on ATS performance, recording significantly reduced nutrient removal efficiency at night.

Issues associated with mixing and oxygen gradients can also be overcome through the use of thin layer reactors, developed in the 1960s (Doucha and Lívanský, 2006). In comparison with open ponds, thin layer reactors typically operate at shallower water depths (<0.05 m) which promotes photosynthetic and respiratory efficiencies, leading to higher biomass productivity (up to 55 g/m^2.day) (Masojídek *et al.*, 2011) and lower hydraulic retention times (3–5 days) (Acién *et al.*, 2016).

The theoretical maximum photosynthetic efficiency[1] of microalgae is 9–10% (Vecchi *et al.*, 2020) compared to terrestrial plants of 1–2% (Peccia *et al.*, 2013). By using thin-layer reactors, a photosynthetic efficiency of 7% has been reported in outdoor conditions (Doucha and Lívanský, 2006), although maximum efficiencies of 9% are thought possible in such scenarios (Doucha and Lívanský, 2009; Morales-Amaral del Mar *et al.*, 2015a). One of the main features of these reactors is the slight slope applied to the surface ($<3\%$) which prevents biomass settling and avoids the use of mixing equipment – reducing energy consumption (Acién Fernández *et al.*, 2013).

Other 'thin layer' adaptations of algae-based technology include algal rotating disk (ARD) and rotating algal biofilm reactor (RABR) systems. The RABR comprises rotating flexible belts that improve CO_2 diffusion and ensure uniform light exposure. One pilot scale study (8,000 L reactor) for photoautotrophic tertiary wastewater treatment reported average total nitrogen and total phosphorus removals of 14.1 g/m^2.day and 2.1 g/m^2.day, respectively, with biomass productivity (both microalgae and bacteria) of 31 g/m^2.day (Christenson and Sims, 2012).

Photobioreactors also perform well when compared with open pond systems, with biomass productivity of up to 47 g/m^2.day (Brennan and Owende, 2010). However, such reactors can have relatively high operational costs. Efficient mixing is necessary to minimise biomass attachment to vessel walls, and to ensure good nutrient and light distribution. This leads to energy demands that can be 15 times higher than required for mixing in open pond systems (Ozkan *et al.*, 2012). Other important factors include manufacturing/construction costs and system maintenance which will vary between technologies and applications.

1. That is, the proportion of energy in intercepted light that is converted to chemical energy via photosynthesis.

6.2.2 Microalgae and Tertiary Wastewater Treatment

Conventional (centralised) wastewater treatment technologies are increasingly energy intensive, and significant contributors of the GHG emissions within the water industry as a whole (Mamais *et al.*, 2015). Of the total energy consumption required for wastewater treatment, aeration is the most energy-intensive process, accounting for about 40–60% (Chae and Kang, 2013; Gu *et al.*, 2017). Furthermore, about 50% of the organic carbon from wastewater loadings is 'lost' as CO_2 during aerobic treatment (Muñoz and Guieysse, 2006) and there is growing concern around N_2O and CH_4 emissions from these processes.

Reliance on aeration during conventional wastewater treatment serves to signpost one of the key attractions of combined microalgal–bacterial systems, in which photoautotrophic metabolism can serve as an oxygen source for bacterial biomass. Photosynthesis ensures oxygen saturation of more than 100% during the light phase, and dissolved oxygen remains at between 30% and 50% saturation even during dark periods (Tiron *et al.*, 2015). Photosynthesis also acts as a temporary sink for CO_2, which can be accumulated by microalgae at an annual rate of around 2 kg CO_2/kg biomass (Xiaogang *et al.*, 2020).

Depending on influent loading, microalgae–bacteria biomass concentrations of up to 1 kg/m^3 wastewater can develop, around five times greater than could be expected in a conventional activated sludge process (Acién *et al.*, 2016). This leads to an excess of microalgal–bacterial biomass that can serve as a feedstock for other processes – whether bioenergy (through fermentative conversion to gaseous or liquid biofuels) or higher value products, supporting the commercial development of microalgae-based technologies and the wider bioeconomy.

The nutrient-accumulating characteristics of microalgae are another primary consideration and can be leveraged in tertiary wastewater treatment systems to reduce nutrient concentrations in final effluent more efficiently than conventional chemically or biologically-mediated approaches. Liu *et al.* (2017) report total nitrogen concentrations of less than 10 mg/L and total phosphorus concentrations of less than 1 mg/L in the effluent from a microalgal/bacterial system. Luxury uptake of phosphorus by microalgae has also been demonstrated, suggesting that lower discharge concentrations might also be possible, mirroring the pattern seen in phosphate-accumulating bacteria (Khanzada, 2020). Through techno-economic analysis, Chalivendra (2014) showed that the annual costs required for inorganic nutrient (N/P) and heavy metal (Cr and Cd) removal by conventional activated sludge processes could be decreased 7- and 26-fold, respectively, by using microalgae.

To date, the potential for microalgae-mediated treatment has been tested on a wide range of water and wastewater sources, including municipal wastewater;

industrial effluents from food processing, textile manufacture, aquaculture and livestock farming; acid mine drainage; centrate from anaerobic digestion; contaminated groundwater and contaminated surface waters (Wang *et al.*, 2012; Van Den Hende *et al.*, 2014; Garcia *et al.*, 2018; Zerrouki and Henni, 2019). Studies (mostly based on laboratory-scale work) have provided evidence for the utility of microalgae for pollutant/nutrient removal; the microalgal–bacterial interactions during wastewater treatment; economic advantages that could be obtained and the wider economic potential for using wastewater to produce microalgae as a feedstock for various high-value compounds and other commercial, industrial or agricultural purposes.

Microalgal–bacterial biomass can be an efficient method for removal of cationic heavy metals and specific toxic compounds (phenols, cresols, nitrophenols, etc.) from wastewater (Surkatti and Al-Zuhair, 2018). Removal of dyes, polycyclic aromatic hydrocarbons and endocrine-disrupting compounds has been shown with microalgae (Zhuang *et al.*, 2020). According to a 2006 literature review (Muñoz and Guieysse, 2006) microalgae cells can accumulate significant concentrations of heavy metals (up to 192 $mg/g_{biomass}$) and sustain a high removal efficiency by adsorption mechanisms (up to 114.2 $mg/g_{biomass}{\cdot}d$), depending on species and the metal being targeted. For example, the microalga *Ulothrix* spp. was tested for heavy metal removal efficiency from acid mine wastewaters on a photo-rotating biological contactor with a 24 hour hydraulic residence time (Orandi *et al.*, 2012). Initial metal concentrations ranged between 80 and 100 mg/L (Cu), 2–3 mg/L (Ni), 35–45 mg/L (Mn), 18–20 mg/L (Zn), 0.005–0.007 mg/L (Sb), 0.03–0.04 mg/L (Se), 0.3–0.5 mg/L (Co) and 0.07–0.09 mg/L (Al). The study reported removal efficiencies ranging between 20% and 50% with the following selectivity Cu>Ni>Mn>Zn>Sb>Se>Co>Al. The tolerance of other species (*Chlorella* spp., *Scenedesmus* spp., *Oscillatoria* spp. and *Nitzchia* spp.) to heavy metals has also been demonstrated (Acién *et al.*, 2016).

Studies have also demonstrated that operational conditions impacted by photoautotrophy (such as the increase in pH, temperature and oxygen values) can prompt a decrease in populations of undesirable pathogens such as faecal coliforms (Ansa *et al.*, 2012). Replacing mechanical with 'biological' oxygenation also decreases risks of pollutant volatilisation (Muñoz and Guieysse, 2006).

6.2.3 Major Barriers to Market Adoption

Despite the many potential benefits of utilising microalgae in wastewater treatment, scaling-up has faced several technical and economic challenges. Cost and energy input linked to microalgae harvesting remains a significant issue – the economic impact of this downstream process, for open ponds, being about 20% of total cultivation costs (Davis *et al.*, 2011; Fasaei *et al.*, 2018).

The harvesting problem comes from microalgae cell particularities, the most commonly used species having a cell diameter lower than 30 μm and a cellular density similar to that of water (Granados *et al.*, 2012; Wang *et al.*, 2013). Centrifugation, chemical flocculation, filtration and dissolved air flotation are some of the most frequently applied harvesting techniques, each with their own advantages and disadvantages. However, even where two or more harvesting techniques are combined, microalgae removal efficiency rarely exceeds 95% (Tiron *et al.*, 2017). Furthermore, the remaining microalgae cells can impact effluent quality and compromise system functionality. Moreover, high levels of retained moisture in harvested biomass present challenges to downstream processing (Polizzi *et al.*, 2017; Khan *et al.*, 2018), and commercial microalgal applications tend to focus on high value functional characteristics or compounds (such as use in dietary supplements or pharmaceutical formulation) that preclude their cultivation in wastewater for reasons of perception, safety or quality management.

6.3 Microalgal–Bacterial Interactions

6.3.1 Overview

Microalgal–bacterial interactions within biofilms have gained significant recent research attention. In a wastewater treatment context, photoautotrophic microalgae, through photosynthesis, provide the necessary oxygen supply for aerobic processes (mainly organic matter degradation and nitrification). In turn, the macronutrients in the medium and/or supplied by bacteria (such as CO_2, NH_4^+, PO_4^{3-} and NO_3^-) are used by microalgae for cellular growth (Fig. 6.2). Inorganic nitrogen can also be provided for microalgae by nitrogen-fixing bacteria, while bacteria such as *Pseudomonas* spp., *Bacillus* spp. and *E. coli* are known to be excellent sources of inorganic phosphate for microalgae (Zhang *et al.*, 2020).

In addition to the bilateral exchange of macronutrients, growth-stimulatory compounds can also be exchanged. For instance, bacteria can provide essential microalgae cell growth compounds such as vitamins and hormones – while microalgae are producers of vitamins and hormones (Kiseleva *et al.*, 2012; Kim *et al.*, 2014) that are important for bacteria. Given that over half of the known microalgae species cannot synthesise essential vitamins required for their own cell development (Fulbright *et al.*, 2018), this interspecies relationship is essential. Indeed, relationships within the microalgal–bacterial phycosphere can be unidirectional (commensalism), bidirectional (mutualism) and/or parasitic (Yao *et al.*, 2019), exploiting three main pathways: stimulation/inhibition of growth; quorum sensing communication and gene transfer (Amin *et al.*, 2012; Kouzama *et al.*, 2015).

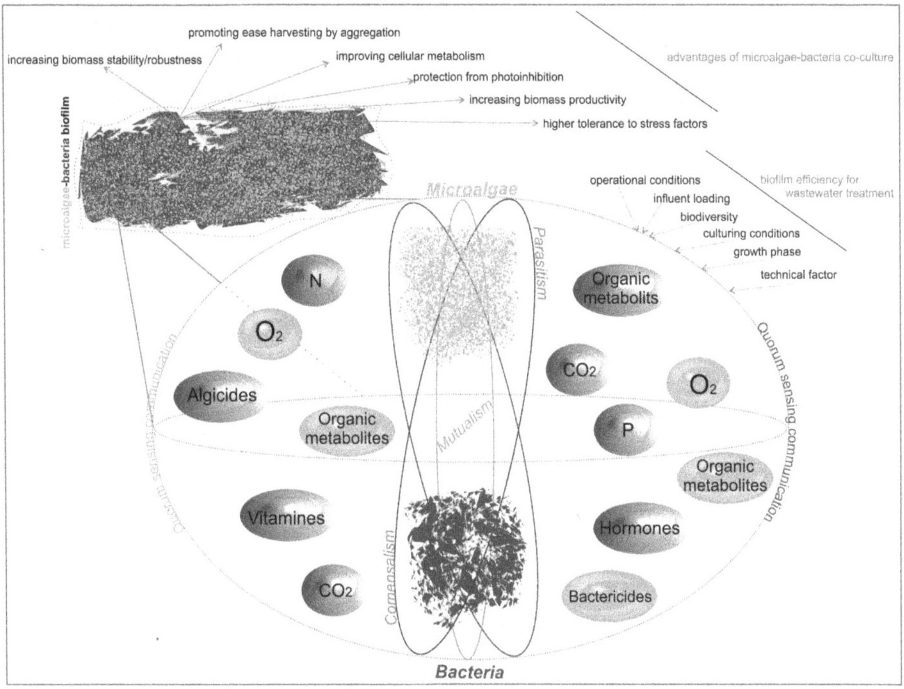

Figure 6.2. Possible microalgae-bacteria interactions occurring in the biofilm structure, the resulting effect on biofilm performance and stability, and factors influencing biofilm efficiency during wastewater treatment

These interactions and interdependencies can benefit growth rates of both bacterial and algal populations. For example, the presence of 'microalgae-growth-promoting bacteria' can increase microalgal productivity by up to 70% (and vice versa) (Ramanan *et al.*, 2016). Kim *et al.* (2014) recorded an increase in the biomass productivity of microalgae *Chlorella vulgaris*, *C. reinhardtii*, *Scenedesmus* spp. and *Botryococcus braunii* by 70.3, 64.5, 92.7 and 59.6%, respectively, in co-culture with *Rhizobium* spp. In turn, the presence of green microalgae increased development of *Rhizobium* spp. by up to 7.8-fold. This positive effect is also seen in wastewater treatment. By using *Auxenonochlorella protothecoides* and *C. sorokoniana* in winery wastewater treatment, Higgins *et al.* (2018) recorded an up to 6-fold increase in bacterial productivity. Toyama *et al.* (2018) reported an increase in productivity of the microalgae *C. reinhardtii*, *C. vulgaris* and *Euglena gracilis* by 1.5, 1.8–2.8 and 2.1-fold, respectively, after 7 days of co-culturing with indigenous bacteria from a swine wastewater effluent. The counterpoint to this is that there is also a range of competitive effects between the different populations. These range from simple competition for essential micronutrients through to parasitism and even the secretion of algaecides or bactericides (Zhang *et al.*, 2020).

Other interactions that can occur during microalgae–bacteria system development – for example cell-to-cell interactions through quorum sensing communication and gene transfers – are less studied. Quorum sensing communication refers to intercellular communication sustained by an exchange of signalling compounds (such as lipid-based molecules, bacterial signalling molecules n-Acyl-homoserine lactones (AHLs) and microalgae secondary metabolites (allelochemicals)) which influence specific gene expression (Zhang *et al.*, 2020) and have a stimulatory, regulatory or inhibitory effect (Gross, 2009). This type of intercellular communication influences a population's abundance and richness – and impacts on practical operational aspects relevant to such systems. For example, while Ramanan *et al.* (2016) reported uncertainties around the mechanisms by which bacteria promote microalgal bioflocculation, Zhou *et al.* (2017) subsequently observed that in the presence of the bacterial signalling molecules AHLs (extracted from activated sludge), the microalga *Chlorophyta* spp. was stimulated to secrete specific aromatic proteins which promoted self-aggregation of the biomass in flocs and increased the efficiency of biomass settling by up to 41%.

Another possible interaction that could arise between microalgae and bacteria populations is horizontal gene transfer. This kind of exchange is more likely under stressful conditions, with transfers occurring from bacteria to microalgae. The stability of such transfers varies tremendously, along with the eventual location of transferred material and mechanism of expression within the receiving cell, but there is evidence that such transfers have allowed eukaryotes to adapt to changing environments – for example, where nutrients are limiting Husnik and McCutcheon (2018). Although they provide insights into the adaptability mechanisms of microalgae and bacteria, the practical implications and applicability of such transfers are as yet unknown, but could eventually support niche wastewater treatment applications.

6.3.2 Advantages of Attached Growth Communities

Algae-based biofilms are characterised by dense, multi-layer biological structures comprising a mixed population of microalgae and bacteria. Biofilm development mainly occurs in two steps. The first step consists of biomass attachment onto a support material through physico-chemical interactions (hydrophobic/hydrophilic, acid–base interactions, etc.), the second step is characterised by the intervention of secreted polymeric substances. The presence of bacteria increases the rate of microalgal adherence through these two mechanisms, decreasing biofilm establishment times and improving the stability of the biomass structure.

Attachment of microalgae–bacteria systems (defined as immobilisation on the surface of the support material) leads to more complex biologic and metabolic

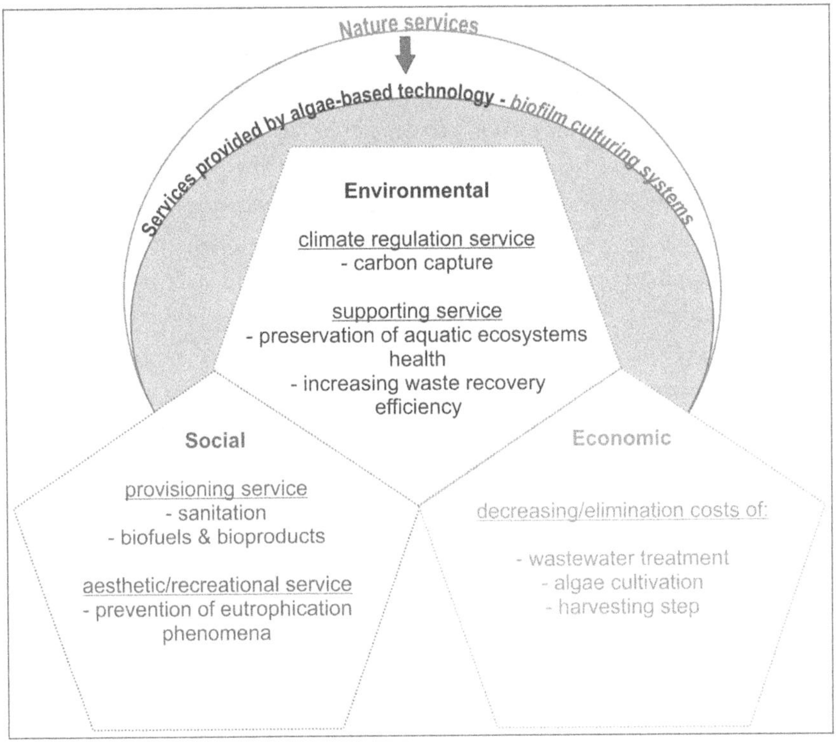

Figure 6.3. Services provided by microalgae-based biofilm culturing systems.

networks compared with suspended biomass. Community composition, extracellular physico-chemical attachment mechanisms and a wide range of operational factors (technical, influent loading, laboratory/outdoor conditions, etc.), lead to systems with different behaviours than those observed in suspended systems with similar community composition. Hence, there is a need for advanced research on interspecies relations, the influence of excreted metabolites on biomass functional activities and treatment performance (involving physico-chemical mechanisms that in turn sustain other biological functions). It is believed that quorum sensing communication could have a high influence on biofilm development and stability, even from the start-up stage, influencing both population size and species richness (Irie and Parsek, 2008). Difficulties in comparing behaviours across previous studies are further compounded by system design and operational differences that also influence these parameters. However, it is unquestionable that attached-growth microalgal–bacterial biofilms can deliver a number of services (Fig. 6.3), including wastewater treatment, which are considered hereunder.

Chlorella spp. and *Scenedesmus* spp. are the most studied microalgae in biofilm systems, being noted from around 40% of published studies in a recent review (Zhuang *et al.*, 2018). Compared with suspended growth cultures, the density of attached growth cultures can significantly decrease the operating land area required (Christenson and Sims, 2012). This has important impacts on the economic analysis of biofilm reactors over (often more easily) operated (low-cost) open pond systems. As an example, in a study on microalgae cultivation Morales-Amaral del Mar *et al.* (2015a) emphasised that even though costs for the operation of thin layer attached reactors are almost four times higher than raceway suspended-growth reactors, the difference between reactor productivity (45 g/m^2.day vs. 24 g/m^2.day) substantially decreased the price for biomass cultured in thin layer attached reactors. Biomass productivity per plan surface area can be further optimised by using multiple layer reactors (Roostaei *et al.*, 2018), applying different operational approaches or different support material configurations. For example, Gross and Wen (2014) obtained biofilm productivity of up to 18.9 g/m^2.day by using a pilot-scale RAB cultivation system in a vertical configuration.

Increases in biomass density can also improve photosynthetic efficiencies (particularly in the surface layers of a mixed community biofilm). When comparing photosynthetic efficiencies of the microalga *Scenedesmus* spp. cultivated in diluted centrate from anaerobic digestion under similar operational conditions, maximum values of 9% were demonstrated in the thin-film attached-growth reactor, compared with 5% in the suspended-growth raceway reactor (Morales-Amaral del Mar *et al.*, 2015a).

Harvested biofilm biomass also has lower water content than harvested suspended biomass. Polizzi *et al.* (2017) report moisture levels of between 80% and 90% in scraped biofilm, which is comparable to that of centrifuged suspended culture. This point was also highlighted by Johnson and Wen (2010), who harvested *Chlorella* spp. biofilm from a dairy wastewater treatment system and determined its water content at around 94% – which was sufficient to avoid the use of a centrifuge for preliminary dewatering. In culturing a *Botryococcus braunii* biofilm, Ozkan *et al.* (2012) with a biomass productivity of 0.71 g/m^2.day and a biomass density of 96.4 kg/m^3, achieved a decrease in required dewatering energy requirements of 99.7% during harvesting and post-harvesting steps.

Harvesting frequency is another important factor, with Johnson and Wen (2010) reporting increased productivity and nutrient removal with shorter harvest intervals (Table 6.1). They treated dairy manure wastewater with *Chlorella* spp. biomass in laboratory conditions, with light irradiance of between 110 and 120 $\mu mol/s.m^2$.

In terms of performance obtained for wastewater treatment, organic matter and N and P removal efficiencies vary depending on factors such as the reactor design, microalgae and bacteria species, influent loadings, harvesting period, etc. Although it has been suggested that attached-growth microalgae–bacteria systems can sustain higher wastewater treatment efficiencies when compared to suspended cultures (Zhang *et al.*, 2020), a recent review highlighted that only 10% of relevant

Table 6.1. Biomass and nutrient removals at various harvest intervals, as reported by Johnson and Wen (2010).

Harvest Intervals	Days	6	10	15
Nitrogen (N)	$g/m^2.d$	0.77	0.59	0.39
Ammonium (NH_4-N)	$g/m^2.d$	0.74	0.45	0.3
Phosphorus (P)	$g/m^2.d$	1.45	0.8	0.47
Biomass	$g/m^2.d$	3.5	3	2

publications included performance data from operations outside laboratory environments (Zhuang *et al.*, 2018). According to Zerrouki and Henni (2019), several pilot-scale systems (only) using attached microalgae technology for wastewater treatment are known to have been successfully implemented worldwide.

According to Acién *et al.* (2016), the theoretical maximum mean nitrogen removal rate that can be achieved by microalgae biomass is 3.5 $g/m^2.d$ with maximum mean biomass productivity of 50 g/m^2.day. Boelee *et al.* (2014a) recorded a removal rate of 3.2 g NH_4-N/m^2.day, 0.41 g PO_4-P/m^2.day and 43 g COD/m^2.day at an HRT of 4.5 hours when treating synthetic municipal wastewater. Treating municipal wastewater in a 32 m^2 pilot-scale thin-layer cascade photobioreactor (0.02 m water depth), Sánchez Zurano *et al.* (2020) recorded daily removal rates of between 15 and 30.6 mg/L NH_4-N, 1.8 and 5.6 PO_4-P mg/L, and 81 to 178.3 mg/L COD depending on seasonal variations; biomass productivity ranged between 28.3 and 47.3 g/m^2.day. This equates to removals of 1.0–2.0 g NH_4-N/m^2.day, 0.12–0.37 g PO_4-P/m^2.day and 5.4–11.9 g COD/m^2.day.

Various supporting matrices have been trialled for algal biofilm wastewater treatment processes – Adey *et al.* (1993) reported highest microalgae productivity (15–27g/m^2.d) when treating agricultural run-off wastewater with biomass attached to plastic screens – while Zhuang *et al.* (2018) found that cotton, polycarbonate and cellulose acetate were the most commonly used supports. Melo *et al.* (2018) tested borosilicate glass, polyurethane foam, polyvinyl chloride, stainless steel, polyethylene and polypropylene, reporting that PVC was the most appropriate for culturing *C. vulgaris* on a rotating flat plate photobioreactor (RFPPB).

6.4 Operational Aspects

6.4.1 Biofilm Community Structures

Biodiversity (referring to community share and species richness) is an important functional parameter that helps define biofilm characteristics and performance.

A high microbiological diversity is associated with complex metabolic and interspecies relations at the cellular level, and can increase biomass value, wastewater treatment performance outcomes and economic and environmental impacts. In other cases selected or engineered biodiversity (e.g., through inoculation) can enhance removals of targeted pollutants, increase biomass value (for further use as a feedstock for energy production or co-product extraction), or even harvesting efficiency (by using self-aggregating/colonial species or large cell size microalgae). As is the case in suspended microalgae–bacteria cultures, filamentous microalgae play an important role in biomass attachment efficiency, density and harvesting – and are specifically targeted for implementation of ATS technologies (Adey *et al.*, 2011).

Identifying factors that can limit the presence of undesirable communities is important in ensuring biofilm efficiency. These factors include (Doucha and Lívanský, 2006):

- Pathogenic viruses and bacteria
- Fungi
- Grazers (protozoa and rotifers) and other predators
- Undesirable microalgae species and some cyanobacteria

Compared to suspended-growth microalgae–bacteria cultures, high physical density within biofilm structures can decrease the risk of culture contamination with unwanted communities (Doucha and Lívanský, 2006). In general, it is far more complex and challenging to maintain 'desired' populations in uncontrolled large-scale applications, as community structures undergo significant changes over time due to the impacts of wastewater physico-chemical and biological characteristics, and wider environmental conditions (Carney *et al.*, 2014).

Diverse biofilm structures contain microalgae species with different trophic patterns (photoautotrophic, heterotrophic and mixotrophic) (Roostaei *et al.*, 2018). Depending on insolation levels, conditions suited to autotrophic and mixotrophic microalgae can occur. Under such conditions, the carbon source (e.g., wastewater) can be assimilated by both heterotrophic bacteria and mixotrophic microalgae, potentially increasing wastewater treatment efficiency. Mixotrophic growth also has a positive effect on microalgae cell lipid content, which can be important when this is a target for downstream processing/valorisation (Zhan *et al.*, 2017). Mixotrophic growth has been found in a large number of microalgae, including *C. vulgaris*, *C. regularis*, *Spirulina platensis*, *Haematococcus pluvialis*, *E. gracilis*, *Nannochloropsis* spp., *Arthrospira* spp., *Synechococcus* spp., *Anabaena* spp., *Phaeodactylum* spp., *Botryococcus braunii*, *Tetraselmis* spp., *Scenedesmus* spp. and *Desmodesmus* spp.

It is also important to note that some microalgae (e.g., *Chlorella* spp.) are capable of nitrification (Gerardi, 2002). Between ammonium and nitrate nitrogen sources, it is assumed that microalgae will assimilate ammonium rather than nitrate compounds due to lower energy requirements for synthesis. However, the preference for one or another nitrogen source varies even within the same genus (Liu and Chen, 2016).

Several methods are commonly used to identify and quantify microorganisms (such as microscopy, fluorescence and PCR/qPCR). In recent years, alternative methods have been proposed for rapid *in vivo* assessment of microalgae communities (and their competitors/predators), such as spectroradiometric monitoring (Reichardt *et al.*, 2020). These emerging techniques should allow biofilm structure and characteristics to be more carefully monitored and controlled, allowing wastewater treatment processes to be further optimised.

6.4.2 Light Irradiance

Alongside other operational parameters, light represents an important driving factor in autotrophic cultivation as it controls photosynthetic activity, and thus oxygen supply – and is closely linked to biofilm productivity.

Photosynthetically active radiation varies with season and latitude but an average of around 1800 $\mu mol/m^2.s$ reaches the surface of the earth on sunny days (Masojídek *et al.*, 2014). However, photosynthetic activity only increases with light intensity to a certain level (light saturation level) which is around 1/10th of maximum irradiance (Torzillo *et al.*, 2010). Prolonged exposure to excessive irradiance can lead to photoinhibition, through damage to photosynthetic structures within algae (Nikolaou *et al.*, 2015). Different microalgal species have different strategies to counter photoinhibition – including lipid accumulation and secretion of carbohydrates – allowing them to become photo-acclimatised (Nikolaou *et al.*, 2016) and (Ramanan *et al.*, 2016).

Determining mechanisms of photo-acclimatisation in biofilm structures is complicated by the physical arrangement of mixed microalgal and bacterial communities – which can lead to physical shading of algae by bacteria (Schnurr *et al.*, 2016), although the mechanisms of light distribution within biofilms require further research (Wang *et al.*, 2015). Although provision of uniform irradiance across and within biofilms might be perceived as useful for maximising photosynthetic productivity, this may not be required or desirable in wastewater treatment applications where chemolithotrophic bacteria use nitrate as an energy source. Indeed, relationships between irradiance, photoperiod and biofilm community and physical structure must all be considered in the context of the wastewater being treated – and the objectives of that treatment.

6.4.3 Flow Velocity and Turbulence

Another important parameter to be considered during biofilm development is the flow velocity and turbulence as this influences shear stress, diffusion processes and biofilm stability. The impacts of flow velocity and turbulence can be technology-specific but in one study which looked at this issue González *et al.* (2008) reported that a flow velocity of 0.4 m/s resulted in biofilm disintegration and compromised treatment performance, particularly in terms of COD removal efficiency. In this study a flow velocity of less than 0.1 m/s was required to ensure biofilm stability.

6.4.4 pH

One of the challenges that arises during microalgae cultivation is control of pH, which is impacted by photosynthetic phenomena (as well as biochemical wastewater treatment processes). In general, organic carbon degradation and nitrification processes will reduce pH through destruction of alkalinity and production of CO_2 whereas denitrification processes or nitrate uptake by microalgae will restore some of this lost alkalinity (for example, through simultaneous cellular OH^- release). However, nitrification may be reduced where microalgae compete with nitrifying bacteria for ammonium.

The impacts of pH vary between microalgae cultivation systems. For example, axenic microalgal cultures can sustain a high increase in pH value due to the absence of bacterial activity. Meanwhile, the presence of bacterial populations can have a strong buffering effect on pH, limiting its increase during microalgae activity. A high pH value (mainly higher than 9) can lead to increased ammonium removal through volatilisation, while higher pH values also favour phosphate precipitation with Ca/Mg or autoflocculation phenomena (Muñoz and Guieysse, 2006).

6.4.5 Wastewater Nutrient Loads

One of the factors which influences nutrient removal efficiency is the ratio between nitrogen and phosphorus concentrations (N:P) in the influent. Although developed during studies on marine phytoplankton, the Redfield ratio (C:N:P – 106:16:1) is commonly used as the basis for microalgal cultivation (Smith *et al.*, 2017). However, the N:P ratio of microalgae biomass is species-specific, and ranges from 8N:1P to 45N:1P (Hecky *et al.*, 1993). Wastewater influent can comprise wide variations in nutrient concentration (both within treatment plants and between various wastewater types) with 'ideal' compositions rarely occurring. This drives the general adoption of nutrient-specific processes. Microalgae have been shown to alter their nutrient composition at a cellular level in response to varying nutrient concentrations in their host environment (Whitton *et al.*, 2016), a feature which can be leveraged in wastewater treatment applications. Where the composition of influent is stable, correlations between nitrogen and phosphorus removal efficiency and biomass productivity can be established (Morales-Amaral del Mar *et al.*, 2015a). However, if the target of a tertiary treatment process is to deliver final effluent of particular nutrient characteristics (for example, to meet regulatory limits), then tailoring of the influent nutrient load and/or optimisation of the biofilm community structure may be required (Sadatshojaei *et al.*, 2020).

Zhuang *et al.* (2020) observed that attached microalgae biomass responds less to the presence of nitrate sources than to organic carbon, phosphate and ammonium (they also reported removal efficiencies ranging between 78.2% and 93.2% for all of the measured parameters: COD, TN, TP, NH_4-N and PO_4-P). At a certain concentration (usually above 100 mg NH_4-N/L, although the precise response is species-specific), ammonium-nitrogen can be toxic to microalgae biomass. For example, Morales-Amaral del Mar *et al.* (2015) found concentrations above 192 mg NH_4-N/L decreased *Scenedesmus* spp. productivity when cultured in diluted centrate from anaerobic digestion.

6.4.6 Environmental Conditions

The performance of attached microalgae–bacteria cultures used for wastewater treatment under laboratory conditions will generally be different from onsite pilot or full-scale tests and thus it is being important to test the biofilm behaviour in environmental conditions that are as similar as possible to the intended final application. For instance, decreases in nutrient removal efficiency (of between 1- and 3-fold) and biomass productivity (of between 10- and 13-fold) as well as modification of the community structure were noted by Van Den Hende *et al.* (2014) when upscaling a novel wastewater treatment technology from indoor lab-scale reactors to outdoor conditions. Boelee *et al.* (2014a) also noted differences in process efficiency when municipal wastewater was treated in a pilot-scale biofilm reactor under real conditions, as compared with treatment at bench scale. They speculated that light and temperature were the possible limiting factors. Bacterial communities can also undergo changes during scaling-up, even between small to medium and large systems operated under the same outdoor conditions (Fulbright *et al.*, 2018).

Even where inoculated with target biomass, biofilm community diversity can undergo multiple changes during the start-up stage. As a result, it is encouraged to use microalgae species with a high tolerance to stressful operational factors such as *Dunaliella salina*, which is found frequently in open pond systems (Xiaogang *et al.*, 2020). Similar shifts in community structure and diversity have also been noted from systems inoculated with native biomass (Sekar *et al.*, 2004) significantly increasing start-up periods (Liu *et al.*, 2017).

Recently, a tool based on photo-respirometry models has been proposed which can enable estimation of algae–bacterial growth rates in different environmental conditions (Rossi *et al.*, 2020). The method proposed is designed to allow for fast and reliable calibration of algae–bacterial growth models as a function of environmental conditions, and optimal growth conditions can be identified for different algae strains. The tool leverages a standard photo-respirometric model calibration protocol and has been validated in an HRAP system treating digestates.

6.4.7 Harvesting Frequency

In contrast to the various methods available for harvesting biomass from suspended-growth systems, published literature suggest that a simple scraping approach is the only applicable method for biofilm harvesting. Questions then arise as to optimal harvest frequency. If biofilm harvesting occurs after the exponential growth phase, treatment performance can decrease significantly as a result of senescence within the film. Dead cells and a destabilised phycosphere can contribute to increased organic carbon and phosphorus concentrations (Jiang *et al.*, 2007). Other consequences of late biomass harvesting are an increase in biofilm thickness and density with a negative effect on photosynthetic activity, increased ash content, biofilm detachment that can negatively impact effluent quality (by increasing turbidity) and the immigration of predators, with their direct impacts on biofilm biodiversity and stability. Furthermore, bacteria have different growth rates to microalgae, meaning that in a mixed culture system – even where microalgae have entered a stationary or decline phase – a significant increase of the bacterial growth rate can still occur through consumption of metabolites and other nutrients released from the microalgal biomass. This can lead, in turn, to destabilisation of the trophic network and compromise system functionality. On the other hand, harvesting immature biomass that has yet to reach its metabolic peak will reduce treatment performance.

Although optimum biomass harvesting frequency is species-dependent, the usual rule of thumb is to use an interval of between 7 and 14 days (Siville *et al.*, 2020).

6.5 INNOQUA Microalgae-based Module

Within the INNOQUA system, the microalgae-based bio-solar purification (BSP) unit is the module designed for polishing lumbrifilter (primary and secondary treated) effluent in mild climates with high insolation. It can provide an addition or an alternative to other tertiary treatment technologies, before disinfection and/or wastewater discharge and reuse.

The focus of the INNOQUA project was decentralised wastewater treatment applications, and thus cost and maintenance efficiency were key considerations. The module developed was an open thin layer cascading photobioreactor, whereby the influent was fed to the reactor during daytime and a recirculation pump was employed to ensure the required exposure of wastewater to sunlight with an overflow mechanism for effluent from the recirculation tank.

6.5.1 Laboratory Scale Testing

This concept was developed and tested at laboratory scale in the Environmental Technology department of ECOIND (a public research and development organisation based in Romania). The laboratory systems were developed to:

(i) Assess the influence of platform design (cascade vs. single platform – Figs. 6.4 and 6.5) on performance;
(ii) Assess the influence of water depth (1 cm and 5 cm) on reactor performance;
(iii) Establish stable operation and assess potential treatment performances of this treatment step prior to pilot scale development and
(iv) Evaluate biomass-specific growth and its impact on effluent suspended solids.

6.5.1.1 Materials and methods

The experimental apparatus comprised two photobioreactor configurations (with one configuration run simultaneously at two different water levels) each with the same total platform area and capacity of 10 L. All reactors were run in duplicate and all run in parallel:

- Double cascade reactors with two platforms each with an average water layer depth on the platform of approximately 5 cm (Fig. 6.5a);
- Thin layer cascade reactors with two platforms each with an average water layer depth on the platform of approximately 1 cm (Fig. 6.5b) and
- Single platform reactors with an average water layer depth of approximately 5 cm (Fig. 6.5c).

All experiments were performed at laboratory scale and used synthetic wastewater designed to replicate secondary treated municipal or domestic wastewater. The reactors were fitted with individual feeding pump and recirculation pumps and illuminated artificially using LED photosynthetic light sources. The experiment used photoperiodicity of 12 hours light and 12 hours darkness at 12,000 lumens/m^2 and a hydraulic loading rate of 250 L/m^2.day. All experiments were performed at room temperature ($22 \pm 5°C$).

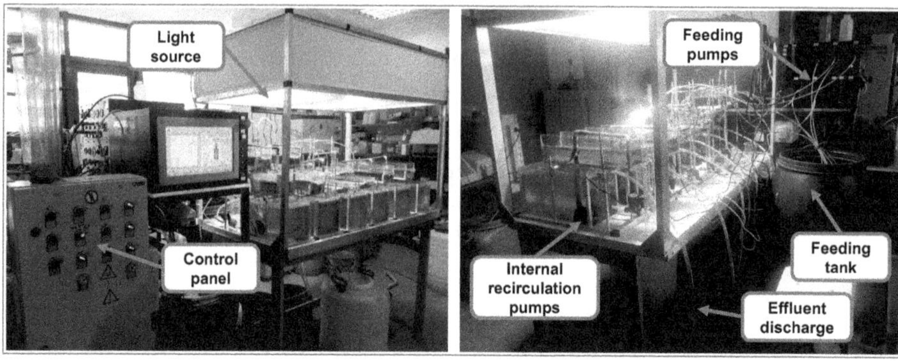

Figure 6.4. Laboratory set-up (front view - left; back view - right).

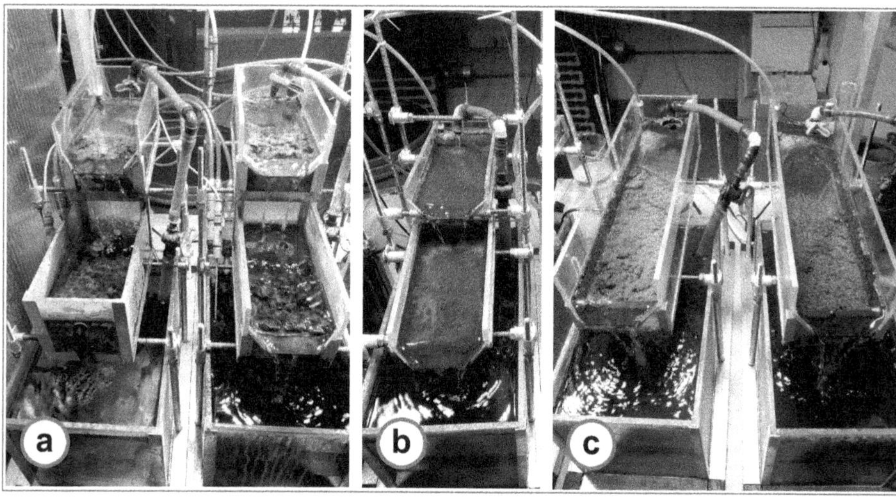

Figure 6.5. Laboratory reactors (a) double cascade photobioreactors, (b) thin layer cascade photobioreactors and (c) single platform photobioreactors.

6.5.1.2 Results and discussion

During steady-state operation, removal efficiencies averaged 70% COD, 80% TN and 60% TP with limited influence of platform design. However, the average specific surface removal performances varied between each reactor type and are summarised in Table 6.2. In all cases the remaining effluent COD was mainly associated with biomass washout (i.e., related to effluent TSS) as the residual COD in filtered samples was below 20 mg/L.

As shown in Fig. 6.6, during the first two weeks the effluent COD concentrations were relatively high which corresponds to suspended growth of microalgae and partial biomass washout in the effluent – reflected as particulate COD. However, after these two weeks, the biomass developed as a mixed microalgae–bacteria biofilm on the platforms and effluent TSS concentrations were relatively low (Fig. 6.7).

Table 6.2. Specific average mass removal of main contaminants (g/m^2.day).

Total Nitrogen Removal			Total Phosphorus Removal			COD Removal		
Double cascade	Thin layer cascade	Single platform	Double cascade	Thin layer cascade	Single platform	Double cascade	Thin layer cascade	Single platform
12.0	12.5	13.4	1.1	1.2	1.3	30.0	29.2	31.2

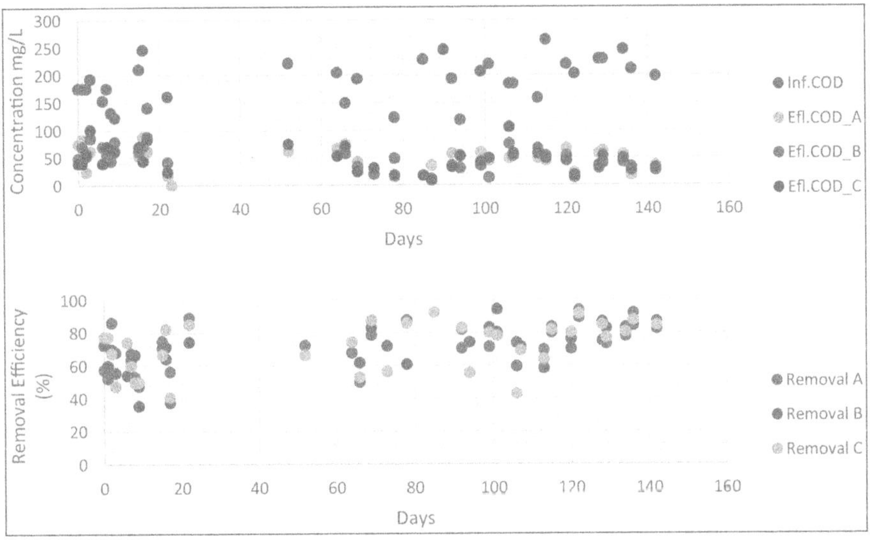

Figure 6.6. COD concentrations and removal performances for each of the tested platforms designs. Note: A is the average data from the double cascade photobioreactors, B the thin layer cascade photobioreactors and C the single platform photobioreactors.

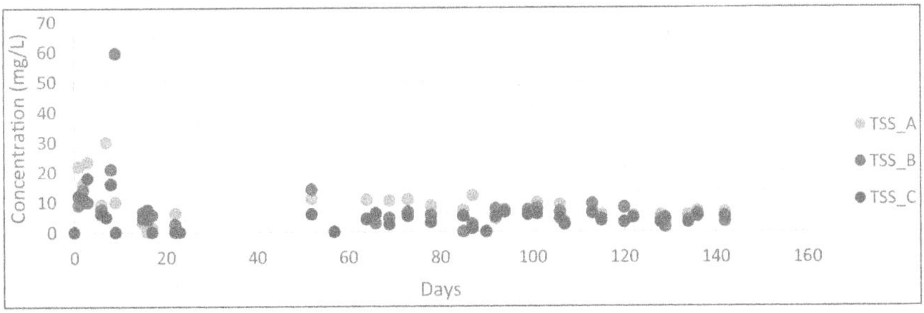

Figure 6.7. TSS concentrations in the effluent of laboratory scale units. Note: A is the average data from the double cascade photobioreactors, B the thin layer cascade photobioreactors and C the single platform photobioreactors.

In general, the cascading platform approach with recirculation was conducive to conditions for good biomass growth and good treatment performance. However, in these configurations biomass accumulation requires the need for regular harvesting and maintenance. For the laboratory scale experiments we observed that

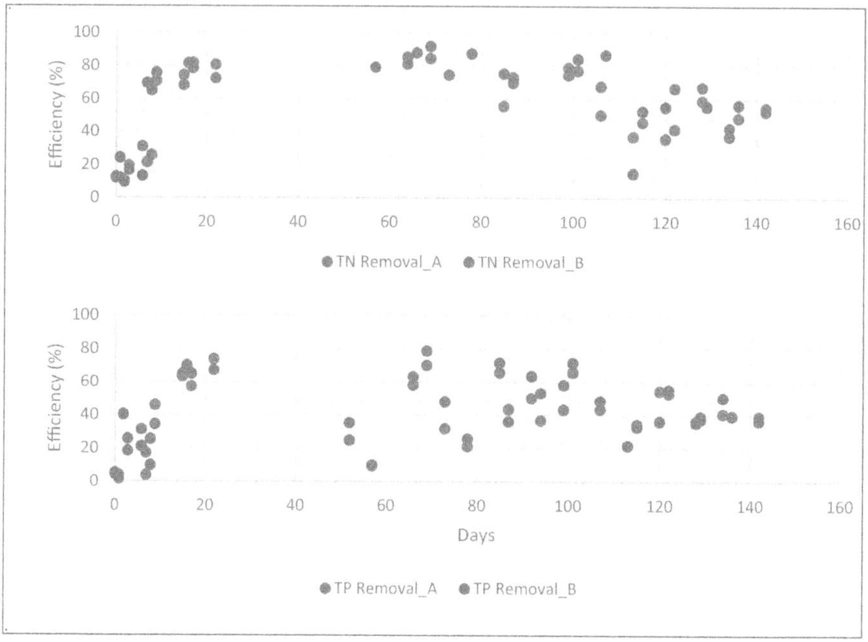

Figure 6.8. Nitrogen and phosphorus removal efficiencies in the lab-scale BSP units. A = double cascade photobioreactors; B = thin layer cascade photobioreactors.

the excess biofilm which detached and accumulated on the platforms needed to be removed periodically to avoid biomass decay and nutrient re-solubilisation within the system (Fig. 6.8). The frequency of biofilm detachment and biomass harvesting was dependent on biomass growth rate, which was in turn dependent on the quality of influent (quantities of nutrients introduced). Other authors have suggested that in phototrophic biofilm photobioreactors used for effluent polishing as part of wastewater treatment, the average biomass production rate is approximately 7 g dry weight m^{-2} day^{-1} while the harvesting frequency should be at least twice a month – as the biofilm starts to spontaneously detach after two weeks (Boelee *et al.*, 2014).

Phosphorus removal performance was most affected by biofilm accumulation and decay – and harvesting twice a month had to be considered to avoid excess biomass accumulation resulting in decay and unstable phosphorus removal performance (Fig. 6.8).

6.5.2 Pilot-scale Testing

Based on lab-scale performance data, the BSP system was designed as a cascading photobioreactor. In addition to performance, the choice of the double platform system also had the advantage of simple and rapid manufacturing (allowing local materials to be used to create an easily handled system that was simple to maintain and adapt).

Pilot-scale BSP units were tested in Spain by the University of Girona, in India by BORDA (Bengaluru) and in Peru by the Catholic University of Santa Maria (Arequipa) (Fig. 6.9).

Each installed system comprised two platforms (each of which had a 2 m^2 surface area) in a cascade sequence. Each system comprised the platforms (constructed from polypropylene), a recirculation tank, a feeding pump and a corresponding recirculation pump. The module was designed to treat 1 m^3 of effluent from a lumbrifilter per day. The modules were self-inoculated with local microflora – either microalgae from ponds/lakes/rivers or local cultures. The start-up duration for each site varied from 1 week to 1 month depending on inoculation technique and amount of inoculum used.

Figure 6.9. Pilot-scale BSP modules of the INNOQUA system installed in India (left) and Peru during inoculation (right).

Table 6.3. Average concentrations of Lumbrifilter influent and effluent and BSP effluent, and the process efficiencies for each treatment step and global efficiency of an integrated LF+BSP system.

	TSS	BOD	COD	NH_4-N	TP
	Average concentrations (mg/L)				
LF influent	2,190	1,165	2,242	104	23.8*
LF effluent	271	90	371	15.2	15.1*
BSP effluent	224	30	183	3.2	6.48
	Process efficiency (%)				
LF	87.63	92.27	83.45	85.38	36.55
BSP	17.34	66.67	50.67	78.95	57.09
LF+BSP	89.77	97.42	91.84	96.92	72.77

*Composite sample on the 8th of September 2020.

In each case, the removal rates for key parameters by lumbrifiltration averaged about 80% for TSS, COD, BOD and NH_4^+-N (Table 6.3). The addition of the

BSP polishing step increased average performances close to 90% for COD and BOD, and above 90% for TSS and ammonium nitrogen. At these sites, excess biomass removal was required approximately once every two weeks.

As a general conclusion, the BSP module as developed within the INNOQUA system requires regular biomass harvesting and maintenance, making it unlikely to be suitable for low-intervention decentralised applications. However, the concept as developed here could be readily adapted for tertiary treatment at a suitably staffed centralised or semi-centralised wastewater treatment facility.

6.6 Conclusions

For more than half a century, microalgal biomass has been investigated as a potential feedstock for advanced generation of biofuels and high-value compounds, and also as an important contributor to sustainable solutions for emerging problems derived from human activities such as climate change, ecosystem pollution and sanitation. During this period, microalgae-based technology has undergone a constant evolution, with improved knowledge of the performance of different cultivation approaches and selection of target species, as well as identification of those economic sectors with high environmental impact and energy consumption (such as wastewater treatment) where microalgae could make an important contribution. However, technological limits – such as those related to costs, biomass harvesting and cultivation – and the fact that most studies have been conducted at the laboratory level, mean that research remains necessary.

The characteristics of photoautotrophic microalgae make them suitable for a number of wastewater applications – their ability to utilise dissolved nutrients supports rapid development of biomass while reducing nutrient loads in final effluents, and their productivity generates oxygen that can be utilised by bacteria to break down dissolved organic pollutants. When combined, these attributes can (potentially) lead to significant reductions in energy and chemical usage in wastewater treatment, whilst simultaneously delivering valuable ecosystem services, increasing sustainability and (even) produce biomass with significant potential in the wider bioeconomy.

Both suspended and attached-growth microalgal–bacterial cultures have been examined, with the latter offering simpler/cheaper opportunities for biomass harvesting that could lend themselves to decentralised wastewater treatment approaches. Microalgal biofilms were at the heart of the INNOQUA BSP module, targeted at treatment of secondary wastewaters from lumbrifilter systems. The BSP module has been demonstrated to be a feasible polishing step for this effluent at both laboratory and pilot scales, using real wastewater with site-specific characteristics. However, before it can be considered ready for commercial deployment, the BSP module requires further optimisation and field testing to better understand maintenance requirements and thoroughly assess its suitability for use in decentralised applications. Although developed within INNOQUA as a tertiary treatment solution, microalgal–bacterial communities also have potential for delivery of primary or secondary wastewater treatment. Current research is exploring the fundamental aspects to support such applications.

References

Acién Fernández G. F., Fernández Sevilla M. J., Molina Grima E., 2013. Photobioreactors for the production of microalgae. Reviews in Environmental Science and Bio/Technology, 12, 131–151. doi: 10.1007/s11157-012-9307-6

Acién, F. G., Gómez-Serrano, C., Morales-Amaral, M. M., Fernández-Sevilla, J. M., Molina-Grima, E., 2016. Wastewater treatment using microalgae: how realistic a contribution might it be to significant urban wastewater treatment? Appl. Microbiol. Biotechnol. https://doi.org/10.1007/s00253-016-7835-7

Adey W. H., Kangas P. C., Mulbry W., 2011. Algal turf scrubbing: Cleaning surface waters with solar energy while producing a biofuel. Bioscience, 61(6), 434–441. doi: 10.1525/bio.2011.61.6.5

Adey W., Luckett C., Jensen K., 1993. Phosphorus removal from natural waters using controlled algal production. Restoration Ecology, 1(1), 29–39. doi: 10.1111/j.1526-100X.1993.tb00006.x

Al-Muyeed, A., 2017. Technical guidelines for designing a decentralised waste water treatment system.

Amin A. S., Parker S. M., Armbrust E. M., 2012. Interactions between diatoms and bacteria. Microbiology and Molecular Biology Reviews, 76(3), 667–684. doi: 10.1128/MMBR.00007-12

Ansa O. D. E., Lubberding J. H., Gijzen J. H., 2012. The effect of algal biomass on the removal of faecal coliform from domestic wastewater. Applied Water Science, 2, 87–94. doi: 10.1007/s13201-011-0025-y

Boelee C. N., Janssen M., Temmik H., Shrestha R., Buisman N. J. C., Wijffels H. R., 2014a. Nutrient removal and biomass production in an outdoor pilot-scale phototrophic biofilm reactor for effluent polishing. Applied Biochemistry and Biotechnology, 172, 405–422. doi: 10.1007/s12010-013-0478-6

Boelee C. N., Temmink H., Janssen M., Buisman N. J. C., Wijffels H. R., 2014. Balancing the organic load and light supply in symbiotic microalgal-bacterial biofilm reactors treating synthetic municipal wastewater. Ecological Engineering, 64, 213–221. doi: 10.1016/j.ecoleng.2013.12.035

Boelee, N. C., Janssen M., Temmink H., Taparavičiūtė L., Khiewwijit R., Jánoska Á., Buisman C. J. N. & Wijffels R. H., 2014. The effect of harvesting on biomass production and nutrient removal in phototrophic biofilm reactors for effluent polishing. Journal of Applied Phycology, 26, 1439–1452.

Brennan L., Owende P., 2010. Biofuels from microalgae – a review of technologies for production, processing, and extractions of biofuels and co-products. Renewable and Sustainable Energy Reviews, 14, 557–577. doi: 10.1016/j.rser.2009.10.009

Bunce, J. T., Ndam, E., Ofiteru, I. D., Moore, A., Graham, D. W., 2018. A Review of Phosphorus Removal Technologies and Their Applicability to Small-Scale Domestic Wastewater Treatment Systems. Front. Environ. Sci. 6, 8. https://doi.org/10.3389/fenvs.2018.00008

Burlew S. J., 1953. Current status of the large-scale culture of Algae. In: Algal culture. From laboratory to pilot plant. Burlew S. J. (Ed.), Carnegie Institution of Washington Publication 600, Washington, 3–23.

Capodaglio, A., 2017. Integrated, Decentralized Wastewater Management for Resource Recovery in Rural and Peri-Urban Areas. Resources 6, 22. https://doi.org/10.3390/resources6020022

Carney L. T., Reinsch S. S., Lane P. D., Solberg O. D., Jansen L. S., Williams K. P., Trent D. J., Lane, T. W., 2014. Microbiome analysis of a microalgal mass culture growing in municipal wastewater in a prototype OMEGA photobioreactor. Algal Research, 4, 52–61. doi: 10.1016/j.algal.2013.11.006

Chae K.-J., Kang J., 2013. Estimating the energy independence of a municipal wastewater treatment plant incorporating green energy resources. Energy Conversion and Management, 75, 664–672. doi: 10.1016/j.enconman.2013.08.028

Chalivendra S., 2014. Bioremediation of wastewater using microalgae. Electronic Thesis. University of Dayton. http://rave.ohiolink.edu/etdc/view?acc_num=dayton1418994496

Chowdhury H., Loganathan B., 2019. Third-generation biofuels from microalgae: a review. Current opinion Green and Sustainable Chemistry, 20, 39–44. doi: 10.1016/j.cogsc.2019.09.003

Christenson L. B., Sims R. C., 2012. Rotating algal biofilm reactor and spool harvester for wastewater treatment with biofuels by-products. Biotechnology and Bioengineering, 109(7), 1674–1684. doi: 10.1002/bit.24451

Cook P. M., 1951. Chemical engineering problems in large-scale culture of algae. Industrial & Engineering Chemistry, 43(10), 2385–2389. doi: 10.1021/ie50502a056

Craggs J. R., Adey H. W., Jenson R. K., John S. ST. M., Green F. B., Oswald J. W., 1996. Phosphorus removal from wastewater using an algal turf scrubber. Water

Science and Technology, 33(7), 191–198. doi: 10.1016/0273-1223(96)00354-X

Davis R., Aden A., Pienkos T. P., 2011. Techno-economic analysis of autotrophic microalgae for fuel production. Applied Energy, 88, 3524–3531. doi: 10.1016/j.apenergy.2011.04.018

de Carvalho J. C., Magalhães A. I., de Melo Pereira G. V., Medeiros A. B. P., Sydney E. B., Rodrigues C., *et al.*, 2020. Microalgal biomass pretreatment for integrated processing into biofuels, food, and feed. Bioresource Technology, 300, 122719. doi: 10.1016/j.biortech.2019.122719

Doucha J., Lívanský K., 2006. Productivity, CO_2/O_2 exchange and hydraulics in outdoor open high density microalgal (Chlorella sp.) photobioreactors operated in a Middle and Southern European climate. Journal of Applied Phycology, 18, 811–826. doi: 10.1007/s10811-006-9100-4

Doucha J., Lívanský K., 2009. Outdoor open thin-layer microalgal photobioreactor: potential productivity. Journal of Applied Phycology, 21, 111–117. doi: 10.1007/s10811-008-9336-2

Fasaei F., Bitter H. J., Slegers M. P., van Boxtel B. J. A., 2018. Techno-economic evaluation of microalgae harvesting and dewatering systems. Algal Research, 31, 347–362. doi: 10.1016/j.algal.2017.11.038

Fulbright P. S., Robbinson-Pianka A., Berg-Lyons D., Knight R., Reardon F. K., Chisholm T. S., 2018. Bacterial community changes in an industrial algae production system. Algal Research, 31, 147–156. doi: 10.1016/j.algal.2017.09.010

Garcia D., Posadas E., Blanco S., Acién G., García-Encina P., Bolado S., Muñoz R., 2018. Evaluation of the dynamics of microalgae population structure and process performance during piggery wastewater treatment in algal-bacterial photobioreactors. Bioresource Technology, 248, 120–126. doi: 10.1016/j.biortech.2017.06.079

Gerardi M. H., 2002. Introduction to nitrification. In: Nitrification and denitrification in the activated sludge process. John Wiley & Sons Inc., New York, 37–41.

Goldman C. J., 1979. Outdoor algal mass cultures – I. Applications. Water Research, 13, 1–19. doi: 10.1016/0043-1354(79)90249-5

González, C., Marciniak, J., Villaverde, S., León, C., García, P. A., Muñoz, R., 2008. Efficient nutrient removal from swine manure in a tubular biofilm photobioreactor using algae-bacteria consortia. Water Sci. Technol. 58, 95–102. https://doi.org/10.2166/wst.2008.655

Granados, M. R., Acien, F. G., Gomez, C., Fernandez-Sevilla, J. M. & Grima Molina, E., 2012. Evaluation of flocculants for the recovery of freshwater microalgae. Bioresource Technology, 118, 102–110. doi: 10.1016/j.biortech.2012.05.018

Gross M. E., 2009. Allelochemical reactions, in: Encyclopedia of Inland Waters, Likens E. G. (Ed.), Academic Press, 715–726. doi: 10.1016/B978-012370626-3.00106-X

Gross M., Wen Z., 2014. Yearlong evaluation of performance and durability of a pilot-scale Revolving Algal Biofilm (RAB) cultivation system. Bioresource Technology, 171, 50–58. doi: 10.1016/j.biortech.2014.08.052

Gu Y., Li Y., Li X., Luo P., Wang H., Wang X., Wu J., Li F., 2017. Energy self-sufficient wastewater treatment plants: feasibilities and challenges. Energy Procedia, 105, 3741–3751. doi: 10.1016/j.egypro.2017.03.868

Haarich, S., Kirchmayr-Novak, S., Fontenla, A., Toptsidou, M., Hans, S., 2017. Bioeconomy development in EU regions. Mapping of EU Member States'/regions' Research and Innovation plans & Strategies for Smart Specialisation (RIS3) on Bioeconomy for 2014–2020. DG Research and Innovation, European Commission, Brussels.

Hayes M., Skomedal H., Mazur-Marzec H., Torunska-Sitarz A., Catala M., Isleten Hosoglu M., Garcia-Vaquero M., 2017. Microalgal proteins for feed, food and health. In: Microalgae-based biofuels and bioproducts. From feedstock cultivation to end-products. Gonzalez-Fernandez C., Muñoz R. (Eds.), Woodhead Publishing, 347–368. doi: 10.1016/B978-0-08-101023-5.00015-7

Hecky E. R., Campbell P., Hendzel L. L., 1993. The stoichiometry of carbon, nitrogen, and phosphorus in particulate matter of lakes and oceans. Limnology and Oceanography, 38(4), 709–724. doi: 10.4319/lo.1993.38.4.0709

Higgins T. B., Gennity I., Fitzgerald S. P., Shannon J. C., Fiehn O., VanderGheynst S. J., 2018. Algal-bacterial synergy in treatment of winery wastewater. Clean Water, 1, 6. doi: 10.1038/s41545-018-0005-y

Hoffmann P. J., 1998. Wastewater treatment with suspended and nonsuspended algae. Journal of Phycology, 34(5), 757–763. doi: 10.1046/j.1529-8817.1998.340757.x

Husnik F., McCutcheon J. P., 2018. Functional horizontal gene transfer from bacteria to eukaryotes. Nature Reviews Microbiology. doi: 10.1038/nrmicro.2017.137

Irie Y., Parsek R. M., 2008. Quorum sensing and microbial biofilms. Current topics in Microbiology and Immunology, 322, 67–84. doi: 10.1007/978-3-540-75418-3_4

Javed F., Aslam M., Rashid N., Shamair Z., Khan A. L., Yasin M., *et al.*, 2019. Microalgae-based biofuels, resource recovery and wastewater treatment: A pathway towards sustainable biorefinery. Fuel, 255, 115826. doi: 10.1016/j.fuel.2019.115826

Jiang L., Yang L., Xiao L., Shi X., Gao G., Qin B., 2007. Quantitative studies on phosphorus transference occurring between *Microcystis aeruginosa* and its attached bacterium (*Pseudomonas* sp.). Hydrobiologia, 581, 161–165. doi: 10.1007/s10750-006-0518-0

Johnson B. M., Wen Z., 2010. Development of an attached microalgal growth system for biofuel production. Applied Microbiology and Biotechnology, 85, 525–534. doi: 10.1007/s00253-009-2133-2

Jorgensen J., Convit C., 1953. Cultivation of complexes of algae with other fresh-water microorganisms in the tropics. In: Algal culture. From laboratory to pilot plant. Burlew S. J. (Ed.), Carnegie Institution of Washington Publication 600, Washington, 190–196.

Kehrein, P., Van Loosdrecht, M., Osseweijer, P., Garfí, M., Dewulf, J., Posada, J., 2020. A critical review of resource recovery from municipal wastewater treatment plants-market supply potentials, technologies and bottlenecks, Environmental Science: Water Research and Technology. Royal Society of Chemistry. https://doi.org/10.1039/c9ew00905a

Khan M. I., Shin J. H., Kim J. D., 2018. The promising future of microalgae: Current status, challenges, and optimization of a sustainable and renewable industry for biofuels, feed, and other products. Microbial Cell Factories, 17, 36. doi: 10.1186/s12934-018-0879-x

Khanzada T. Z., 2020. Phosphorus removal from landfill leachate by microalgae. Biotechnology Reports, 25, e00419. doi: 10.1016/j.btre.2020.e00419

Kim B.-H., Ramanan R., Cho D.-H., Oh H.-M., Kim H.-S., 2014. Role of Rhizobium, a plant growth promoting bacterium, in enhancing algal biomass through mutualistic interaction. Biomass and Bioenergy, 69, 95–105. doi: 10.1016/j.biombioe.2014.07.015

Kiseleva A. A., Tarachovskaya R. E., Shishova F. M., 2012. Biosynthesis of phytohormones in algae. Russian Journal of Plant Physiology, 59(5), 595–610.

Kouzama A., Watanabe K., 2015. Exploring the potential of algae/bacteria interactions. Current Opinion in Biotechnology, 33, 125–129. doi: 10.1016/j.copbio.2015.02.007

Li K., Liu Q., Fang F., Luo R., Lu Q., Zhou W., Huo S., Cheng P. *et al.*, 2019. Microalgae-based wastewater treatment for nutrients recovery: a review. Bioresource Technology, 291, 121934. doi: 10.1016/j.biortech.2019.121934

Liu J., Chen F., 2016. Biology and industrial applications of *Chlorella*: advances and prospects. Advances in Biochemical Engineering/Biotechnology, 153, 1–35. doi: 10.1007/10_2014_286

Liu J., Wu Y., Wu C., Muylaert K., Vyverman W., Yu H.-Q., Muñoz R., Rittman B., 2017. Advanced nutrient removal from surface water by a consortium of attached microalgae and bacteria: a review. Bioresource Technology, 241, 1127–1137. doi: 10.1016/j.biortech.2017.06.054

Maity P. J., Bundschuch J., Chen C.-Y., Bhattacharya P., 2014. Microalgae for third generation biofuel production, mitigation of greenhouse gas emissions and wastewater treatment: present and future perspectives – a mini review. Energy, 78, 104–113. doi: 10.1016/j.energy.2014.04.003

Mamais D., Noutsopoulos C., Dimopoulou A., Stasinakis A., Lekkas D. T., 2015. Wastewater treatment process impact on energy savings and greenhouse gas emissions. Water Science and Technology, 71(2), 303–308. doi: 10.2166/wst.2014.521

Masojídek J., Koblížek M., Torzillo G., 2014. Photosynthesis in microalgae. In: Handbook of Microalgal Biotechnology and Applied Phycology, Richmond A. (ed.), Blackwell Publishing Ltd, 20–39.

Masojídek J., Kopecký J., Giannelli L., Torzillo G., 2011. Productivity correlated to photobiochemical performance of *Chlorella* mass cultures grown outdoors in thin-layer cascades. Journal of Industrial Microbiology & Biotechnology, 38, 307–317. doi: 10.1007/s10295-010-0774-x

Matassa, S., Batstone, D.J., Hülsen, T., Schnoor, J., Verstraete, W., 2015. Can direct conversion of used nitrogen to new feed and protein help feed the world? Environ. Sci. Technol. 49, 5247–5254. https://doi.org/10.1021/es505432w

Mehrabadi A., Craggs R., Farid M. M., 2015. Wastewater treatment high rate algal ponds (WWT HRAP) for low-cost biofuel production. Bioresource Technology, 184, 202–214. doi: 10.1016/j.biortech.2014.11.004

Meier R. L., 1955. Biological cycles in the transformation of solar energy into useful fuels. In F. Daniels F. & Duffie J. A. (Eds.), Solar energy research Madison, WI, USA: University of Wiscons in Press, 179–183.

Melo M., Fernandes S., Caetano N., Borges T. M., 2018. *Chlorella vulgaris* (SAG 211-12) biofilm formation capacity and proposal of a rotating flat plate photobioreactor for more sustainable biomass production. Journal of Applied Phycology, 30, 887–899. doi: 10.1007/s10811-017-1290-4

Morales-Amaral del Mar M., Gómez-Serrano C., Acién G. F., Fernandez-Sevilla M. J., Molina-Grima E., 2015. Production of microalgae using centrate from anaerobic digestion as the nutrient source. Algal Research, 9, 297–305. doi: 10.1016/j.algal.2015.03.018

Morales-Amaral del Mar M., Gómez-Serrano C., Acién G. F., Fernandez-Sevilla M. J., Molina-Grima E., 2015a. Outdoor production of *Scenedesmus* sp. in thin-layer raceway reactors using centrate from anaerobic digestion as the sole nutrient source. Algal Research, 12, 99–108. doi: 10.1016/j.algal.2015.08.020

Muñoz R., Guieysse B., 2006. Algal-bacterial processes for the treatment of hazardous contaminants: a review. Water Research, 40, 2799–2815. doi: 10.1016/j.watres.2006.06.011

Nagarajan D., Lee Duu-Jong, Chen Chun-Yen, Chang Jo-Shu, 2020. Resource recovery from wastewaters using microalgae-based approaches: a circular bioeconomy perspective. Bioresource Technology, 302, 122817. doi: 10.1016/j.biortech.2020.122817

Nikolaou A., Bernardi A., Meneghesso A., Bezzo F., Morosinotto T., Chachuat B., 2015. A model of chlorophyll fluorescence in microalgae integrating photoproduction, photoinhibition and photoregulation. Journal of Biotechnology, 194, 91–99. doi: 10.1016/j.jbiotec.2014.12.001

Nikolaou A., Hartmann P., Sciandra A., Chachuat B., Bernard O., 2016. Dynamic coupling of photoacclimation and photoinhibition in a model of microalgae growth. Journal of Theoretical Biology, 390, 61–72. doi: 10.1016/j.jtbi.2015.11.004

Orandi S., Lewis M. D., Moheimani R. N., 2012. Biofilm establishment and heavy metal removal capacity of an indigenous mining algal-microbial consortium in a photo-rotating biological contactor. Journal of Industrial Microbiology & Biotechnology, 39, 1321–1331. doi: 10.1007/s10295-012-1142-9

Oswald J. W., Golueke G. C., 1960. Biological transformation of solar energy. Advances in Applied Microbiology, 2, 223–262. doi: 10.1016/S0065-2164(08)70127-8

Oswald J. W., Asce A. M., Gotaas B. H., Asce M., 1955. Photosynthesis in sewage treatment. American Society of Civil Engineers, 2849, 73–105.

Ozkan A., Kinney K., Katz L., Berberoglu H., 2012. Reduction of water and energy requirement of algae cultivation using an algae biofilm photobioreactor. Bioresource Technology, 114, 542–548. doi: 10.1016/j.biortech.2012.03.055

Park K. B. J., Craggs J. R., Shilton N. A., 2011. Wastewater treatment high rate algal ponds for biofuel production. Bioresource Technology, 102, 35–42. doi: 10.1016/j.biortech.2010.06.158

Patil L., Kaliwal B. B., 2019. Microalga *Scenedesmus bajacalifornicus* BBKLP-07, a new source of bioactive compounds with in vitro pharmacological applications. Bioprocess and Biosystems Engineering, 42, 979–994. doi: 10.1007/s00449-019-02099-5

Peccia J., Haznedaroglu B., Gutierrez J., Zimmerman B. J., 2013. Nitrogen supply is an important driver of sustainable microalgae biofuel production. Trends in Biotechnology, 31(3), 134–138. doi: 10.1016/j.tibtech.2013.01.010.

Polizzi B., Bernard O., Ribot M., 2017. A time-space model for the growth of microalgae biofilms for biofuel production, Journal of Theoretical Biology, 432, 55–79. doi: 10.1016/j.jtbi.2017.08.017

Posadas E., Alcantara C., Garcia-Encina A. P., Gouveia L., Guieysse B., Norvill Z., Acien G. F., Markou G., Congestri R., Koreiviene J., Muñoz R., 2017. Microalgae cultivation in wastewater. In: Microalgae-based biofuels and bioproducts.

From feedstock cultivation to end-products. Gonzalez-Fernandez C., Muñoz R. (Eds.), Woodhead Publishing, 67–91. doi: 10.1016/B978-0-08-101023-5.00003-0

Ramanan R., Kim B.-H., Cho Dae-Hyun, Oh Hee-Mock, Kim Hee-Sik, 2016. Algae-bacteria interactions: evolution, ecology and emerging applications. Biotechnology Advances, 34, 14–29. doi: 10.1016/j.biotechadv.2015.12.003

Reichardt A. T., Maes D., Jensen J. T., Dempster A. T., McGowen A. J., Poorey K., Curtis J. D., Lane W. T., Timlin A. J., 2020. Spectroradiometric detection of competitor diatoms and the grazer Poteriochromonas in algal cultures. Algal Research, 51, 102020. doi: 10.1016/j.algal.2020.102020

Roostaei J., Zhang Y., Gopalakrishnan K., Ochocki J. A., 2018. Mixotrophic microalgae biofilm: a novel algae cultivation strategy for improved productivity and cost-efficiency of biofuel-feedstock production. Scientific Reports, 8, 12528. doi: 10.1038/s41598-018-31016-1

Rossi S., Casagli F., Mantovani M., Mezzanotte V., Ficara E., 2020. Selection of photosynthesis and respiration models to assess the effect of environmental conditions on mixed microalgae consortia grown on wastewater. Bioresource Technology, 305, 122995. doi: 10.1016/j.biortech.2020.122995

Sadatshojaei, E., Mowla, D., Wood, D. A., 2020. Review of Progress in Microalgal Biotechnology Applied to Wastewater Treatment. Springer, Cham, pp. 539–557. https://doi.org/10.1007/978-3-030-42284-4_19

Sánchez Zurano A., Garrido Cárdenas A. J., Gómez Serrano C., Morales Amaral M., Acién-Fernández G. F., Fernández Sevilla M. J., Molina Grima E., 2020. Year-long assessment of a pilot scale thin-layer reactor for microalgae wastewater treatment. Variation in the microalgae-bacteria consortium and the impact of environmental conditions. Algal Research 50, 101983. doi: 10.1016/j.algal.2020.101983

Schnurr J. P., Molenda O., Edwards E., Espie S. G., Allen G. D., 2016. Improved biomass productivity in algal biofilms through synergistic interactions between photon flux density and carbon dioxide concentration. Bioresource Technology, 219, 72–79. doi: 10.1016/j.biortech.2016.06.129

Sekar R., Venugopalan V. P., Nandakumar K., Nair K. V. K., Rao V. N. R., 2004. Early stages of biofilm succession in a lentic freshwater environment. Hydrobiologia, 512, 97–108. doi: 10.1023/B:HYDR.0000020314.69538.2c

Siville B., Boeing J. W., 2020. Optimization of algal turf scrubber (ATS) technology through targeted harvest rate. Bioresource Technology Reports, 9, 100360. doi: 10.1016/j.biteb.2019.100360

Smith R. D., Jarvie P. H., Bowes J. M., 2017. Carbon, nitrogen, and phosphorus stoichiometry and eutrophication in river Thames Tributaries, UK. Agricultural & Environmental Letters, 2, 170020. doi: 10.2134/ael2017.06.0020

Surkatti R., Al-Zuhair S., 2018. Microalgae cultivation for phenolic compounds removal. Environmental Science and Pollution Research, 25, 33936–33956. doi: 10.1007/s11356-018-3450-8

Tang D. Y. Y., Khoo K. S., Chew K. W., Tao Y., Ho S.-H., Show P. L., 2020. Potential utilization of bioproducts from microalgae for the quality enhancement of natural products. Bioresource Technology, 304, 122997. doi: 10.1016/j.biortech.2020.122997

Tiron O., Bumbac C., Manea E., Stefanescu M., Nita Lazar M., 2017. Overcoming microalgae harvesting barrier by activated algae granules. Scientific Reports, 7, 4646. doi: 10.1038/s41598-017-050

Tiron O., Bumbac C., Patroescu I. V., Badescu V. R., Postolache C., 2015. Granular activated algae for wastewater treatment. Water Science & Technology, 71(6), 832–839. doi: 10.2166/wst.2015.010

Torzillo G., Giannelli L., Martínez-Roldan A. J., Verdone N., De Filippis P., Scarsella M., Bravi M., 2010. Microalgae culturing in thin-layer photobioreactors, Chemical Engineering Transactions, 20, 265–270. doi: 10.3303/CET1020045

Toyama T., Kasuya M., Hanaoka T., Kobayashi N., Tanaka Y., Inoue D., Sei K., Morikawa M., Mori K., 2018. Growth promotion of three microalgae, *Chlamydomonas reinhardtii*, *Chlorella vulgaris* and *Euglena gracilis*, by in situ indigenous bacteria in wastewater effluent. Biotechnology for Biofuels, 11, 176. doi: 10.1186/s13068-018-1174-0

Ubando, B. A., Felix, B. C., Chen W.-H., 2020. Biorefineries in circular economy: a comprehensive review. Bioresource Technology, 299, 122585. doi: 10.1016/j.biortech.2019.122585

Van Den Hende S., Beelen V., Bore G., Boon N., Vervaeren H., 2014. Up-scaling aquaculture wastewater treatment by microalgal bacterial flocs: from lab reactors to an outdoor raceway pond. Bioresource Technology, 159, 342–354. doi: 10.1016/j.biortech.2014.02.113

Vecchi V., Barera S., Bassi R., Dall'Osto, 2020. Potential and challenges of improving photosynthesis in algae. Plants, 9, 67. doi: 10.3390/plants9010067

Wang H., Xiong H., Hui Z., Zeng X., 2012. Mixotrophic cultivation of *Chlorella pyrenoidosa* with diluted primary piggery wastewater to produce lipids. Bioresource Technology, 104, 215–220. doi: 10.1016/j.biortech.2011.11.020

Wang J., Liu J., Liu T., 2015. The difference in effective light penetration may explain the superiority in photosynthetic efficiency of attached cultivation over the conventional open pond for microalgae. Biotechnology for Biofuels, 8, 49. doi: 10.1186/s13068-015-0240-0

Wang, H., Gao, L., Chen, L., Guo, F. & Liu, T., 2013. Integration process of biodiesel production from filamentous oleaginous microalgae *Tribonema minus*. Bioresource Technology, 142, 39–44. doi: 10.1016/j.biortech.2013.05.058

Whitton R., Le Mevel A., Pidou M., Ometto F., Villa R., Jefferson B., 2016. Influence of microalgal N and P composition on wastewater nutrient remediation. Water Research, 91, 371–378. doi: 10.1016/j.watres.2015.12.054

Xiaogang Hu, Jalalah M., Jingyuan Wu, Zheng Y., Li X., Salama El-Sayed, 2020. Microalgal growth coupled with wastewater treatment in open and closed systems for advanced biofuel generation. Biomass Conversion and Biorefinery. doi: 10.1007/s13399-020-01061-w

Yao S., Lyu S., An Y., Lu J., Gjermansen C., Schramm A., 2019. Microalgae–bacteria symbiosis in microalgal growth and biofuel production: a review. Journal of Applied Microbiology, 126, 359–368. doi: 10.1111/jam.14095

Zerrouki D., Henni A., 2019. Outdoor microalgae cultivation for wastewater treatment. In: Application of Microalgae in Wastewater Treatment, Gupta K. S., Bux F. (Eds.), Springer Nature, 81–98. doi: 10.1007/978-3-030-13913-1_5

Zhan J., Rong J., Wang Q., 2017. Mixotrophic cultivation, a preferable microalgae cultivation mode for biomass/bioenergy production, and bioremediation, advances and prospect. International Journal of Hydrogen Energy, 42, 8505–8517. doi: 10.1016/j.ijhydene.2016.12.021

Zhang B., Li W., Guo Y., Zhang Z., Shi W., Cui F., Lens N. L. P., Tay H. J., 2020. Microalgal-bacterial consortia: from interspecies interactions to biotechnological applications. Renewable and Sustainable Energy Reviews, 118, 109563. doi: 10.1016/j.rser.2019.109563

Zhou D., Zhang C., Fu L., Xu L., Cui X., Li Q., Crittenden C. J., 2017. Responses of the microalga Chlorophyta sp. to bacterial quorum sensing molecules (*N*-acylhomoserine lactones): aromatic protein-induced self-aggregation. Environmental Science and Technology, 51(6), 3490–3498. doi: 10.1021/acs.est.7b00355

Zhuang L.-L., Li M., Ngo H. H., 2020. Non-suspended microalgae cultivation for wastewater refinery and biomass production. Bioresource Technology, 308, 123320. doi: 10.1016/j.biortech.2020.123320.

Zhuang L.-L., Yu D., Zhang J., Liu F.-F., Wu Yin-Hu, Zhang T.-Y., Dao G.-H., Hu H.-Y., 2018. The characteristics and influencing factors of the attached microalgae cultivation: a review. Renewable and Sustainable Energy Reviews, 94, 1110–1119. doi: 10.1016/j.rser.2018.06.006

DOI: 10.1561/9781680837810.ch7

Chapter 7

Disinfection Options for Decentralised Wastewater Treatment Introduction

By Ziye Dai

Removal of pathogenic microorganisms can represent a key challenge for decentralised wastewater treatment solutions (Naughton and Mihelcic, 2017). Although conventional mechanical and biological treatments that aim for chemical contaminant removal can also reduce pathogen content, they are not normally designed to reduce pathogen loads to a safe level for treated wastewater discharge to water bodies and other sensitive areas, use in irrigation, recreation, or reuse as drinking water (Momba *et al.*, 2019; UK Government, 2013; USEPA, 2012). In such cases, dedicated microorganism removal or inactivation treatments are required to adequately disinfect treated wastewater. Disinfection is the killing of infectious agents outside the body through direct exposure to chemical or physical agents. In a wastewater context this is typically via the use of ultraviolet light (UV), chlorine, ozone and other methods. Decentralised wastewater treatment systems are often deliberately deployed where resources for operation, monitoring and maintenance are lacking – which means that disinfection options for decentralised wastewater treatment need to be low in technological complexity, energy demand, cost and maintenance (Naughton and Mihelcic, 2017).

This chapter gives a background to the key pathogens in wastewater and associated discharge standards (Section 7.1). Section 7.2 introduces conventional disinfection methods used in wastewater treatment, while Section 7.3 examines the economic feasibility of two possible approaches to disinfection in decentralised wastewater treatment. Finally, Section 7.4 considers the disinfection approach and performance demonstrated by the INNOQUA project.

7.1 Introduction

Wastewater commonly contains a range of hazards that can lead to problems in the receiving environment. These hazards include chemical substances (such as phosphorus that can lead to harm through eutrophication, and a wide range of organic compound contaminants that can directly impact flora and fauna) and pathogenic microorganisms (including enteric bacteria, viruses and protozoan cysts) that are associated with human diseases (Chahal *et al.*, 2016; USEPA, 2003). Removing or inactivating pathogenic microorganisms helps to lower public health and environmental impacts from discharged wastewater and furthermore helps protect local potable water sources. Table 7.1 shows the infectious microorganisms potentially present in untreated domestic wastewater, and the diseases that can result from contact with humans (e.g., through the environment, water or food pathways). Table 7.2 shows typical concentrations of various organisms that can be expected in raw domestic wastewaters. At a household level, black water (containing faecal material from toilet flushing) has higher pathogen loadings than grey water. The latter may also contain disinfectants that reduce pathogen loads (Bogler *et al.*, 2020; Eregno *et al.*, 2018).

Expressing population levels and removal rates for microorganisms requires different mathematical units than those used for chemical contaminants. Colony-forming units (CFU) or most probable number (MPN) counts are typically used to enumerate populations (or 'concentrations'), (Beckley *et al.*, 2014). To measure CFU, serial dilutions of the sample under test are plated onto an appropriate growth medium such as agar which is then incubated under defined conditions for a defined duration. The number of separate colonies is then counted and (depending on the initial dilution) back-calculated to give the starting population. For MPN measurement, serial dilutions of the sample are added to a liquid medium (e.g., multiple

Table 7.1. Infectious microorganisms potentially present in untreated domestic wastewater. Adapted from USEPA (1999a) and Chahal *et al.* (2016).

Types	Organisms	Diseases Caused
Bacteria	*Campylobacter jejuni*	Gastroenteritis
	E. coli	Gastroenteritis
	Salmonella spp.	Salmonellosis, typhoid, paratyphoid
	Vibrio cholerae	Cholera
Protozoa	*Balantidium coli*	Balantidiasis
	Cryptosporidium parvum	Cryptosporidiosis
	Giardia lamblia	Giardiasis
	Toxoplasma gondii	Toxoplasmosis
Viruses	Adenovirus	Upper respiratory infection and gastroenteritis
	Astrovirus	Gastroenteritis
	Coxsackie virus	Meningitis, pneumonia, fever
	Echovirus	Meningitis, paralysis, encephalitis, fever
	Hepatitis A virus	Infectious hepatitis
	Hepatitis E virus	Infectious hepatitis, miscarriage and death
	Human calicivirus	Epidemic gastroenteritis with severe diarrhoea
	Rotavirus	Gastroenteritis

Table 7.2. Typical concentrations of various organisms expected in raw domestic wastewater.

Total Coliforms $\log_{10}$ cfu/100 mL	Faecal Coliforms $\log_{10}$ cfu/100 mL	*E. coli* $\log_{10}$ cfu/100 mL	Enterococci $\log_{10}$ cfu/100 mL	References
7.50–7.70	7.10–7.30	N/A	5.62–6.36	Kay *et al.* (2008)
N/A	7.10–8.00	7.00–7.49	6.71–7.18	Contreras *et al.* (2017)
7.00–10.00	6.00–9.00	7.00–8.00	5.00–7.00	Momba *et al.* (2019)

tube fermentation/Colilert), which is then incubated under defined conditions for a defined duration. Positive wells/tubes are then counted, and a statistical conversion is applied to derive the estimated starting population. Although derived in different ways, MPN and CFU data can be considered broadly equivalent. Since the number of microorganisms in a sample is normally significant, counts and removal are usually presented in terms of log base 10 (FAO/WHO, 2016). Thus a starting population of 1 million CFU would be presented as 1×10^6 CFU. Likewise a million-fold removal (from 10^9 to 10^3 or 99.9999%) would be presented as a 1×10^6 reduction.

7.1.1 Indicator Bacteria

As shown in Table 7.1, wastewater can contain various types of bacteria, viruses and protozoa. It is not usually practical to analyse all of them to develop a detailed microbiological profile of the wastewater. Standards, regulations and guidance instead tend to use faecal indicator bacteria (FIB) to provide an overall view of the presence of pathogens (Momba *et al.*, 2019). These bacteria are typically present in the gastrointestinal tract of humans and animals, and contaminate the environment in faecal material. Commonly-used FIB are described in the following sections.

Total coliforms and faecal coliforms

Total and faecal coliforms are Gram-negative facultative anaerobic bacteria that are rod-shaped and non-spore-forming. The two categories are distinguished from one another by their response under laboratory conditions, with total coliforms being capable of fermenting lactose to acid and gas within 48 hours at 35°C while faecal coliforms comprise a thermo-tolerant subset, capable of fermenting lactose within 24 hours at 44°C (Harwood *et al.*, 2017). Although there are requirements to measure total coliform populations in various wastewater related standards, regulations and guidance, it should be noted that there is no quantifiable relationship between this group of bacteria and pathogens in wastewater – as these organisms can be found in non-faecally-contaminated water and soils (Von Sperling and De Lemos Chernicharo, 2005). Faecal coliforms are associated with wastes of human and animal origin, but may also be found in non-faecal material (Cabral, 2010). Thus total and faecal coliform measurements should only be taken to infer possible faecal contamination of the sample under test and are only broadly suggestive of the presence of pathogenic organisms (Harwood *et al.*, 2017). These uncertainties have prompted some countries and organisations, such as New Zealand and the European Union, to substitute Enterococci and *E. coli* for total and faecal coliform measurements as the preferred FIBs (Harwood *et al.*, 2017).

Escherichia coli

E. coli is a faecal coliform. The majority of *E. coli* strains are harmless, but some are pathogenic, for example, *E. coli* O157 H7 can cause intestinal and urinary tract infections (Scheutz and Strockbine, 2015). Compared to total and faecal coliforms which can originate from non-faecally-contaminated sources, *E. coli* is a more specific indicator as it originates exclusively from humans and other warm-blooded animals (DWFA, 1996). It has been estimated that more than 90% of the coliforms in human and animal faeces are *E. coli* ((Hurst *et al.*, 2007). Thus, *E. coli* is a good indicator of faecal pollution and, furthermore, the detection of *E. coli* is relatively easy. For these reasons, key organisations such as the United States Environmental Protection Agency (USEPA), European Union (EU) and World Health Organization (WHO) use *E. coli* for regulatory and assessment purposes (EEC, 2006; USEPA, 2012; WHO, 2006a).

Although a robust indicator, *E. coli* can have drawbacks under certain conditions. Good FIBs need to have a consistent relationship with pathogens at any time, after any treatment – but in tropical environments *E. coli* can tolerate or even reproduce substantially at high temperatures while pathogens of concern might not (Desmarais *et al.*, 2002; Harwood *et al.*, 2017; Solo-Gabriele *et al.*, 2000; Winfield and Groisman, 2003). *E. coli* are also more sensitive than other pathogenic bacteria, viruses and protozoa to inactivation via disinfection processes (Harwood *et al.*, 2017; Sinclair *et al.*, 2009). Hence, in some cases, the actual pathogen level post-disinfection could be higher than the value as indicated by *E. coli*. Therefore, while *E. coli* is generally robust as an FIB, data should always be treated with caution and confirmatory checks made with other organisms where required.

Intestinal enterococci

Intestinal enterococci are those members of the genus *Enterococcus* found in the human intestine, most commonly *Enterococcus faecalis* (Harwood *et al.*, 2017; Schleifer and Kilpper-Bälz, 1984). Intestinal enterococci are not as sensitive as *E. coli* to inactivation via disinfection processes such as chlorination, and even though they may also originate from sediments and soils, they are considered as a reliable FIB alongside *E. coli* for indicating the level of faecal pollution and the efficacy of disinfection (Harwood *et al.*, 2017).

FIBs used in different countries and organisations

Different organisations and countries use different FIBs in their standards for regulatory control or guidance relating to the microbial quality of treated wastewater and (in some cases) the receiving environment. Some of these are shown in Table 7.3.

7.1.2 Standards, Regulations and Guidance for Recreational Waters and Irrigation

Regulations and guidance regarding the reuse of treated wastewater for discharge to recreational waters or use in irrigation have been set by various organisations,

Table 7.3. FIB used in different standards, regulations and guidance of the microbial quality of the treated wastewater and the surface wastewater. Adapted from Harwood *et al.* (2017).

Areas	Regulatory Uses	FIB Used
El Salvador	Wastewater discharged to the environment	Total coliforms; Faecal coliforms
Kenya	Effluent discharge to the environment; wastewater use in agriculture; recreational waters	Total coliforms; Faecal coliforms; *E. coli*
Turkey	Treated wastewater	Total coliforms; Faecal coliforms
Sri Lanka	Treated wastewater	Faecal coliforms
EU	Bathing water	Enterococci; *E. coli*
USA	Recreational water	Enterococci; *E. coli*

including the WHO, USEPA and the EU. In the context of decentralised wastewater treatment systems wastewater is commonly used for irrigation, and may also be discharged into surface water bodies that are used for recreational purposes such as swimming (Edokpayi *et al.*, 2017).

Irrigation

Untreated, partially treated or adequately treated wastewater contains nutrients such as nitrogen and phosphorus (DEFRA, 2012; WHO, 2006b). Reusing the wastewater for agricultural purposes can reduce the consumption of and thereby expenditure on fertilisers, and is a common practice in rural communities (Adegoke *et al.*, 2018; WHO, 2006b). The EU recently developed regulations to allow the reuse of treated wastewater in a number of irrigation applications (Table 7.4). These were based (*inter alia*) on a review of pre-existing national standards in six EU member states, which identified that FIB limits ranged between 5 and 100,000 cfu/100 mL for *E. coli*, depending on the intended end use (European Commission, 2016). This has been reflected in the regulations, with tighter limits where irrigated crops are expected to be consumed raw and more relaxed limits where irrigated crops are not intended for human consumption (Table 7.4). A similar pattern can also be observed in a sub-set of INNOQUA partner countries (Table 7.5).

As can be seen from Table 7.4, regulations also account for the impacts of irrigation methods. For example, the permitted agricultural uses of EU Class B and Class C reclaimed water are identical, but Class C has a higher *E. coli* limit and manages risk by requiring that a different irrigation method is used when compared with Class B. Thus, it is sometimes possible to relax the requirements for disinfection

Table 7.4. Classes and quality of reclaimed water and permitted agricultural use and irrigation method. Adapted from Regulation (EU) 2020/741 of the European Parliament and of the Council of 25 May 2020 on minimum requirements for water reuse.

Classes	*E.* coli cfu or mpn/100 mL	Permitted Agricultural Uses	Permitted Irrigation Methods
A	≤10	All food crops consumed raw where the edible part is in direct contact with reclaimed water and the root crops are consumed raw.	All irrigation methods
B	≤100	Food crops consumed raw where the edible part is produced above ground and is not in direct contact with reclaimed water, processed food crops and non-food crops including crops used to feed milk- or meat-producing animals.	All irrigation methods
C	≤1,000	Food crops consumed raw where the edible part is produced above ground and is not in direct contact with reclaimed water, processed food crops and non-food crops including crops used to feed milk- or meat-producing animals.	Drip irrigation (*) or other irrigation method that avoids direct contact with the edible part of the crop
D	≤10,000	Industrial, energy and seeded crops	All irrigation methods

*Drip irrigation (also called trickle irrigation) is a micro-irrigation system capable of delivering water drops or tiny streams to the plants and involves dripping water onto the soil or directly under its surface at very low rates (2–20 litres/hour) from a system of small-diameter plastic pipes fitted with outlets called emitters or drippers.

where irrigation measures are more controlled and contact with edible parts of the crop is limited. The effective pathogen reductions delivered by a number of different irrigation techniques is shown in Table 7.6.

The standards, regulation, and guidelines for wastewater reuse in agriculture aim to not only minimise consumers' exposure to pathogens, but also impact on farmers – who can be directly exposed to the wastewater (Adegoke *et al.*, 2018; WHO, 2006b). Some classes of reclaimed wastewater have a higher allowable content of *E. coli* than, for example, the minimum standards in the EU Bathing Water Directive (BWD) or the USEPA Recreational Water Quality Criteria (RWQC) which regulates recreational water to which humans are directly exposed. This means that reclaimed wastewater can pose health risks to farmers if they are in direct

Table 7.5. Classes and quality of reclaimed water and permitted agricultural uses and irrigation methods for a sub-set of countries involved in the INNOQUA project. Adapted from Harwood *et al.* (2017), MINAM (2017), Turkish Ministry of Environment and Forestry (2010) and Government of Tanzania (2007).

	Parameters, cfu or mpn/100 mL	Permitted Irrigation Methods
Ecuador	Faecal coliforms: 1,000	Restricted irrigation
	Faecal coliforms: 0	Unrestricted irrigation
Peru	Thermotolerent coliforms: 2,000 *E. coli*: 1,000	Restricted irrigation
	Thermotolerent coliforms: 1,000	Unrestricted irrigation
Turkey	Faecal coliforms: 0	Irrigation for rooted herbs (carrots, onions) eaten raw
	Faecal coliforms: 0 for sprinkling irrigation; Faecal coliforms: <200 for drip irrigation	Irrigation for large-leaved plants growing on or near the surface and plants not in contact with the ground
	Faecal coliforms: <200	Irrigations for plants that are peeled before eating and treated before eating, superficial plants processed before eating, plants, cultivation, pastures and pastures not for human consumption
Tanzania	Total coliforms: $\leq 10{,}000$; *E. coli*: ≤ 100; Faecal coliform: $\leq 1{,}000$; Helminth eggs: ≤ 5	Restricted irrigation
	Total coliforms: ≤ 300; *E. coli*: ≤ 10; Faecal coliform: ≤ 200; Helminth eggs: ≤ 1	Unrestricted irrigation

Table 7.6. Pathogen reductions possible through use of different irrigation methods. Adapted from European Commission (2016).

Irrigation Methods	Pathogen Reduction, $\log_{10}$
Drip irrigation	2
Surface drip irrigation	4–6
Spray drift control irrigation	1
Spray buffer zones irrigation	1

Table 7.7. Bathing Water Directive (EEC, 2006) and Recreational Water Quality Criteria (USEPA, 2012).

Regulations	Water Types	Grades	Intestinal Enterococci (cfu/100 mL)[1]	*E. coli* (cfu/100 mL)[1]
BWD	Inland	Excellent[2]	200	500
		Good[2]	400	1,000
		Sufficient [3]	330	900
		Poor[3]	>330	>500
	Coastal	Excellent[2]	100	250
		Good[2]	200	500
		Sufficient [3]	185	500
		Poor[3]	>185	>500
RWQC	Marine and fresh	Illness rate 32/1,000[3]	110	320
		Illness rate 36/1,000[3]	130	410

[1] BWD uses the arithmetic mean, and RWQC uses the geometric mean. CFU stands for colony-forming unit.
[2] 95th percentile.
[3] 90th percentile.

contact with it. Furthermore, some irrigation methods, such as sprinkler irrigation, will produce aerosols that may allow pathogens to drift to adjacent areas (Adegoke *et al.*, 2018). Thus, farmers need to ensure good practice is employed when using reclaimed wastewater, and consistently treating reclaimed wastewater to bathing water standards (or better) might be desirable where possible.

Recreational waters

In the EU, the quality of inland and coastal bathing waters is prescribed by the BWD (EEC, 2006). Similarly, the USEPA uses RWQC to categorise the quality of recreational waters, including rivers, lake and coastal waters (USEPA, 2012). Both of these regimes use *E. coli* and intestinal Enterococci as the FIBs (Table 7.7). While the EU approach provides for four categories of water quality – which allows for individual Member States to develop national targets to improve water quality in specific locations – the US approach provides two options based on the predicted level of gastro-intestinal illness that might be deemed acceptable in each State. Table 7.8 shows the recreational water standards of some of the INNOQUA partner countries. These tend to use a simpler risk tier structure when compared with the EU approach.

Table 7.8. Recreational water standards of some INNOQUA partner countries (CPCB, 2007; Mateus *et al.*, 2019; MINAM, 2017).

	Water Types	Parameters, CFU or MPN/100 mL
India	Outdoor bathing water	Total coliforms: ≤500
Peru	Recreational water	**Primary contact*:** *E. coli*: 0 Intestinal enterococci: ≤200 **Secondary contact**:** *E. coli*: 0 Intestinal enterococci: N/A
Ecuador	Recreational water	Total coliforms: ≤2,000 Faecal coliforms: ≤500

*Primary contact is from activities that could result in the ingestion of water or immersion, such as swimming, kayaking, water skiing.
**Secondary contact is from activities where the majority of participants would have very little direct contact with the water and where ingestion of water is unlikely, such as wading, canoeing, motor boating, fishing.

7.2 Disinfection

Typical centralised wastewater treatment comprises preliminary mechanical processes for removal of large (often inorganic) solids, followed by primary settlement of solids and biological treatment of the clarified effluent. Primary settlement aims to remove larger settleable organic solids while biological treatment relies on microorganisms and the use of electro-mechanical approaches to remove soluble and un-settleable contaminants via biological predation, assimilation and oxidation (Tchobanoglous *et al.*, 2004). In smaller wastewater treatment systems that rely on more nature-based approaches, a septic tank can often act as the primary treatment system to settle the larger particulates, followed by treatment of the supernatant via percolation systems, constructed wetlands, reed beds and lagoons that serve as secondary (or tertiary) treatment to remove soluble and un-settleable contaminants physically and biologically. In either case, both primary and secondary treatment can deliver pathogen removal through the settlement of pathogen-containing particulates, as well as via filtration and biological predation. Table 7.9 shows the typical pathogen removal efficiencies of various wastewater treatment technologies.

For example, decentralised treatment that only uses a septic tank can deliver a 2 $\log_{10}$ removal of indicator bacteria. With the addition of secondary treatment, the removal can be increased to 4–6 $\log_{10}$. Using *E. coli* as an example, if its concentration in raw wastewater is 7 $\log_{10}$, a 5–6 $\log_{10}$ removal would be required for the treated effluent to comply with EU recreational/bathing water standards (Table 7.7) and wastewater reuse in agriculture (Table 7.4). Decentralised

Table 7.9. Typical pathogen removal efficacy of different treatment technologies. Adapted from (Momba *et al.*, 2019).

Treatments	Treatment Methods	Typical log_{10} Removal
Centralised	Sedimentation	<1
	Activated sludge	2–3
	Trickling filter	<1
Decentralised	Septic tank	2
	Filter	~2
	Construct wetlands (reed bed)	2–4
	Lagoon	3–4
	Package plant	2–5[1]

[1]A package plant typically comprises two compartments: (i) a compartment that serves as a primary settlement/septic tank and (ii) a compartment that serves as filter or activated sludge secondary treatment stage.

wastewater treatment solutions comprising only primary and secondary treatment will struggle to meet these requirements. Therefore, additional disinfection measures for pathogen removal are necessary to ensure decentralised treatment can produce an appropriate quality final effluent.

Disinfection uses chemical and/or physical measures to inactivate or destroy pathogens. This is achieved through the following five principal mechanisms (Tchobanoglous *et al.*, 2004):

1. Damage to the cell wall;
2. Altering cell permeability;
3. Disrupting the colloidal nature of the protoplasm;
4. Altering the DNA and RNA of the organism; and
5. Inhibiting enzyme activity.

Both potable water and wastewater treatment facilities widely apply chlorination, ozonation or UV for disinfection (EPA, 2011; USEPA, 2003) – individually or in combination. In order to achieve the required pathogen disinfection standards, contaminants in the wastewater such as suspended solids and organic matter must first be removed. Therefore, disinfection treatment is usually located at the end of the overall treatment process (EPA, 2011).

A wide range of other approaches to disinfection have been trialled in decentralised wastewater treatment applications. Leverenz *et al.* (2006) considered techniques as diverse as peracetic acid dosing and biological filtration before focussing on ozonation, chlorination and UV treatment. Likewise, Fedler *et al.* (2012) compared technologies as diverse as membrane filtration, bromination, potassium permanganate dosing and electrochemical disinfection as alternatives to chlorination – concluding that ozonation and UV treatment were most suited of the prevalent 'market ready' solutions. Membrane filtration is particularly appealing, as it relies on simple physical processes to remove particles of specific sizes from the wastewater flow. Generally, virus particles range in size between 0.01 μm and 0.1 μm, and bacteria between 0.1 μm and 10 μm. Thus, nanofiltration with a pore size of approx. 0.001 μm can completely remove pathogens from water. While theoretically not as effective, microfiltration (pore size ranging between 0.1 μm and 10 μm) and ultrafiltration (pore size ranging between 0.01 μm and 0.1 μm), can still deliver 4–6 $\log_{10}$ removal of bacteria and protozoa (Parsons and Jefferson, 2006). However, the implementation of membrane technology is limited by membrane fouling, and is often associated with high energy demand as well as intensive labour and maintenance requirements (Bodzek *et al.*, 2019). For these reasons it is not considered suitable for decentralised wastewater treatment and is not considered further within this chapter.

7.2.1 Chlorination

Chlorination is the generic term which refers to disinfection processes using chlorine (Cl_2) and chlorine derivatives, such as sodium hypochlorite (NaClO), calcium hypochlorite ($Ca(ClO)_2$) and chlorine dioxide (ClO_2) (EPA, 2011). Elemental chlorine, the hypochlorite anion (ClO^-) and ClO_2 are strong oxidants. They can oxidise the cellular material of the target organism, modify cell wall permeability, precipitate proteins and alter and inactivate enzymes to achieve pathogen inactivation (Tchobanoglous *et al.*, 2004).

Historically, chlorination was mainly conducted via dosing Cl_2 gas into water. When Cl_2 reacts with water, it forms elemental chlorine and hypochlorite anions (Tchobanoglous *et al.*, 2004). However, Cl_2 gas is toxic, and hence its storage and dosage present challenges in terms of handling and associated health and safety risk (EPA, 2011; USEPA, 1999a). Hypochlorite salts (e.g., NaClO or $Ca(ClO)_2$), are considered preferable alternatives because they are safer and easier to handle. Figure 7.1 shows the schematic of a typical NaClO disinfection system (EPA, 2011).

NaClO is generally supplied as an aqueous solution. NaClO can also be produced *in situ* via electrolysis of diluted high purity sodium chloride solution – this method can be applied where it is desirable to produce the NaClO locally (EPA, 2011). For smaller applications, commercially available calcium hypochlorite tablets can be a good alternative as they require no onsite production and calcium hypochlorite is relatively safe to store and handle. As shown in Fig. 7.2, tablets are stacked within a dedicated tablet feeder which allows contact with the water flow

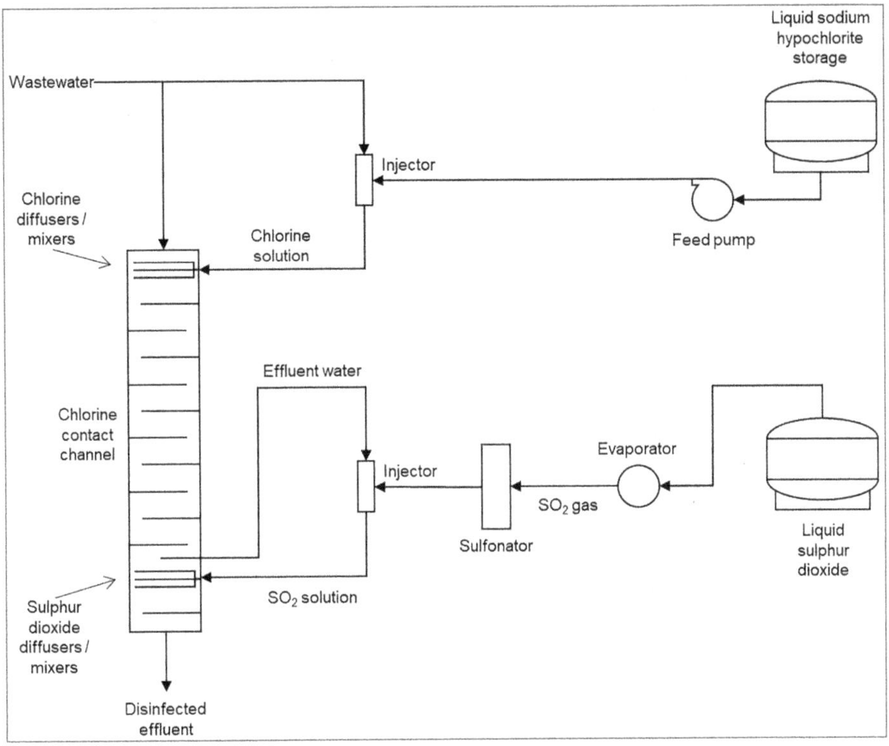

Figure 7.1. Schematic of a typical sodium hypochlorite disinfection system (re-drawn from Tchobanoglous *et al.* (2004)).

at the base of the stack. Contact and tablet release are flow-dependent, allowing the dose to be controlled on a volumetric basis, assuming consistent dissolution of the tablets. Water passes from the tablet stack into a contact tank which provides an hydraulic buffer to improve treatment efficacy (Norweco *et al.*, 2006a).

The ClO^- ion and its associated acid – hypochlorous acid (HClO) – are both effective disinfectants. At low pH, HClO is the dominating species while ClO^- dominates at high pH, as shown in Fig. 7.3. HClO is a much stronger oxidant and hence a much stronger disinfectant than ClO^-. Therefore, hypochlorite disinfection is more effective in neutral to acidic conditions than in alkaline conditions (EPA, 2011).

In designing chlorine-based disinfection systems it is important to consider that chlorine-related residues in the treated wastewater may have toxic effects on aquatic organisms in receiving waters. Therefore, reductants, such as sulphur dioxide, sodium sulphite and sodium bisulphite may be dosed into treated wastewater to react with the residual oxidative disinfectant to form the non-harmful chloride ion (EPA, 2011; Tchobanoglous *et al.*, 2004; USEPA, 2003). A typical sulphur dioxide post-dosing flow is shown in Fig. 7.1.

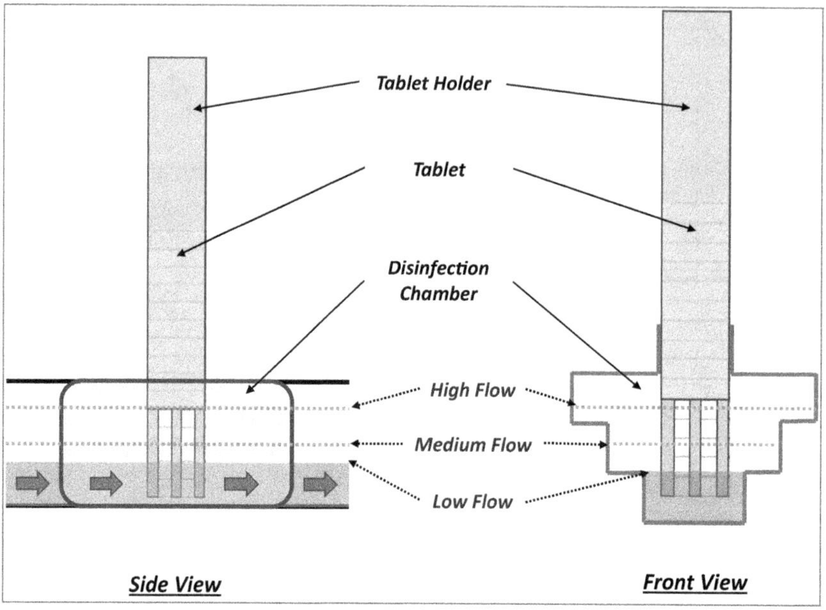

Figure 7.2. Calcium hypochlorite tablet disinfection (redrawn from Norweco *et al.* (2006a)).

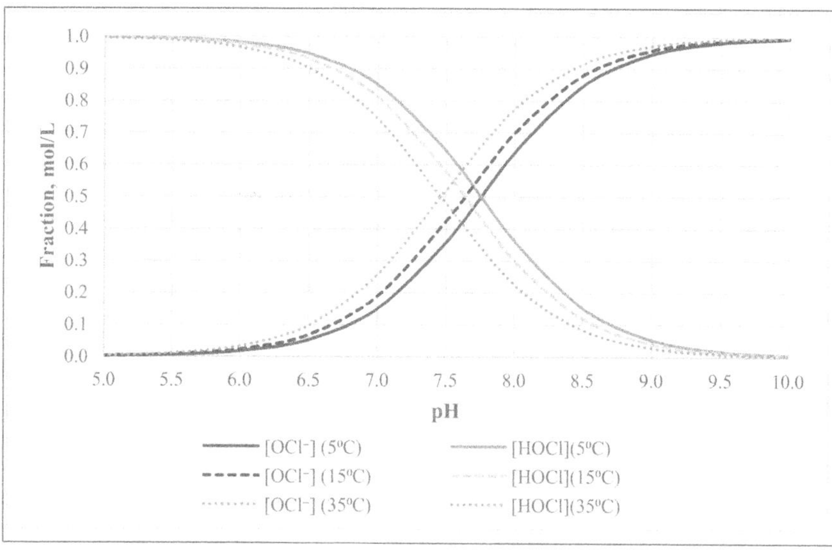

Figure 7.3. HOCl-OCl⁻ equilibrium at different pH and temperature (redrawn from EPA, 2011).

7.2.2 Ozonation

When the oxygen molecule, O_2, is dissociated by an energy source, oxygen atoms are formed. If the oxygen atom collides with an oxygen molecule, then ozone (O_3) is generated. In nature, lightning can produce ozone in the atmosphere.

When ozone is dosed into water a number of reactions occur that produce free radicals with strong oxidation capabilities. Equations 7.1–7.4 show the reactions that occur to produce these free radicals (hydroperoxyl: HO_2, and hydroxyl: HO) (Tchobanoglous *et al.*, 2004):

$$O_3 + H_2O \rightarrow HO_3^+ + OH^- \tag{7.1}$$

$$HO_3^+ + OH^- \rightarrow 2HO_2 \tag{7.2}$$

$$O_3 + HO_2 \rightarrow HO + 2O_2 \tag{7.3}$$

$$HO + HO_2 \rightarrow H_2O + O_2 \tag{7.4}$$

HO_2 and HO are free radicals with significant oxidising abilities and are considered the active disinfectants in the ozonation process. They can directly oxidise or destroy the cell wall causing leakage of cellular constituents from the cell, damage to purines and pyrimidines which are the constituents of the nucleic acid, and depolymerisation through breaking of carbon–nitrogen bonds (Tchobanoglous *et al.*, 2004).

Ozone is an unstable gas whose decomposition will happen shortly after generation – it must therefore be generated locally to where it will be used. It is commonly produced by inducing a high voltage (6–20 kV) alternating current across a dielectric discharge gap in an oxygen-containing gas. The oxygen-containing gas can be either air or high-purity oxygen (EPA, 2011; Tchobanoglous *et al.*, 2004). A schematic of ozone generation and disinfection is shown in Fig. 7.4.

Since ozone is also an irritating and toxic gas, residual ozone in the off-gas must be destroyed to prevent its discharge to atmosphere. The destruction product is oxygen, and hence, in some cases, the treated off-gas can be recycled for use in the ozone generator (Tchobanoglous *et al.*, 2004).

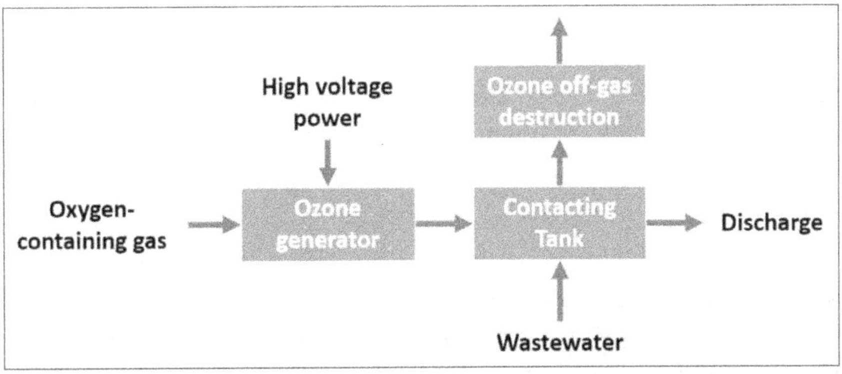

Figure 7.4. Schematic of ozone disinfection (adapted from Tchobanoglous *et al.*, 2004).

7.2.3 Ultraviolet

UV is a form of electromagnetic radiation with wavelengths ranging from 10 nm to 400 nm (EPA, 2011; Tchobanoglous *et al.*, 2004). Within this spectrum, the optimum wavelength to effectively inactivate microorganisms is generally considered to be in the range of 220–320, or more precisely between 250 nm and 270 nm; mainly covered by the UV-C spectrum (200–280 nm) (EPA, 2011; Tchobanoglous *et al.*, 2004; USEPA, 1999b). UV radiation can damage DNA and RNA, which has the highest UV radiation absorbance at a wavelength of approximately 260 nm. This damage prevents microorganisms from growing or reproducing (Tchobanoglous *et al.*, 2004).

Low-pressure UV lamps are the most widely used disinfection technology, producing monochromatic light at a wavelength of 253.7 nm via electrical discharge through mercury vapour (USEPA, 1999b). Depending on the internal mercury vapour pressure and the discharge current, low-pressure UV lamps can be characterised as low intensity (low output) or high intensity (high output). At higher intensities, fewer lamps are needed to supply the same UV dose (Tchobanoglous *et al.*, 2004).

Unlike ozonation and chlorination, UV disinfection is a physical rather than a chemical treatment method. It requires no disinfectant preparation, storage or handling of hazardous chemicals (USEPA, 1999b). As shown in Fig. 7.5, the disinfection process takes place as wastewater passes over the surface of a quartz sleeve, within which the UV lamp is housed. UV radiation passes through the sleeve and into the wastewater, where disinfection takes place. Contact time can often be less than a minute, which is less than required for ozonation or chlorination (USEPA, 1999b). Furthermore, UV disinfection leaves no toxic residues in the treated wastewater and requires no further disinfectant destruction steps. Due to its simplicity and safety, UV disinfection is often considered a user-friendly system (USEPA, 1999b).

However, UV disinfection can be significantly inhibited by the presence of suspended solids and organic carbon in the wastewater as these will 'block' UV radiation and prevent it from reaching the microorganisms. Similarly, the quartz sleeve – which isolates the UV lamp and has direct contact with the water – needs regular cleaning to ensure high levels of UV light transmittance.

7.2.4 Technology Comparison

When designing disinfection systems, a key metric is the 'Ct' value – i.e., a measurement of the dose. For ozone and chlorine this can be calculated by multiplying the concentration of the disinfectant by the contact time with the water/wastewater to

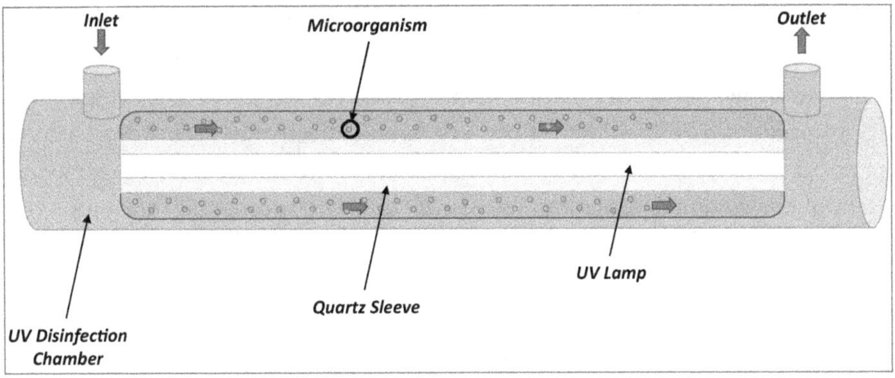

Figure 7.5. Schematic of UV disinfection (redrawn from Alfaa UV, 2018).

Table 7.10. Required Ct values for inactivation of different microorganisms by Cl_2, O_3 and UV radiation (EPA, 2011) and (Tchobanoglous *et al.*, 2004).

Microorganisms	Log_{10} Removal	Ct Values of Cl_2, mg/L Per Minute	Ct Values of O_3, mg/L Per Minute	Ct Values of UV, mJ/cm^2
Cryptosporidium (protozoa)	0.5	N/A	4.9	1.6
	3	N/A	30.0	12.0
Giardia (protozoa)	0.5	17.0	0.23	1.50
	2	104.0	1.43	5.20
Viruses	2	2.5–3.5	0.5	20–30
	4	6.0–7.0	1.0	70–90
Bacteria	2	0.4–0.8	3.0–4.0	30–60

be treated (generally given in units of mg/L per minute). For UV systems this refers to the UV dosage (generally given in mW/cm^2 multiplied by the residence time (seconds) of the water/wastewater in the lamp module (or tank where the UV lamps are placed)) and is generally denoted in units of mJ/cm^2 (EPA, 2011). For each of chlorine, ozone and UV systems there are guideline Ct values for achieving certain levels of inactivation for specified pathogens. In an example of chlorination disinfection for bacteria, if a 2 $\log_{10}$ removal requires a Ct of 15 mg/L per minute, the process can be operated at 0.5 mg/L chlorine for 30 minutes of contact time, or 1 mg/L chlorine for 15 minutes. Table 7.10 lists examples of the required Ct for inactivation of protozoa, virus and bacteria by Cl_2, O_3 and UV radiation.

As might be expected, organisms which are more resistant to disinfectants have higher Ct values. As shown in Table 7.10, chlorination is effective at removing bacteria and viruses but will not readily inactivate protozoa such as *Cryptosporidium*

(EPA, 2011; USEPA, 2003). Ozone is a stronger bactericide and virucide than chlorine and its derivatives (EPA, 2011; USEPA, 1999c) meaning that the Ct value of ozonation is lower than chlorination in most cases. UV light is effective at inactivating protozoa and viruses. Furthermore, UV light is effective at inactivating bacteria, but not at inactivating bacterial spores. For example, the Ct values for 2 $\log_{10}$ inactivation of *E. coli* (non-sporulating) and *Bacillus subtilus* (sporulating) are reported as 4.8 and 39, respectively. Moreover, microorganisms with certain enzymatic systems are able to repair damage caused by UV (EPA, 2011; USEPA, 1999b).

In general, ozone can be considered as being an effective disinfectant for the majority of waterborne microorganisms. Nevertheless, it comes with challenges such as technology costs, the requirement for high voltage power supplies, and the need for corrosion-resistant equipment (USEPA, 1999c). For these reasons ozonation is not as widely applied as chlorination and UV in decentralised systems (EPA, 2011; USEPA, 1999b; Zyara *et al.*, 2016). While both UV and chlorination systems can be challenged by certain pathogens, they can be used effectively in combination where required as each can target pathogens which may be resistant to the other. In such scenarios chlorination is typically applied as primary disinfection, with UV serving as the secondary step (Zyara *et al.*, 2016).

Since some pathogens are more resistant to disinfection processes than others it is critical to design systems in a targeted way accounting for both the FIBs which are to be measured for compliance purposes and any pathogens of concern that might be more resistant to disinfection than the FIBs.

7.3 Economic Study: Calcium Hypochlorite Disinfection vs. UV Disinfection

As discussed in Section 7.1.2, delivering water quality that complies with the higher end of irrigation standards and the EU bathing/recreational water standards requires approximately 5–6 $\log_{10}$ removal of pathogens from wastewater. While primary and secondary treatment systems can deliver up to 4–6 $\log_{10}$ removal, disinfection is still required to deliver the final 2 $\log_{10}$ removal. Thus, in the context of decentralised applications, a suitable disinfection measure should:

- Be effective at delivering a 2 $\log_{10}$ pathogen removal,
- Comprise simple to moderate technological complexity,
- Require minimal operation and maintenance, and
- Be low cost.

Among the conventional disinfection measures introduced in Section 7.2, hypochlorite and UV disinfection can be considered options that can meet

Table 7.11. Applicability of hypochlorite tablet disinfection and UV disinfection. Adapted from USEPA (1999a).

Considerations	Hypochlorite Tablet	UV
Bactericidal	Good	Good
Virucidal	Moderate	Moderate
Cysticidal	Poor	May be effective against oocyst
Fish toxicity	Potentially toxic	Nontoxic
Hazardous by-product	Yes	No
pH dependent	Yes	No
Corrosive	Yes	No
Contact time (related to footprint)	Long	Short
Electricity requirement	No	Yes

these criteria. Both are good in terms of bacteria removal capability and moderate in terms of virus removal capability (Table 7.11). UV disinfection has advantages in that it produces no hazardous by-products and will thus not negatively impact receiving waters. However, traditional UV disinfection systems require a steady electrical supply, which can be challenging in some locations.

Cost is also a determining factor for the implementation and selection of decentralised disinfection systems. The following sections provide a basic economic comparison between UV and hypochlorite (tablets) for disinfection.

7.3.1 Economic Assessment of Technology Options (Case-study Example)

In this case-study a rural property is considered with the following characteristics:

- has a 3 m^3/day (125 L/h) flow of wastewater,
- has no restrictions in terms of footprint and electricity access, and
- requires a minimum 2 $\log_{10}$ removal of bacteria.

The appraisal of the two disinfection options focuses on capital expenditure (CAPEX), operational expenditure (OPEX) and total expenditure (TOTEX). The following caveats should be noted:

- Whilst CAPEX estimates have been sourced directly from suppliers or from supplier websites, accurate costing would be highly site-specific. As a result, these CAPEX estimates should only be considered as indicative for the purposes of this comparison.

- The CAPEX does not include the costs of installation which are site-specific and highly variable.
- The OPEX considers two key issues: (i) ongoing consumable costs and (ii) maintenance. The former includes electricity and/or chemicals consumed for operations, while the latter involves cleaning and changes of components, etc.
- This study considers the electricity consumption within the disinfection treatment, but excludes any electricity consumed as a result of pumping to or between different stages of treatment. It can be assumed, in this case, that all wastewater transfers will be by gravity.
- An annual OPEX has been estimated over a 20-year asset life. No compounding has been applied to these annual costs to account for inflation or other cost increases (i.e., net present value is not considered).
- The TOTEX is the sum of the CAPEX and the OPEX for a 20-year asset life.

Since the costs of equipment and consumables can be vendor or country-specific, it is important to understand how such uncertainty impacts the TOTEX estimation. Furthermore, determining influential factors can help to identify areas where cost improvements can be made. A sensitivity analysis examines ±20% changes on CAPEX, annual maintenance OPEX and annual consumable OPEX.

7.3.2 Estimation of CAPEX and OPEX

Tablet chlorination disinfection

The process uses calcium hypochlorite tablets to treat secondary treated effluent with relatively low suspended solids, organic carbon and nutrient concentrations. The process involves two major steps:

1. The treated effluent flows through the tablet feeder where calcium hypochlorite is dissolved into the liquid phase.
2. The wastewater containing the calcium hypochlorite will flow to a contact tank where the hypochlorite will inactivate the microorganisms.

The CAPEX includes the purchase of a tablet feeder, a contact tank and other components such as pipework, joints and pumps. The OPEX for consumables considers purchase of calcium hypochlorite, while maintenance is expected to comprise routine cleaning and repairs.

The USEPA suggests that a residual chlorine concentration of 0.5 mg/L after 15 minutes contact is likely to result in satisfactory disinfection of secondary wastewater effluent (Norweco *et al.*, 2006a). For secondary (activated sludge or trickling filter) treated effluent, Shammas and Wang (2015) reported that an initial chlorine concentration of 3–9 mg/L is required to produce a residual chlorine concentration of 0.5 mg/L after 15 minutes. Leverenz *et al.* (2006) reported higher chlorine doses (ranging from 10 mg/L to 25 mg/L) while USEPA (2002) cites starting doses as high as 65 mg/L in wastewater at pH=8. This study assumes a chlorine dosage of 17.5 mg/L.

We have considered two similar products, Bio-Sanitizer® produced by Norweco and Accu-tab® produced by Axiall Corporation, to estimate the cost of hypochlorite. The specifications are shown in Table 7.12. To meet the 17.5 mg/L chlorine dosage, daily tablet consumption is approx. 1/2 tablet.

Table 7.12. Comparison of the market available calcium hypochlorite tablets, Bio-Sanitizer and Accu-Tab.

	Bio-Sanitizer®	Accu-Tab®
Supply package, kg[1]	20.4	20.4
Weight, g	140	145[2]
Chlorine content%[1]	73%	70%[3]
Chlorine content per tablet, g	102.2	101.5
Price per tablet[1,4], €	1.00	1.23

1: The information presented in this table is drawn largely from www.usabluewater.com, which cites lower prices than other sites.
2: The number of tablets is estimated from a 25 lb package that has 75 tablets.
3: The purity is reported varying from 65% to 76%.
4: The price for 45 lb packages of Bio-Sanitizer® and Accu-Tab® are USD 161.95 and 187.95, respectively. They are converted to Euro (€) at a ratio of 0.89.

To populate our case study, we selected the LF2000 tablet feeder from Norweco. The lowest cost was ~€ 200 (October 2020). Since the price and usage of contact tank, pipes and joints are site-specific, this study assumed the cost of this is the same as the tablet feeder and the system is gravity fed. Thus, the total CAPEX is ~€ 400.

The annual OPEX for consumable chemicals is 0.51 tablets × €1/tablet × 365 days = € 187.50.

The annual OPEX for maintenance is estimated at € 120.00, based on:

- The labour cost is estimated as € 10/hour, (ILOSTAT, 2021), and the annual labour cost is assumed as € 10/hour × 1 visit × 8 hour/visit = € 80
- The annual material cost, i.e., for parts replacement due to hypochlorite corrosion (Leverenz *et al.*, 2006), is assumed as 50% of the labour cost and is € 40

UV disinfection

The costs of UV disinfection have been based on the VIQUA Model S2Q-PA/2B produced by VIQUA (a division of Trojan Technologies Group). Both technical and cost information used were obtained from the UK supplier KK Water. Costs have been converted from GBP to EUR at a ratio of £:€ = 1.09.

The CAPEX of the VIQUA Model S2Q-PA/2B is quoted as € 307.30.

Annual OPEX for consumables is estimated at €183.10, based on:

- The power consumed by the VIQUA Model S2Q-PA/2B is 22 W. Assuming a 24-hour operation throughout the year, the annual electricity consumption is 192.7 kWh. Based on a domestic electricity price of € 0.146 per kWh, the annual electricity cost is € 28.30.
- The lifespan of the UV lamp is approximately one year (9,000 hours), with a replacement cost of € 58.70.
- It is recommended that the quartz sleeve be replaced every 2–3 years, at a cost of € 32.10. We have assumed a biennial replacement cycle, giving annual costs of € 32.10÷2 = € 16.05.
- We assume an annual one day inspection and maintenance visit at € 80, in line with the estimate for the hypochlorite dosing system.

7.3.3 Result: TOTEX and Sensitivity

TOTEX

The CAPEX, OPEX and TOTEX over a 20 year asset life for the two disinfection options are shown Table 7.13. Both options are OPEX-dominated, and since UV disinfection has around 40% less OPEX and less TOTEX than tablet chlorination disinfection it can be considered the more cost-effective option in this scenario.

Table 7.13. CAPEX, OPEX and TOTEX of tablet chlorination disinfection and UV disinfection.

	Total CAPEX	Annual Maintenance OPEX	Annual Consumable OPEX	Annual OPEX	Total OPEX Over Asset Life	TOTEX
Tablet chlorination	€ 398.70	€ 120.00	€ 187.50	€ 307.50	€ 6,150	€ 6,549
UV disinfection	€ 307.30	€ 154.80	€ 28.30	€ 183.10	€ 3,661	€ 3,969

Sensitivity study

Spider plots for both disinfection options are shown in Fig. 7.6. Because both options are OPEX-dominated (Table 7.13), the CAPEX on the tablet feeder and the UV system have limited influence on the TOTEX. For the tablet chlorination, the most influential factor in the TOTEX is the annual consumable OPEX, which accounts for >60% of the annual OPEX (Table 7.13). For UV disinfection, the annual maintenance OPEX (>80% of total OPEX) is the most influential factor.

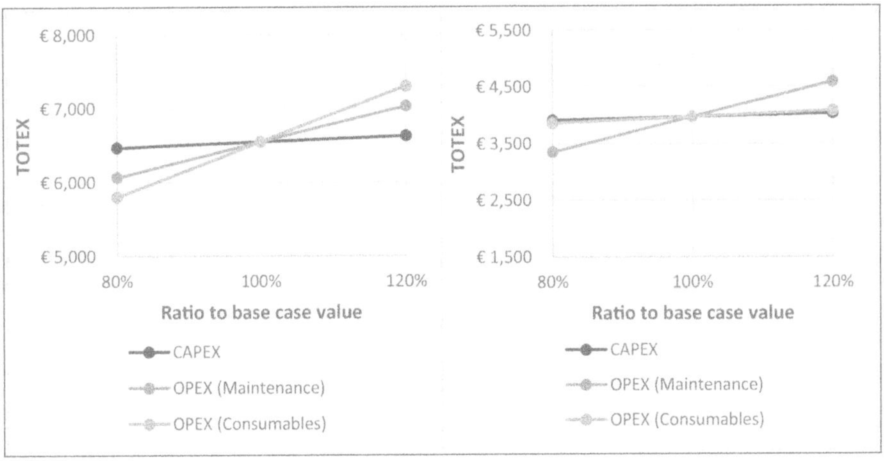

Figure 7.6. Spider plot of the two disinfection options, (A) tablet chlorination disinfection and (B) UV disinfection.

In the case of tablet chlorination disinfection, both the dose and unit price of the calcium hypochlorite tablet significantly impact the consumable OPEX. Given that both the influent pathogen level and the required contact time impact on the dose, if improved pathogen removal was achieved upstream (e.g., via constructed wetlands or lagoon) the influent pathogen level could be lower and the dose could be reduced. We used a chlorine dosage of 17.5 mg/L, but actual requirements could be as low as 5–10 mg/L (Leverenz *et al.*, 2006). If the dosage were reduced to 7.5 mg/L, TOTEX for the chlorination system would be €4,406, which is comparable to the UV system.

In the case of UV disinfection, both the replacement frequencies and costs of the UV lamp and quartz sleeve affect the maintenance OPEX. Dynamic control of the UV system, i.e., switching off the UV system when there is no flow, would be essential in most systems to prevent overheating. This would add complexity but (potentially) increase the lifespan of the UV lamp and therefore makes the replacement less frequent. Alternatively, a bespoke UV dosing system could be fabricated in which the lamp was positioned within a final buffer/contact tank.

These conclusions are in line with those of previous studies. For example, USEPA (2002) suggested similar CAPEX costs for UV and chlorination units of ~USD2,000 for units, installation and housing. Annual OPEX (maintenance and consumables) were assumed to be similar (~USD180), due to the low cost of replacement chlorination tablets, estimated at just USD50 per year compared with our example of USD188 per year. Leverenz *et al.* (2006) suggest CAPEX of up to USD600 and annual consumable costs of USD100 for chlorination systems, compared with CAPEX of up to USD1000 and annual lamp replacement costs of up to USD80 for UV treatment.

7.4 Disinfection and INNOQUA

Chapter 6 has considered the potential for a nature-based decentralised tertiary treatment system to remove nutrients and (to a lesser extent) pathogens from wastewater subjected to primary and secondary treatment in other nature-based systems. Since Bio-Solar Purification relies on high levels of natural insolation, it was not deemed suitable for providing disinfection at all locations where the INNOQUA solutions might eventually be deployed. To add flexibility, UV disinfection was therefore trialled at a number of the demonstration sites (Table 7.14). In most cases UV units were installed to provide disinfection of effluent from the Daphniafilters, but in Italy disinfection direct from the Lumbrifilter was required. In two cases (India and Tanzania) microbiological data are not presented here, since they were recorded for the systems as a whole (the 'global' efficiency) at those sites, rather than for the UV unit in isolation. Results are summarised in Table 7.15.

Although it varied from site to site, overall performance of the UV system was satisfactory. Across the whole fleet of demonstration sites, total ('global') removal of pathogens through the INNOQUA systems was around $5\log_{10}$ – more than adequate to deliver compliance with local norms. It should be noted that although turbidity is normally considered to negatively impact UV transmission and disinfection efficacy, the data in Table 7.15 don't fully reflect this (for example, compare faecal coliform removals and turbidity data at the Turkish and Peruvian sites). Recent studies indicate that although UV performance is impacted by turbidity, aspects such as suspended solids and organic carbon are also important (Fitzhenry *et al.*, 2016), highlighting the importance of establishing an appropriate testing suite before seeking to demonstrate the efficacy of new treatment approaches.

The demonstration site managers noted a number of operational aspects with respect to UV operation, including:

1. The importance of sizing the UV system according to the anticipated wastewater flow from any preliminary treatment unit;

2. The importance of ensuring adequate warm-up time for the UV lamp before re-commencing flows and
3. The importance of weekly cleaning of the outer surface of the quartz sleeve to maintain transmissivity.

Table 7.14. Information on the INNOQUA demonstration sites equipped with UV disinfection.

Locations	Design Capacity, m^3/day	Source of Wastewater	Pre-UV Treatment
India (Bengaluru)	1.50	Toilets and washing rooms	1. Settlement tank 2. Lumbrifilter 3. Daphniafilter
Italy (Vasto)	2.00	Toilets and kitchens	1. Settlement tank 2. Lumbrifilter
Peru (Arequipa)	1.00	Toilets, kitchen, laboratories, veterinary and agronomic faculties	1. Settlement tank 2. Lumbrifilter 3. Daphniafilter
Tanzania (Dar-es-Salaam)	1.50	Toilets and shower	1. Septic tank 2. Lumbrifilter 3. Daphniafilter
Turkey (Sinop)	3.00	Toilets, bathrooms and kitchens	1. Equalisation tank 2. Lumbrifilter 3. Daphniafilter

Table 7.15. Reduction of various FIB at INNOQUA demonstration sites (average $\log_{10}$ reduction (±standard deviation)), together with average turbidity of influent wastewater in Nephelometric Turbidity Units (NTU).

Locations	NTU	*E. coli*	Faecal Coliforms	Enterococci
Italy	9.9	1.59 (±0.79)	1.86 (±0.74)	1.10 (±0.95)
Peru	2.8	0.77 (±0.63)	0.62 (±0.58)	–
Turkey	26.3	–	1.6 (±0.83)	–

7.5 Summary

Disinfection is an increasingly important aspect of wastewater treatment design and operation. It can be required in scenarios where wastewater reuse is required or desirable and where there is a need to protect receiving water bodies (such as bathing waters and potable water sources). Disinfection can be achieved with various techniques, but in decentralised scenarios it is important that the selected technology is robust, easily maintained and suited to specific local conditions including wastewater flows, quality and temperature. Furthermore, the selection and design of the system needs to be targeted at the pathogens to be removed and cognisant of both upstream wastewater quality and regulatory requirements.

Total coliforms, faecal coliforms, *E. coli* and Enterococci are commonly used as FIBs to predict pathogen levels in water, and regulators use one or more FIBs to categorise the water quality. Typically, 4–6 $\log_{10}$ FIB removal from raw wastewater is required for the treated wastewater to be suitable for discharge or irrigation reuse under European and WHO standards.

Chlorination is effective in removing bacteria and viruses but protozoa can be highly resistant to it. UV is effective in inactivating protozoa, bacteria and some (but not all) viruses. Ozone has been shown to be effective for the majority of water-borne pathogens, but its implementation is more technologically complicated and comes with specific operational and maintenance challenges if used in decentralised applications.

Irrespective of the solution selected, appropriate design, installation, monitoring and maintenance are necessary to ensure that any disinfection system operates effectively. When comparing chlorination and UV treatment systems for use in decentralised applications, Leverenz *et al.* (2006) made the following specific observations:

- Tablet chlorination systems are susceptible to episodic failure due to non-uniform erosion of tablets, while UV systems are subject to progressive failure as fouling occurs on the lamp housing. Frequent (brief) inspection and maintenance will be required to address these issues.
- The chlorine dose from dissolution of calcium hypochlorite tablets is difficult to predict and is not related to water quality. Similarly, the rate of dissolution can be variable if flow equalization is not used.
- UV systems are sensitive to water mineral content, periods of no flow while the lamp remains on, flow rate through the unit, and reliability of the pre-treatment system to provide adequate water quality.
- UV systems designed for disinfection of drinking water can be used successfully for wastewater, but should be tested using wastewater to determine capacity.
- All disinfection systems should be used in conjunction with flow equalization to minimise the peak flows expected from small wastewater systems.

Careful attention to system operation and maintenance is also highlighted by USEPA (2002), who suggest that domestic users do not necessarily possess the skills needed to perform proper servicing of chlorination or UV treatment units, and that long-term management through service contracts or local management programmes is an important aspect of successful long-term operation. Nonetheless, both tablet chlorination and UV treatment are cheap and simple to operate when compared with other disinfection approaches, and our assessment suggests that costs for both are OPEX-dominated over a typical 20-year life cycle. While UV proved effective at the INNOQUA demonstration sites, these required a constant electrical supply and frequent maintenance to minimise fouling of the internal quartz sleeve. This is manageable under an experimental setting, but may not be realistic under genuine market conditions. None of the disinfection options is maintenance-free.

References

Adegoke, A.A., Amoah, I.D., Stenström, T.A., Verbyla, M.E., Mihelcic, J.R., 2018. Epidemiological evidence and health risks associated with agricultural reuse of partially treated and untreated wastewater: a review. Front. Public Heal. 6, 337.

Alfaa UV, 2018. All about UV Disinfection Systems for Water Treatment [WWW Document].

Beckley, J., Herron, E., Stepenuck, K., 2014. How to monitor for Bacteria, in: 9th National Monitoring Conference. National Water Quality Monitoring Council, Cincinnati, USA.

Bodzek, M., Konieczny, K., Rajca, M., 2019. Membranes in water and wastewater disinfection. Arch. Environ. Prot. 45.

Bogler, A., Packman, A., Furman, A., Gross, A., Kushmaro, A., Ronen, A., Dagot, C., Hill, C., Vaizel-Ohayon, D., Morgenroth, E., Bertuzzo, E., Wells, G., Kiperwas, H.R., Horn, H., Negev, I., Zucker, I., Bar-Or, I., Moran-Gilad, J., Balcazar, J.L., Bibby, K., Elimelech, M., Weisbrod, N., Nir, O., Sued, O., Gillor, O., Alvarez, P.J., Crameri, S., Arnon, S., Walker, S., Yaron, S., Nguyen, T.H., Berchenko, Y., Hu, Y., Ronen, Z., Bar-Zeev, E., 2020. Rethinking wastewater risks and monitoring in light of the COVID-19 pandemic. Nat. Sustain. https://doi.org/10.1038/s41893-020-00605-2

Cabral, J.P.S., 2010. Water microbiology. Bacterial pathogens and water. Int. J. Environ. Res. Public Health 7, 3657–3703.

Chahal, C., van den Akker, B., Young, F., Franco, C., Blackbeard, J., Monis, P., 2016. Pathogen and particle associations in wastewater: significance and implications for treatment and disinfection processes. Adv. Appl. Microbiol. 97, 63–119.

Contreras, J.D., Meza, R., Siebe, C., Rodríguez-Dozal, S., López-Vidal, Y.A., Castillo-Rojas, G., Amieva, R.I., Solano-Gálvez, S.G., Mazari-Hiriart, M.,

Silva-Magaña, M.A., 2017. Health risks from exposure to untreated wastewater used for irrigation in the Mezquital Valley, Mexico: A 25-year update. Water Res. 123, 834–850.

CPCB, 2007. Guidelines for Water Quality Monitoring.

DEFRA, 2012. Waste water treatment in the United Kingdom – 2012. London, UK.

Desmarais, T.R., Solo-Gabriele, H.M., Palmer, C.J., 2002. Influence of soil on fecal indicator organisms in a tidally influenced subtropical environment. Appl. Environ. Microbiol. 68, 1165–1172.

DWFA, 1996. South African Water Quality Guidelines. Pretoria, South Africa.

Edokpayi, J.N., Odiyo, J.O., Durowoju, O.S., 2017. Impact of wastewater on surface water quality in developing countries: a case study of South Africa. Water Qual. 401–416.

EEC, 2006. Directive 2006/7/EC of the European Parliament and of the council of 15 February 2006 concerning the management of bathing water quality and repealing Directive 76/160/EEC.

EPA, 2011. Water treatment manual: disinfection. Wexford, Ireland.

Eregno, F.E., Tryland, I., Myrmel, M., Wennberg, A., Oliinyk, A., Khatri, M., Heistad, A., 2018. Decay rate of virus and faecal indicator bacteria (FIB) in seawater and the concentration of FIBs in different wastewater systems. Microb. Risk Anal. 8, 14–21.

European Commission, 2016. EU-level instruments on water reuse Final report to support the Commission's Impact Assessment. Luxembourg.

Fedler, C.B., Francis, R., Parekh, D., Blanchet, S., 2012. Review of Potential Onsite Wastewater Disinfection Technologies.

Fitzhenry, K., Barrett, M., O'Flaherty, V., Dore, W., Cormican, M., Rowan, N., Clifford, E., 2016. The effect of wastewater treatment processes, in particular ultraviolet light treatment, on pathogenic virus removal: EPA Research Report 171.

Friedler, E., F. Chavez, D., Alfiya, Y., Gilboa, Y., Gross, A., 2021. Impact of Suspended Solids and Organic Matter on Chlorine and UV Disinfection Efficiency of Greywater. Water 13, 214. https://doi.org/10.3390/w13020214

Government of Tanzania, 2007. The Environmental Management (Water Quality Standards) Regulations.

Harwood, V., Shanks, O., Koraijkic, A., Verbyla, M., Ahmed, W., Iriate, M., 2017. General and host-associated bacterial indicators of faecal pollution, in: Rose, J.B., Jiménez-Cisneros, B. (Eds.), Global Water Pathogen Project. https://doi.org/https://doi.org/10.14321/waterpathogens.6

Hurst, C.J., Crawford, R.L., Garland, J.L., Lipson, D.A., 2007. Manual of environmental microbiology. American Society for Microbiology Press.

ILOSTAT, 2021. Statistics on labour costs [WWW Document].

Kay, D., Crowther, J., Stapleton, C.M., Wyer, M.D., Fewtrell, L., Edwards, A., Francis, C.A., McDonald, A.T., Watkins, J., Wilkinson, J., 2008. Faecal indicator organism concentrations in sewage and treated effluents. Water Res. 42, 442–454. https://doi.org/10.1016/j.watres.2007.07.036

Leverenz, H., Darby, J., Tchobanoglous, G., 2006. Evaluation of Disinfection Units for Onsite Wastewater Treatment Systems.

Mateus, C., Guerrero, C.A., Quezada, G., Lara, D., Ochoa-Herrera, V., 2019. An integrated approach for evaluating water quality between 2007–2015 in Santa Cruz Island in the Galapagos Archipelago. Water 11, 937.

Mills, N., Pearce, P., Farrow, J., Thorpe, R.B., Kirkby, N.F., 2014. Environmental & economic life cycle assessment of current & future sewage sludge to energy technologies. Waste Manag. 34, 185–195. https://doi.org/10.1016/j.wasman.2013.08.024

MINAM, 2017. DECRETO SUPREMO N° 004-2017-MINAM.

Momba, M., Edbon, J., Kamika, I., Verbyla, M., 2019. Using indicators to assess microbial treatment and disinfection efficacy, in: Rose, J.B., Jiménez-Cisneros, B. (Eds.), Global Water Pathogen Project. Michigan State University, E. Lansing, MI, UNESCO. https://doi.org/https://doi.org/10.14321/waterpathogens.9

Naughton, C., Mihelcic, J., 2017. Introduction to the Importance of Sanitation, in: Jiménez-Cisneros, J.B.R. and B. (Ed.), Global Water Pathogen Project. Michigan State University, E. Lansing, MI, UNESCO. https://doi.org/https://doi.org/10.14321/waterpathogens.1

Norweco, 2006a. LF Series 1000, 2000, 3000 Installation And Operation Manual.

Norweco, 2006b. BIO-DYNAMIC® DRY CHEMICAL TABLET FEEDERS AND TABLETS.

Parsons, S.A., Jefferson, B., 2006. Introduction to potable water treatment processes. Blackwell publishing.

Scheutz, F., Strockbine, N.A., 2015. Escherichia, in: Bergey's Manual of Systematics of Archaea and Bacteria. Bergey's Manual Trust.

Schleifer, K.H., Kilpper-Bälz, R., 1984. Transfer of Streptococcus faecalis and Streptococcus faecium to the genus Enterococcus nom. rev. as Enterococcus faecalis comb. nov. and Enterococcus faecium comb. nov. Int. J. Syst. Evol. Microbiol. 34, 31–34.

Shammas, N.K., Wang, L.K., 2015. Water engineering: hydraulics, distribution and treatment. John Wiley & Sons.

Sinclair, R.G., Jones, E.L., Gerba, C.P., 2009. Viruses in recreational water-borne disease outbreaks: A review. J. Appl. Microbiol. 107, 1769–1780.

Solo-Gabriele, H.M., Wolfert, M.A., Desmarais, T.R., Palmer, C.J., 2000. Sources of Escherichia coli in a coastal subtropical environment. Appl. Environ. Microbiol. 66, 230–237.

Tchobanoglous, G., Burton, F.L., Stensel, H.D., 2004. Wastewater engineering: treatment, disposal, and reuse, 4th ed. McGraw-Hill New York.

Turkish Ministry of Environment and Forestry, 2010. Başbakanlık Mevzuatı Geliştirme ve Yayın Genel Müdürlüğü [WWW Document]. URL https://www.resmigazete.gov.tr/eskiler/2010/03/20100320-7.htm (accessed 4.26.21).

UK Government, 2013. The Bathing Water Regulations.

USEPA, 2012. 2012 Recreational Water Quality Criteria [WWW Document]. URL https://www.epa.gov/wqc/2012-recreational-water-quality-criteria (accessed 7.6.20).

USEPA, 2003. Wastewater Technology Fact Sheet Disinfection for Small Systems.

USEPA, 2002. Onsite Wastewater Treatment Systems Technology Fact Sheet 4 Effluent Disinfection Processes.

USEPA, 1999a. Wastewater Technology Fact Sheet: Chlorine Disinfection.

USEPA, 1999b. Wastewater Technology Fact Sheet: Ultraviolet Disinfection.

USEPA, 1999c. Wastewater Technology Fact Sheet: Ozone Disinfection.

Von Sperling, M., De Lemos Chernicharo, C.A., 2005. Biological Wastewater Treatment in Warm Climate Regions Volume II. IWA Publishing, London, UK.

Weaver, R., n.d. Onsite Wastewater Treatment Systems: Tablet Chlorination [WWW Document]. URL https://agrilifeextension.tamu.edu/library/water/onsite-wastewater-treatment-systems-tablet-chlorination/ (accessed 4.27.21).

WHO, 2006a. WHO guidelines for the safe use of wastewater, excreta and greywater, Volume 1 policy and regulatory aspects.

WHO, 2006b. Guidelines for the safe use of wastewater, excreta and greywater Volume 2: Wastewater use in agriculture. Geneva, Switzerland.

Winfield, M.D., Groisman, E.A., 2003. Role of nonhost environments in the lifestyles of Salmonella and Escherichia coli. Appl. Environ. Microbiol. 69, 3687–3694.

Zyara, A.M., Torvinen, E., Veijalainen, A.-M., Heinonen-Tanski, H., 2016. The effect of UV and combined chlorine/UV treatment on coliphages in drinking water disinfection. Water 8, 130.

DOI: 10.1561/9781680837810.ch8

Chapter 8

Tools for Appraising and Supporting the Adoption of Nature-based Systems

By Louise Hannon and Eoghan Clifford

8.1 Introduction

Despite evidence that nature-based solutions often represent more efficient and cost-effective solutions to climate change threats than traditional approaches (European Commission, 2015, 2020c), key knowledge gaps exist that may affect the wide scale adoption and implementation of such solutions. Knowledge gaps concerning the relationships between nature-based solutions and society and, more specifically, the stakeholder involvement and impact of human-nature interactions in forming or altering lifestyles, beliefs, and preferences have been explored by Kabisch *et al.* (2017) while Istenic *et al.* (2015) studied the status of decentralised wastewater treatment systems and barriers for the implementation of nature-based solutions for these systems in central and eastern Europe, determining that these solutions were rarely used. Potentially linked to this they found that promoters of nature-based solutions were mainly found within specific-interest stakeholder groups, including ecological engineers, environmental non-governmental organisations (NGOs) and "green" movements.

In ecosystem-based approaches to climate change adaptation and mitigation, different drivers for the implementation of nature-based solutions can be identified at the project level. These range from policy and strategic objectives at national or supra-national level and/or local policies as well as specific community needs and stakeholder motivations. In some instances it may be a combination of these factors such as in the case of the INNOQUA project demonstration installation in the village of Littlemill in the Scottish Highlands. The motivation to explore and implement a wastewater treatment system that employs natural systems in this rural village was directly linked to the commitment of Scottish Water (the utility company providing potable and wastewater services to the people of Scotland) to develop innovative solutions for water and wastewater treatment and supply. In this case, among other needs, it met the aspirations of Scottish Water to be socially sustainable in addressing Scotland's water and wastewater requirements. The commitment of a utility is, in most cases, key to bringing such solutions from submarginal or rarely used to commonplace. In addition, the residential community whose wastewater systems were to be connected to the INNOQUA demonstration installation were supportive of the proposal following consultation and discussion with the utility. The local community's endorsement of a low impact nature-based wastewater treatment solution is consistent with previous experience in Scotland, where rural communities often exhibit high levels of connection and guardianship with their local natural environment. While the main drivers for nature-based solutions and the wider field of green infrastructure are often to meet biodiversity conservation and other environmental objectives, the potential contribution of such solutions to wider sustainability adaptation and mitigation efforts can be overlooked (Naumann *et al.*, 2011). The European Commission recognises that nature-based solutions have the potential to not only address societal challenges in sustainable ways but can also provide co-benefits for health, the economy, society and the environment (European Commission, 2015, 2020c). These co-benefits represent a significant opportunity to increase Society's motivation and support for the adoption of nature-based solutions.

Given the increasing and indisputable evidence of the effects of anthropogenic activities on the planetary environment, and with high profile activists and media efforts increasing society's environmental knowledge (Dunn *et al.*, 2020) it is perhaps unsurprising that concerns about our environment and the effects of climate change are increasingly prevalent in the minds of many citizens. A 2020

Eurobarometer survey found that protecting the environment was important to 94% of citizens in EU Member States and that 91% of citizens considered that climate change was a serious problem in the EU (European Commission, 2020a). Such concerns and anxieties are likely to stimulate more interest in less impactful and more sustainable solutions to resource provision such as those that are nature-based. To translate such interest into adoption, a robust evidence base is required to address any knowledge gaps that might otherwise hinder the uptake of such novel or non-traditional technologies or practices.

This chapter leverages the INNOQUA Project's experience in adapting the proposed INNOQUA technologies to local conditions, with an emphasis on overcoming barriers to the adoption of a novel nature-based wastewater treatment system. The importance of early and extensive stakeholder engagement is highlighted.

8.2 Implementation of Wastewater Infrastructure

In trying to anticipate and address potential issues with the implementation of nature-based sanitation solutions, we can first look to experiences gained during the planning and installation of existing on-site sanitation infrastructure. The World Health Organisation recognises that in the case of countries in the Global South, while some of these 'grey infrastructure' type sanitation projects follow a detailed planning and approval process, others move forward as self-builds with little involvement from any external parties. The reality for most sanitation projects and programmes lies somewhere between these extremes (Franceys *et al.*, 1992).

8.2.1 Key Project Phases

Figure 8.1 summarises the key phases in the planning and implementation of decentralised sanitation projects. An issue that is consistently important in each phase is the consideration of perspectives of key interest groups i.e. stakeholders. The concept of stakeholders is taken from corporate management theory where it was originally proposed as a novel approach to strategic management but has since become a critical step in most project delivery. The classic definition of a stakeholder is based on the work of Freeman where, in the context of strategic management of firms and such organisations, a stakeholder was defined as any group or individual who can effect or be affected by the achievement of the firm's objectives (Freeman, 1984).

The first phase in the process is the identification of the needs or drivers for a solution to a sanitation problem. Depending on the project, initial motivation and problem recognition may come from an individual, a community, a legislative requirement, a local government initiative etc. At this early stage it is possible to

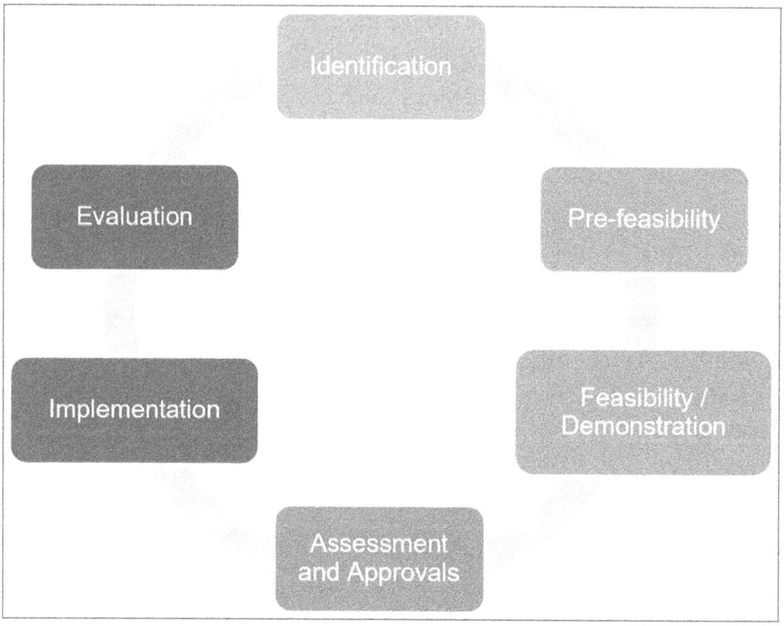

Figure 8.1. Key phases in traditional decentralised sanitation projects (Franceys *et al.*, 1992).

identify key stakeholders and to begin to understand what role they might play in any subsequent project and begin to determine to what extent engagement with these groups or individuals should be planned. As a minimum the key stakeholders will be the promoter of the project, the implementing body or agency and the community, group or individual whose sanitation needs will be addressed by the project.

Phase 2 allows feasible solutions to the identified problem to be considered and should engage all the key stakeholders – to identify their support, objections, concerns and motivations towards the project. Phase 3 seeks to identify a preferred solution that satisfies technical, social, environmental and economic criteria. Stakeholder buy-in is critical at this point. Phase 4 encompasses the final approvals of the project promoter and funder, and allows the project to move to securing technical and planning consents. The implementation phase (Phase 5) then requires training, demonstration and promotion to facilitate acceptance, proper usage and correct maintenance/operation of the sanitation solution into the future. Phase 6 is often under-appreciated, but lessons learned from any evaluation form a key step in the project lifecycle which provides data and feedback on which to base future improvements of the system. User testimony collected at this stage can also be used to promote the sanitation solution and encourage wider adoption.

The phases outlined above are equally applicable to the planning and implementation of nature-based sanitation solutions. However, given the novelty of some of these solutions – satisfying technical, regulatory and legislative criteria in Phases 1 and 2, as well as the pre-feasibility and feasibility requirements in Phases 2 and 3, and appraisal and approval requirements in Phase 4, may all be particularly onerous. The suitability of the proposed site can also demand particular attention as nature-based solutions may require more land and/or benefit from bespoke design to ensure local environmental suitability in terms of climate and resources, and to optimise overall project sustainability.

More recently a framework was developed by the WHO water, sanitation and hygiene (WASH) community specifically in relation to neglected tropical diseases (NTD) to encourage, support and develop investment in support of the goals of the WHO 2012 Neglected Tropical Diseases roadmap (World Health Organization, 2012). The BEST Framework was proposed by the NTD NGO Network in 2016 and focuses on four key elements: (i) behaviour, (ii) environment (iii) social inclusion and (iv) treatment and care (NTD NGO Network, 2021). This conceptual framework ensures that all actions needed for control and elimination of NTDs, including water and sanitation, are approached in a sustainable and holistic manner. The framework comprises five stages within each of which relevant tools are made available. While the focus of this toolkit is not strictly sanitation provision, it can, in conjunction with the key phase approach, be adapted to inform the development of any sanitation and wastewater solutions for decentralised applications. The five BEST Framework stages are outlined in Table 8.1.

Truly understanding the position, expectations, attitudes and power or influence of groups and individuals who can affect or be affected by the implementation of a conventional sanitation solution is extremely important – and even more so with respect to the implementation of novel solutions such as nature-based sanitation.

8.2.2 Barriers to the Implementation of Nature-based Solutions for Water Management and Sanitation

It is not unexpected that any relatively "new" or novel approach to water management or sanitation will encounter some barriers to implementation. However, like all such innovations it is useful to consider these barriers in the context of the key drivers for such innovation. In some cases, issues that are currently barriers can also serve in the future as key drivers for innovation (e.g. regulation).

Katsou *et al.* (2020) outlined four main steps for the implementation of nature-based solutions in creating Circular Cities: (i) planning, (ii) design, (iii) assessment and, (iv) communication of results. This approach aligns well to the WHO project phases in sanitation solutions presented in Section 8.2.1, but with the addition of a

Table 8.1. BEST Framework (NTD NGO Network and World Health Organization, 2019).

Stage	Key Questions
1. Setting the programme vision	– What are you trying to achieve? – What will it take? – How is it linked to a broader (e.g. national) agenda?
2. Building partnership	– Why should you collaborate with partners? – How do you collaborate with partners? – How do you get started?
3. Analysing the situation	A situation analysis protocol is suggested which contains the following key headings: – Identify the analysis team – Identify and formally involve key stakeholders – Collect information – Analyse – Recommend – Report
4. Planning and programme design	This approach emphasises the following key areas: – Prioritising long-term policies and strategies rather than long-term targets – Continuously linking planning to implementation, not detailed pre-implementation planning followed by little monitoring – Regular monitoring and evaluation to learn from errors on a continuous basis, rather than periodic external evaluations – Continuous dialogue with intended beneficiaries to adjust activities to their needs Within this stage the project planning phases suggested are: (i) gather, (ii) synthesis, (iii) align, (iv) act, (v) verify and (vi) revisit and realign A variety of advisory documents and a planning tool are provided
5. Implementing and monitoring	This stage focuses on monitoring and evaluation. This includes monitoring of the implemented project but also evaluation of issues such as accountability and cost-benefit analysis. The document proposes the use of a logical framework (logframe) and includes tools to help carry out this stage

further step between steps (ii) and (iii) to include implementation. Barriers will be encountered in each of these phases, but it should be noted that while some may be similar across many regions, others may be specific to a particular country or region. For example, in Europe where there has been increased harmonisation across many sectors, regional differences are still common. In Western Europe national guidelines adopted by the government are available for the design and operation of wastewater treatment wetlands. However, in Central and Eastern Europe, the adoption of such guidelines is varied: some countries have no national guidance and some have standards that are decades old (Kabisch *et al.*, 2017). This is despite treatment wetlands being a well-established technology that is familiar to experts across all countries.

Table 8.2 summarises from literature and the experience of the INNOQUA project team, key barriers to the implementation of nature-based water management and sanitation systems alongside mitigation measures. Many of the barriers and mitigation measures are interlinked, and mitigation measures for one barrier can reduce barriers elsewhere.

8.2.3 Drivers for the Implementation of Nature-based Solutions for Water Management and Sanitation

Unlike barriers, drivers for the implementation of nature-based solutions are most prevalent at the early phases of a project. As described in Section 8.1, advocates for the implementation of nature-based solutions are most often niche groups, ecological engineers, environmental non-governmental organisations (NGOs) and "green movements" (Istenic *et al.*, 2015). So, in considering the current drivers for greater implementation we should recognise that certain elements that may be current barriers have the potential, with some adaptation, to be drivers. Table 8.3 summarises the key drivers for the implementation of nature-based water management and sanitation.

8.3 Acceptance of Nature-based Technologies

An aspect of nature-based biological systems that is particularly relevant to novel solutions is the issue of their acceptance by society. The INNOQUA team recognised this as requiring careful consideration, particularly in the context of the global reach of the project and the additional complexities that this implies in terms of adapting technologies to varying geographical locations, socio-economic groups and cultural norms. By identifying broad stakeholder groups relevant to the specific installations at each site and recognising broader interests in a demonstration site

Table 8.2. Barriers to the implementation of nature-based water management and sanitation systems alongside mitigation measures (Istenic *et al.*, 2015; Kabisch *et al.*, 2017; Katsou *et al.*, 2020; Kisser *et al.*, 2020).

Barriers	Mitigation Measures
Public awareness and social acceptance	• High stakeholder engagement • Improved use of social and digital media tools • Increased communication regarding the benefits of nature-based systems • Integration with educational and schools' programmes • Use of local expertise and services during planning, design, construction and operation
Decision-makers lacking knowledge of the potential impact of nature-based systems in the transition to sustainable and circular systems	• Link policy and innovation to ensure improved dissemination between research, deployment, performance metrics and policy makers • Holistic analysis of costs and benefits to include environmental, economic, ecological and social aspects – i.e. full life cycle analysis • Analysis of project's impacts over larger physical and time scales • Use of multi-scale physical and temporal scenario analyses/modelling to help de-risk decision making
Unsupportive (or lack of) legal/regulatory frameworks Lack of standards (when compared to those for traditional engineered systems Varying legal frameworks in countries from the same geographical region	• Develop legal frameworks that define the roles of nature-based systems in the water sector • Inclusion of nature-based systems as part of the regulatory toolkit to address water and sanitation challenges • Harmonisation of regulatory systems, design standards and policies across boundaries where possible • Focus on sharing of best practice between countries and regions
Financial resourcing	• Policy that puts sustainability at the heart of investment (for example, see European Commission (2019) and GCF (2021)) • Holistic analyses of financial costs and benefits that include multi-functional benefits • Increased analysis of the role of nature-based systems in addressing the costs of climate change

(*Continued*)

Table 8.2. Continued

Barriers	Mitigation Measures
	• Develop procurement policies that prioritise overall life cycle cost and sustainability rather than up-front capital costs. This can be at local, regional and national level but also includes investment banks and other funders.
Need for large scale demonstrations and pilots and lack of technological maturity	• Innovation funding to enable larger demonstrations and longer-term performance measurement
Numerous possibilities for derived end-products can cause competition and also reduce benefits	• Focus on end products can improve commercial outcomes • Reduce internal competition between end products
The role of nature-based solutions in densely populated or historical urban centres	• Innovation between stakeholders such as urban planners, historians, archaeologists, engineers and architects and the public to re-imagine how such systems can respect and compliment sensitive and historical urban centres
Relatively large land area required by many nature-based systems	• Design of nature-based systems with multi-functional uses and integration with existing land-uses (e.g. parks, green areas etc.)

region or country, the project was able to segment relevant groups whose acceptance would be most important to the adoption of the proposed technologies.

8.3.1 Stakeholder Identification

For INNOQUA, stakeholders were first identified in generalised terms in an early project deliverable (D1.2, European Commission (2020b)). The focus of this initial assessment was stakeholder groups relevant to future demonstrator replication and exploitation activities such as early adopters, potential end users and potential partners etc.

A later project deliverable, concerning the Identification and Assessment of Exploitable Results further refined the INNOQUA stakeholder groups in terms of target customers and/or end-users for each exploitable result. These target customer and end-user groups were used as the basis for the identification of the following generalised key stakeholder groups relevant to the planning, implementation and success of the INNOQUA system in the INNOQUA Training and Education Programme Target segments (D7.4, European Commission (2020b)):

Table 8.3. Drivers for the implementation of nature-based water management and sanitation systems.

Driver	Rationale
Research and Innovation Policy	• For example, the ambition of research and innovation policy is to position the European Union as a leader in innovating with nature to achieve more sustainable and resilient societies (European Commission, 2021). This in turn will push the adoption of nature-based solutions
Regional and National Policy Initiatives and/or regulation	• European Green Deal (European Commission, 2019) • Climate Action policies • Sustainable resource usage policies • Environmental protection policy and regulation
Financial Models	• Effective water pricing can stimulate uptake of new innovations if it reflects true financial, environmental and resource costs including operation and maintenance costs (Hrovatin and Bailey, 2001; Barraqué, 2020) • Separate charges for water use and effluent treatment, in particular, can drive industry towards increased efficiency, investment in water treatment innovation and closing of local water cycles
Economic Costs: Implementation, Operation, Maintenance	• Relative cost of implementation of traditional centralised sanitation systems and that of decentralised nature-based solutions • Relative cost of operation and maintenance of traditional centralised systems and that of decentralised nature-based solutions
Community/social stewardship	• Benefits of co-design and citizen engagement in community based green initiatives allow citizens to develop ownership of the local landscape, which may in turn increase the engagement in such projects (Shandas and Messer, 2008)

(*Continued*)

Table 8.3. Continued

Driver	Rationale
Social and environmental responsibility	• Greater social and environmental responsibility at all levels towards responsible water resource usage will influence government and industry – and citizens • The concept of water footprinting provides a quantitative approach to determining impact on water resources (European Commission, 2021)
Awareness and education of nature-based solutions	• Studies have shown that one of the most important factors in the successful implementation of sustainable water management initiatives is public outreach. The Windhoek water reuse project in Namibia points to continued public education campaigns including media campaigns and education of children at public schools that led to the project being embraced and supported by the public to the extent of deriving pride from it (du Pisani, 2006; Lahnsteiner and Lempert, 2007)

- End Users
- Technical Professionals, Installers and Operators
- Promoters, Agents and other interest groups
- Technical Decision makers

Table 8.4 details some of the INNOQUA Stakeholder Groups as they relate to the initial cohort described above. It should be noted that while Table 8.4 comprises an overview of all stakeholder groups across the various INNOQUA demonstration sites, each demonstration site may have involved only particular groups. However, the stakeholder mapping exercise proved very useful in identifying groups in one geography that might not otherwise have been considered in another geography.

8.3.2 INNOQUA Social Acceptance Questionnaire

To understand the likely acceptance of the INNOQUA solution a social acceptance questionnaire was developed (D7.4, European Commission (2020b)) as a key engagement tool. The questionnaire was designed to gather information on likely barriers or issues that could affect the acceptance of respondents of a system

Table 8.4. Key INNOQUA Stakeholder Groups by segment.

Customers & End Users	**Technical Professionals, Installers & Operators**	**Promoters, Agents & Other Interest Groups**	**Technical Decision Makers**
Decentralised communities and housing collectives	Technology providers	Promoters of collective sanitation systems	Planning authorities/bodies
Landowners, self-builders	Utility companies, municipalities and water authorities/bodies	Sustainable Design Professionals	Regulatory authorities/bodies
Hotel Groups, tourism industry	Managers of existing wastewater treatment facilities	NGOs involved in implementation of health solutions in rural areas.	Policy makers – regional, local, national
Decentralised business and commercial operations especially agri-industries	Designers and specifiers, i.e. engineers, architects	Farms, landowners and local communities in close proximity to the demonstration sites	Municipalities interested in water reuse

such as that proposed by the INNOQUA Project. Specifically, the questionnaire was developed with four principle aims, to:

1. Establish the level of knowledge of the respondent in terms of wastewater and wastewater treatment systems
2. Establish the priorities of the respondent in terms of adopting nature-based, environmentally sustainable wastewater treatment solutions
3. Establish the priorities of the respondent in choosing a wastewater treatment solution
4. Establish whether the respondent would consider installation of an innovative nature-based system, such as the INNOQUA solution

Although INNOQUA was a global project, a single questionnaire was developed, without any regional variations (except for local translation), to simplify comparison between different regions. The questionnaire was principally multiple choice to allow for rapid completion by respondents regardless of their level of understanding and interest in wastewater treatment and environmental sustainability. The development of the questionnaire was informed by the work of De Groot and Steg (2009) and was founded in the concept of the norm activation model presented in 1977 (Schwartz, 1977). The INNOQUA questionnaire was divided into five parts, as set out below.

8.3.2.1 Part 1 – General information about the respondent

This section established key demographic and personal information about the respondent relating to gender, age, religious affiliation, education, employment, location and type of dwelling, and current wastewater treatment infrastructure. The section comprised 13 questions (Table 8.5):

- Questions 1 – 5 captured information that could be used to determine particular patterns or consistencies.
- Questions 6 – 8 related to employment and the type of industry, if any, in which the respondent was engaged. This may have a direct impact on interpretation of responses to later questions regarding technical knowledge and the importance or not of adequate wastewater treatment. A question regarding income was included as this can be a significant issue in the prioritisation of sanitation.
- Questions 9 – 11 related to the respondent's living situation as this can have a bearing on the context for answers to later questions.
- Questions 12 and 13 related specifically to the existing treatment of wastewater and established the general knowledge of the respondent of sanitation and wastewater treatment issues.

8.3.2.2 Part 2 – General questions about the respondent's pro-environmental activities

This section was developed to establish the involvement of the respondent in pro-environmental activities as a determinant of likely future adoption of pro-environmental behaviours, based on the work of De Groot and Steg (2009) and Schwartz (1977). The questions established the personal norms of the respondent by asking them to select the frequency, on a five point Likert scale, with which they undertook certain actions (Table 8.6). This type of ranking based on 'vaguely' quantified frequency response options is not favoured by some academics in the field of behavioural research but was considered a useful measure in the context of this study.

Table 8.5. Questions 1 – 13, including multiple choice answers (where suggested).

1. What is your gender?

Male | Female | Other | Don't Want to Say

2. What is your age range?

18–29 | 30–49 | 50–64 | 65 years or over

3. In which of the following regions do you currently reside?

Europe | South America | North America | Africa | Asia | Other

4. Which (if any) is your religious affiliation?

Christian* | Muslim | Buddhist | Hindu | Jewish | Sikh | Not Religious | Other | Don't Want to Say

*(All denominations)

5. What is the highest education that you finished?

No formal education | Primary School | Lower Secondary | Upper Secondary | Third Level* | Third Level Post Grad | Other | Don't Want to Say

*(College or University)

6. What best describes the industry in which you are usually employed?

Agriculture, Forestry & Fishing	Professional, scientific and technical activities
Industry	Administrative and support service activities
Construction	Public administration and defence
Wholesale & Retail Trade	Education
Accommodation and food service activities	Human health and social work activities
Information and communication	Student
Financial, insurance and real estate activities	Other

7. What best describes your job?

Management Staff | Professional | Operational Staff | Technical Staff | Administrative Staff | Other

8. What is your average monthly income?

€500 or Less | € 500–1,000 | € 1,000–2000 | € 2,000–3,000 | € 3000 or above | Would prefer not to specify

9. What best describes your living situation?

I am a homeowner | I rent my home | I live with my Parents | I am in a house/apartment share | Other

10. What best describes your main residence?

Detached House | Semi-Detached House | Apartment/Flat | Other

11. How many people are in your household?

I live alone | Number of Adults | Number of Children

12. Is wastewater arising from your main residence currently connected to a sewerage network?

Yes | No | I Don't Know

13. How is wastewater arising from your main residence currently treated?

Untreated | On-site settlement tank system | On-site septic tank & percolation system | Proprietary Biological or Mechanical on-site treatment system | Connected to a Municipal/ Centralised Treatment | I Don't Know

Table 8.6. Likert scale and questions in Part 2 that used this scale.

Never	Rarely	Sometimes	Often	Always	N/A

1. Do you incorporate pro-environmental activities in your daily routine?
2. Do you take measures to minimise the volume of waste that you generate?
3. Do you favour nature-based solutions when selecting new products or technologies where possible?
4. Do you choose low energy technologies where possible?
5. Do you choose water saving technologies where possible?

8.3.2.3 Part 3 – The respondent's opinion on the treatment of wastewater

In this section, the focus was on establishing the respondent's awareness of the problem of uncontrolled and/or untreated wastewater discharge (Problem Awareness) and respondent's perception of the effectiveness of mitigation measures (Outcome Efficacy) that they might take to address the problem. Specific questions regarding the usual behaviour of respondents are also included to further contribute to establishing "personal norms". This section was again based on the work of De Groot and Steg (2009). Questions 2, 3, 6 and 7 addressed Problem Awareness, questions 8 to 11 addressed Outcome Efficacy and 1, 4 and 5 Personal Norms (Table 8.7). In this part of the questionnaire, a five-point scale similar to Likert was used to gauge the respondent's agreement or not with a series of statements related to wastewater.

8.3.2.4 Part 4 – The respondent's criteria in selecting wastewater treatment systems

In the penultimate section, the questionnaire focused on the importance of various criteria to the respondent in the selection of a wastewater treatment system. In this part of the questionnaire, a five-point scale similar to Likert was used to gauge how important or not a certain criterion was to the respondent (Table 8.8). These questions were important to better understand the key criteria that different categories of respondents would consider important when deciding between various options for wastewater treatment. The results of this aspect of the survey informed the development of the multi-assessment criteria (discussed in Section 8.4.2) that were used to compare various design options for the technology. The results also indicated differences between regions in the level of importance given to each of the criteria below. For example, respondents from Italy and India considered all of the criteria to be very or extremely important (i.e. an average score >4 for each question) whereas respondents from France considered efficiency and performance to be very or extremely important with aesthetics and visual impacts being of moderate importance (an average score of between 3 and 4). These results were similar

Table 8.7. Five-point scale and questions in Part 3 that used this scale.

Completely Disagree	1	2	3	4	5	Completely Agree

1. I am aware of what wastewater is and the various sources of wastewater occurring at the property where I live
2. I have a strong personal obligation to ensure that wastewater generated by me or my household is connected to an effective treatment system
3. I worry about the negative impact of untreated or poorly treated wastewater on the environment
4. I consider that biological treatment systems using earthworms that can treat wastewater to acceptable quality before reuse or discharge are positive solutions
5. I consider that biological treatment systems using crustaceans that can treat wastewater to acceptable quality before reuse or discharge are positive solutions
6. I consider that biological treatment systems using sunlight that can treat wastewater to acceptable quality before reuse or discharge are positive solutions
7. In selecting a new wastewater treatment system or other technology, I'd feel guilty if I chose a less sustainable solution over a more sustainable solution of similar cost
8. The ineffective treatment of wastewater is a problem for society
9. The lack of sustainable and effective wastewater treatment systems has a significant negative environmental impact
10. Making improvements to the treatment of wastewater from my home/property will not have an effect on the overall quality of treated wastewater or the resources
11. Promoting sustainable and pro-environmental wastewater technologies at work /school /college will have a positive effect on people's adoption of such technologies
12. The provision of incentives (such as grants) for installing/adopting sustainable and pro-environmental wastewater technologies would have a positive effect in peoples adoption of such technologies

to responses from Tanzania and Peru, albeit with the importance of each category being rated more equally in Ecuador and aesthetics receiving the lowest average score in Tanzania. Such data is useful to better understand each market but also very useful in understanding the differences between stakeholders within each market. However, as noted later in this chapter, care should be when analysing the results and drawing conclusions.

8.3.2.5 Part 5 – The respondent's views on adopting innovative wastewater treatment systems

In the final section of the questionnaire, the respondent's views and willingness to use a new nature-based, sustainable wastewater treatment system (such as those proposed in the INNOQUA project) were sought. The possible responses were

Table 8.8. Five-point scale and questions in Part 4 that used this scale.

Not Important	1	2	3	4	5	Extremely Important

1. Ease of Installation (E.g. factors such as size, weight, number of ancillary parts, whether this is installed over or underground)
2. Efficiency and Performance (E.g. Ability to produce a very high-quality final effluent)
3. Sustainability and Energy Requirements (E.g. Energy consumption during Operation)
4. Aesthetics and Visual Impacts (E.g. Factors such as size of unit, whether this can be installed over or underground)
5. Initial Purchase and Installation Cost
6. Ease of use and maintenance requirements
7. Noise and Odours

Table 8.9. Likert scale and questions in Part 5 that used this scale.

Definitely Not	Probably Not	Possibly	Probably	Definitely Yes

1. Would you be willing to adopt a nature-based solution incorporating earthworms or other micro-organisms for the treatment of wastewater at your residence?
2. Would you be willing to adopt a nature-based solution incorporating crustaceans or other micro-organisms for the treatment of wastewater at your residence?
3. Would you be willing to use treated wastewater from an onsite treatment system for non-consumable use? (E.g. Irrigation or domestic applications such as toilet)

based on a 5 point Likert scale indicating the likelihood of the respondent to adopt a nature-based sanitation solution (Table 8.9).

Overall, the results indicate a greater acceptance of the reality of nature-based solutions by respondents in countries where a significant proportion of wastewater is currently untreated. This was reflected in higher overall scores for questions 4, 5 and 6 in Table 8.7 (which ask whether a respondee believes that technologies such as the ones proposed in INNOQUA could work) and also in relation to the questions from Table 8.9. Taking the examples of Ecuador, France, India, Italy and Peru – in all cases respondents from France and Italy were less likely to adopt nature-based solutions when compared to respondents from Ecuador, India and Peru. It should be noted that to draw wider conclusions further research it would be necessary to broaden the response base – but it is an area that should be explored. There has been previous work analysing public perceptions on treated wastewater reuse (e.g. Smith *et al.* (2018); Akpan *et al.* (2020)) but very limited work exploring acceptance of: (i) nature-based solutions and (ii) whether public perceptions differ between countries or regions. Such work would be valuable to understand what lessons can be learned from countries with successful applications of nature-based solutions compared to those where more "traditional" electro-mechanical solutions have been deployed.

8.3.3 Training and Education

An INNOQUA Training and Education Programme (D7.4, European Commission (2020b)) developed five flexible modules to address some of the identified barriers, increase knowledge regarding wastewater treatment and provide detailed knowhow regarding the INNOQUA system and its implementation. The education programme was designed with flexibility to serve the needs of the four principle target participant groups, as described in Section 8.3.1, in the diverse geographical locations represented within the project consortium.

Although extensive surveys and comprehensive training and education programmes may not be practical for smaller projects, the concept of using questionnaires or similar methods to identify barriers and drivers – and to generate information that might help develop solutions that mitigate against potential barriers – is relevant irrespective of the scale of development. Furthermore, the development and generation of data from such surveys can be a very useful way to connect with local stakeholders and develop new opportunities such as building on expressions of interest for new collaborations or leveraging pro-environmental propensities of any particular group.

8.4 The INNOQUA Approach

The planning, design, development and installation of the INNOQUA project demonstration sites followed five distinct phases as indicated in Figure 8.2. These were adapted from Figure 8.1 and were aligned with relevant parts of Steps 2, 3 and 4 in the BEST Framework. Issues such as system start-up, commissioning, operation and maintenance are considered elsewhere in this book and thus the focus here is on the planning and advance works.

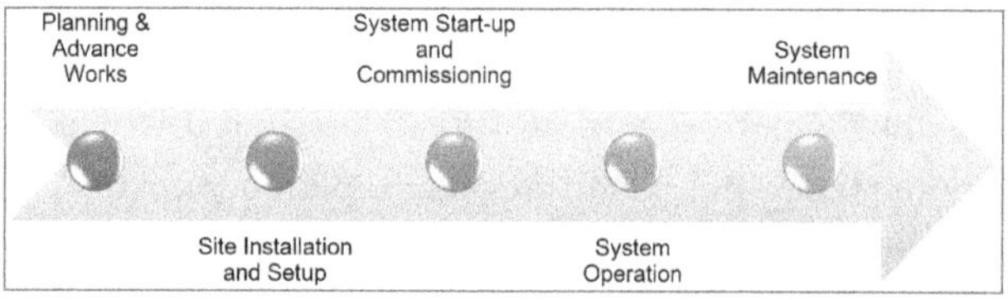

Figure 8.2. INNOQUA Demonstration Site Design, Development and Installation Phases.

This first phase considers system selection and any necessary permissions and approvals that might be required to allow the installation to proceed. In the case of the INNOQUA project a critical early step in this phase was a multi-criteria assessment to determine the suitability of the INNOQUA solution to solve the sanitation issues presented at a potential demonstration site, and to determine the optimal configuration of the modular INNOQUA solution for that site. The multi-criteria assessment comprised two stages: Go-No Go and Impact Assessment. This approach could readily be adapted for other applications, as an aid to the identification of an optimal solution.

8.4.1 Multi Criteria Assessment

8.4.1.1 Go-No Go Criteria

The three key Go-No Go criteria included:

- System alignment with local certification requirements
- Confidence that an expected level of performance will be achievable
- Compatibility with local climatic conditions

In the case of INNOQUA the above criteria gave high level decision outcomes as to technology suitability for each demonstration site. For example, the biosolar purification system would likely not be compatible with local climate conditions in Ireland (or Scotland) and thus would not achieve expected levels of performance. On the other hand the technology was likely to have significant application in India, Spain or Peru. In the case of local certification requirements there were no particular barriers in this project to the implementation of any given technology in a full-scale demonstration capacity – but for commercial deployment local certification would be necessary.

8.4.1.2 Impact assessment

A key goal in the INNOQUA project was to develop an easy to use "scoring" system that would allow comparison of various design options in a comprehensive and easily understood manner. Weightings could be applied to different assessment criteria to favour one criterion over another (e.g. to emphasise sustainability over cost). The impact assessment could be used to compare design options for any given technology (e.g. the use of different materials during manufacturing) or used to compare different technologies as a solution to the challenge. Such an assessment approach could also, for example, be used to compare solutions developed to enable electricity-free operation of a technology versus one that relied on a centralised electricity supply or a solution whereby electricity supply is decentralised (e.g. use of

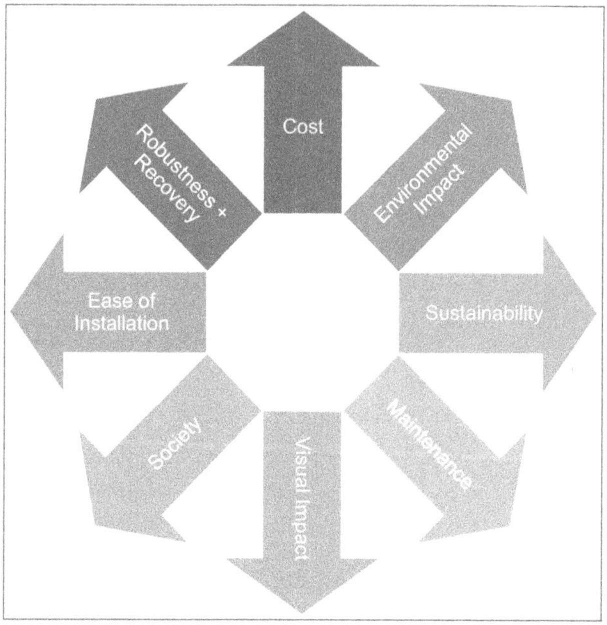

Figure 8.3. Typical project impact assessment criteria.

solar PV with back up battery storage). The various impact assessment criteria are shown in Figure 8.3.

Although many of the criteria are qualitative, for ease of comparison between alternative systems all factors were "quantified" or "scored". It should be noted that the specific scales presented in the following sections to "score" or "rate" different options can be changed for different projects.

In the case of the INNOQUA project the key comparisons being made were mainly between various design choices for each of the technologies being deployed rather than specific technologies. However, the assessment approach outlined was also used to inform later comparison between INNOQUA technologies and other commercially available systems. In the case of this project such an assessment could only be done after data had been collected from various demonstration sites. An example of one such comparison is provided in Chapter 9.

8.4.1.2.1 Cost

The cost of a system was considered as both capital and operational cost incurred over the expected life of the system. In the case of INNOQUA, life cycle costs (LCC) were assessed in accordance with the recommendations of ISO 15686-5:2017 and the European Commission Joint Research Centre (Hauschild *et al.*, 2011; JRC-IES, 2012). It should be noted that LCC considers costs only, while whole life costing (WLC) includes costs as well as benefits over a defined

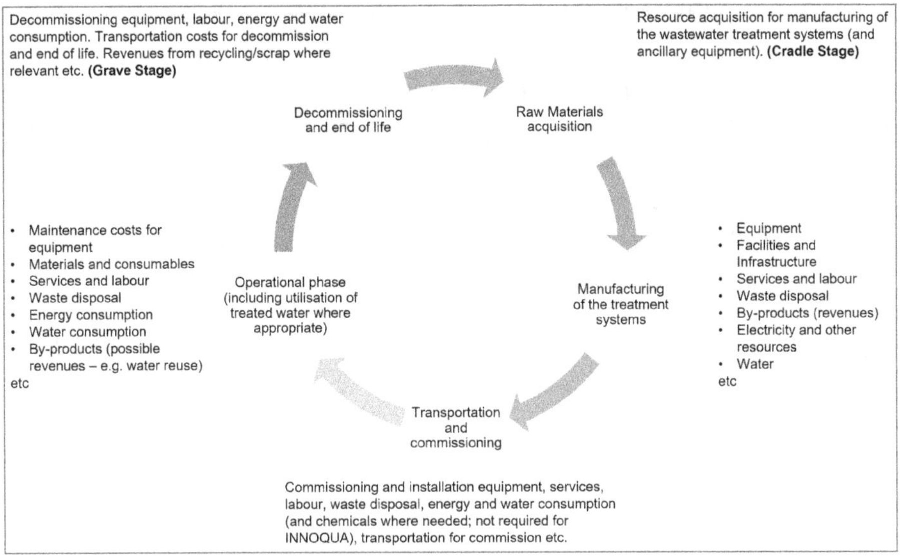

Figure 8.4. Overview of the life cycle stages (from Cradle to Grave) of a wastewater treatment plant as applied in INNOQUA.

period. When assessing options for the design, construction and operation of the INNOQUA technologies, the life cycle costs of an integrated treatment solution were analysed. Figure 8.4 shows the system boundaries and scope of the cost assessment in this case.

Life cycle costs are generally calculated as a net present value (NPV), which allows the value of projected future costs over the system life cycle to be considered and compared in terms of their current value. In this project a simple rating on a scale of 0 to 3 was applied to various costs associated with different design choices – whereby '0' was applied to high cost solutions and '3' applied to low cost solutions. This comprehensive approach may not be practical for small scale installers, but consideration of the operation and maintenance costs of any system should always be made in addition to initial capital investment costs.

8.4.1.2.2 Environmental impact

Life cycle assessment (LCA) can be used to evaluate the environmental impact of the sanitation solution proposed for the demonstration sites. The LCA process is the compilation and evaluation of potential inputs, outputs and environmental impacts of a product, process or service, through its life cycle (bsi, 2020). The initial phases of LCA in which the goals and scope of the assessment are defined and life cycle inventory analysis are carried out, are critical to the robustness of the final outcome (JRC-IES, 2012). This is where the process boundaries and assessment assumptions are confirmed and also where data is collected and interpreted in order to calculate life cycle impacts.

The environmental impact categories adopted for INNOQUA (to assess life cycle impacts of different design choices) were those required by the international reference life cycle data system (ILCD) midpoint method (JRC-IES, 2012). The choice was made to use only midpoint indicators due to their lower uncertainty when compared to endpoint indicators; endpoint indicators are impacts at the level of the areas of protection (e.g. the natural environment, human health or natural resources), while midpoint indicators indicate impacts somewhere between the emission point and the endpoint (Finnveden *et al.*, 2009). The associated list of environmental impact categories is listed below:

- Resource depletion: Mineral, fossils and renewables
- Resource depletion: Water
- Land use
- Acidification
- Climate change
- Ionising radiation ecosystems/human health
- Particulate matter
- Ozone depletion
- Photochemical ozone formation
- Freshwater ecotoxicity
- Human ecotoxicity carcinogens/no-carcinogens
- Freshwater eutrophication
- Terrestrial eutrophication
- Marine eutrophication

A simple rating on a scale of 0 to 3 was again applied, whereby '0' indicated a relatively high and negative environmental impact and "3" indicated a high and positive environmental impact. As with the approach to life cycle costing, this comprehensive LCA approach to the assessment of environmental impact may not be practical for small scale installers or installations but even a rudimentary assessment of the potential positive and negative environmental impacts of installation options is beneficial in determining a suitable solution – and can assist greatly with stakeholder engagement.

8.4.1.2.3 Sustainability

Demonstrating the feasibility of wastewater reclamation and the reuse of treated water was one of the key aims of the INNOQUA project and this was used as an assessment indicator for sustainability. This was done after consultation between the various partners within the project and cognisant of cost and life cycle impacts being accounted for separately. Depending on the specific analysis being undertaken "sustainability" can obviously refer to other factors.

An evaluation of the potential for reuse of treated water was carried out by comparing the regional limits of physical, chemical and biological characteristics for various reuse options with the expected characteristics of water post treatment by the INNOQUA system. Full details of the relevant regulations for water reuse (and indeed other relevant aspects) in each of the demonstration site locations is given in Deliverable D1.1 (European Commission, 2020b); it can be time consuming to gather relevant regulations from various countries and thus this resource should prove useful to other projects.

As before, a rating on a scale of 0 to 3 was used to represent reuse potential. A rating of '0' indicated no potential for reuse without further treatment following discharge from the INNOQUA system. A rating of '3' indicated more than one potential reuse activity following discharge from the INNOQUA system without the necessity for further treatment. In some sites water reuse was not required or not considered desirable and in such scenarios would not be considered or this heading could be given a lower weight compared to others.

8.4.1.2.4 Maintenance

The maintenance of any treatment system can be considered as largely comprising two distinct elements: (i) routine control and inspection e.g. water level, visual aspects, functioning of pumps and (ii) higher intervention tasks, often requiring some specialist technical training e.g. parts replacement or disassembly of components for cleaning or inspection. The frequency of each of the maintenance tasks and the level of expertise required for the higher intervention tasks were used to compare design scenarios for the INNOQUA system. A rating of '0' indicated that a relatively high frequency or complexity of maintenance may be required at numerous points within a system (resulting from design choices) and also indicated a dependency on suitably qualified professionals. A rating of '3' indicated a low frequency and/or complexity of maintenance required at a limited number of points within the system and a limited dependency on suitably qualified professionals.

8.4.1.2.5 Visual impact

The visual impact of different design options for the INNOQUA technologies was evaluated by analysing the overall external volume of the installations. Initially the project footprint on site was calculated by the total surface occupied (in metres squared) multiplied by the height above ground of the highest component of the system. A rating of '0' indicated a significant bulk volume visible aboveground covering a large footprint and/or height aboveground and indicated high potential for visual impact. A rating of '3' indicated low potential for visual impact with significant amounts of the system below ground level and minimal modules of height visible above ground. Aesthetics were not considered and thus design choices in which a system was more visible were considered as less favourable than those where the system was less visible.

8.4.1.2.6 Social acceptance

A key consideration in the success of any novel technology is whether or not it will be adopted by end users. Thus social acceptance can be defined as the use, or adoption of a technology versus just the passive "approval" of a technology, or intention to use it (Mallett, 2007). In the INNOQUA Project, social acceptance was explored using a questionnaire circulated to various stakeholders in different regions to gauge and evaluate their responses, as described in Section 8.3.2. A key part of this evaluation was the degree to which nature-based systems (particularly the Lumbrifilter which contained worms) might be accepted by local communities and professionals.

8.4.1.2.7 Ease of installation

The degree of complexity of installation is a consideration in the selection of any system. Those requiring extensive and/or invasive civil engineering works and the involvement of a variety of technical specialists might be less favoured. The quantity of individual components comprising a system is a good indication of the ease with which installation can be achieved: a component can be a pump station or tank or piped connection. In this case ease of installation was assessed on a scale of 1 to 3 whereby a rating of '1' indicated a number of component parts in excess of 7, a rating of '2' indicated a number of component parts above between 5 and 7 and a rating of '3' indicated a number of component parts below 5.

8.4.1.2.8 Robustness and recovery

These criteria considered the likelihood of the system breaking down during normal operations and its expected recovery after breakdown or normal periods of shut down. System recovery following a breakdown was assessed as the time that the system requires to return back to normal working conditions after a shut-down period – in this case a shut-down period of less than 1 month was assumed. This was considered a key feature in many sites – for example an office or commercial premises may be closed for holidays or a wastewater treatment system in a tourist region may only receive influent wastewater during certain periods of the year. In the case of the INNOQUA project – design choices that may enable a system to recover to normal working conditions with no additional intervention in a matter of days was given a rating of '0'; if a time period of a number of weeks and additional intervention (e.g. seeding with activated sludge or new worms etc) was required a rating of "3" would apply.

8.4.2 Site installation and Setup

8.4.2.1 Planning and location

Once a solution has been identified and its suitability to meet the necessary criteria has been established e.g. using assessments such as those presented above, there are a number of additional aspects to be considered in locating the system on a site. Some of these are linked to securing any necessary permits or planning consents due to the proposed installation e.g. in Ireland, the Environmental Protection Agency (EPA) requires certain minimum distances to be maintained between boundaries, dwellings, roadways and any onsite treatment systems (EPA, 2010). The relevant local legislative or regulatory body should be consulted to determine requirements as these are likely to vary considerably from country to country and even within countries, particularly in areas that have been designated as having particular ecological or environmental significance. The practicalities of transportation and access for delivery, installation and ongoing maintenance of a system are also factors for consideration at this early stage in deciding the final onsite location.

8.4.2.2 Pre-installation site works

In most cases, in the implementation of a new on-site sanitation system (whether nature-based or not), some works are required in advance to prepare the location and to facilitate the installation. Such works can include the provision of electrical power supply, installation of pipework and excavations for below-ground installations.

Pre-installation works were required at the majority of INNOQUA demonstration sites, the most common being the provision of additional pipework, pumping stations and access chambers – as well as electrical supplies and suitable level (generally concrete) plinths to take the installations. In Italy, a dedicated room was designed and constructed in the basement of the demonstration residential property, while in Tanzania an excavation pit and shelter for the daphniafilter was required. The availability of contractors, the temperature extremes that limit working hours and the availability of building materials all posed challenges to the completion of the works at the site in Tanzania. At the demonstration site in Arequipa (Peru), the pre-installation works were complicated by difficulties locating existing power utilities and possible connection points. It is not unusual to encounter such challenges and careful planning and management of pre-installation works will ensure that any impact on a planned installation is minimized.

8.4.3 Stakeholder Engagement

In the development, demonstration and then commercial application of a novel sanitation system, the process of identifying and engaging stakeholders will vary depending on the nature and scale of the proposal e.g. whether it is a single house installation or community-based solution. In this context, stakeholder engagement is the notification and consultation of those who can effect or be affected by an installation and its operation. This is carried out with a view to understanding and addressing any concerns that may be held by such groups and is particularly important in a community setting. Where a site is private property and the sanitation solution concerns a single house installation with limited exposure to the local community, the number of potential key stakeholders who may be directly impacted by an installation is limited. However in the case of novel solutions where the role of early adopters is critical to a solution's wider acceptance, stakeholders may be identified beyond the householder, including (but not limited to) technology promoters, installers, neighbouring property owners and planning/regulatory bodies. The process of identifying stakeholders is often intuitive and guided by local knowledge and/or experience in carrying out similar exercises.

The case for a comprehensive stakeholder engagement strategy was elucidated by the varying experience of the community-based demonstration sites in Scotland, Tanzania and India. In the case of the INNOQUA installation at Little Mill (Scotland), Scottish Water began stakeholder meetings well in advance of installation. Key stakeholders such as the Scottish EPA (SEPA), Scottish Water operations teams the local planning authority and local residents were identified in the early project planning phases. Consultations were then held to identify and understand potential issues. Each of the local residents was contacted individually in the planning stage to discuss the proposed installation and a Scottish Water communications team also met with the local community. Residents were kept updated on project progress by letter communication. These consultations proved invaluable and not only avoided planning objections as the residents' concerns relating to aspects such as negative visual impact were identified and addressed in the progressed scheme, but also promoted a sense of local community stewardship in the project as demonstrated by their participation in a site open day where the science of the technology was demonstrated.

In the case of Tanzania, the complex socio-economic landscape within the informal settlements of Dar es Salaam made early INNOQUA team engagement with local government authorities, officials and local leaders as well as influential members of the community extremely difficult. The inability to identify, contact and work with all relevant local stakeholders led to significant opposition during subsequent system operation, prompted by odours generated during wastewater dosing that were exacerbated by operational errors that led to overflows of raw wastewater into the treatment compound. This highlighted common sensitivities to wastewater management schemes, which can historically be associated with bad odours and illnesses. Subsequent close collaboration with influential local leaders, coupled with relocation of the treatment plant within the compound and improvements to operational practices transformed attitudes and acceptance of the new system, which then had a local champion with a stake in its success.

In the case of the INNOQUA demonstration site in India, efforts to engage with community representatives were made early on in the project, however there were still concerns that led to opposition to the installation of the wastewater treatment system directly within the community where it could have had the most positive impact. Instead the facility had to be re-located nearby.

While early-stage engagements were made with community partners and other stakeholders, our key recommendations to other practitioners include:

1. The use of the stakeholder surveys (outlined in this chapter) could have been more extensive. This type of survey can be performed online, in-person or via virtual meetings and can help highlight any potential issues. In the case of INNOQUA this survey was mainly answered by practitioners and a broader response base would have been preferable. Response numbers also varied between countries and to develop more robust analysis more responses from different stakeholders would be necessary. This should be borne in mind when analysing results from such surveys and drawing conclusions. The advantage of allocating resources to such an activity is the development of a much deeper understanding of local capacity and requirements and thus the ability to tailor a solution for that market.
2. The development of educational and training materials was planned for the second half of the INNOQUA project. The development and deployment of such materials earlier in the project may have enabled improved feedback from stakeholders and again highlighted potential barriers and opportunities.
3. At all demonstration sites the project was committed to engaging with, utilising and learning from local experts and stakeholders in formal (e.g. via the survey and site open days) and informal (e.g. impromptu engagement with stakeholders at demonstration sites) ways. These approaches worked well and enabled demonstration sites to feel "ownership" over the project. Furthermore, this approach led to better design solutions across the project with lessons from experts in India being applicable not just to the demonstration site in India but also to those in other countries.

Overall, a number of established frameworks and formalised assessment approaches are available to help practitioners understand attitudes to new technologies, to contextualise those technologies within a local market landscape and to support their adoption. Although significant weight is commonly given to engineering and other operational aspects when developing new solutions – our experience has highlighted the value of early and extensive stakeholder engagement in securing longer-term success.

References

Akpan, V. E., Omole, D. O. and Bassey, D. E. (2020). 'Assessing the public perceptions of treated wastewater reuse: opportunities and implications for urban communities in developing countries', *Heliyon*. Elsevier Ltd, 6(10), p. e05246. doi: 10.1016/j.heliyon.2020.e05246.

Barraqué, B. (2020). 'Full cost recovery of water services and the 3 T's of OECD', *Utilities Policy*. Elsevier Ltd, 62, p. 100981. doi: 10.1016/j.jup.2019.100981.

bsi (2020). 'BS EN ISO 14040:2006+A1:2020 – Environmental management. Life cycle assessment. Principles and framework'. Available at: https://shop.bsigroup.com/ProductDetail?pid=000000000030379352 (Accessed: 29 April 2021).

De Groot, J. and Steg, L. (2009). 'Morality and prosocial behavior: The role of awareness, responsibility, and norms in the norm activation model', *Journal of Social Psychology*. Heldref, 149(4), pp. 425–449. doi: 10.3200/SOCP.149.4.425-449.

Dunn, M. E., Mills, M. and Veríssimo, D. (2020). 'Evaluating the impact of the documentary series Blue Planet II on viewers' plastic consumption behaviors', *Conservation Science and Practice*. Wiley, 2(10). doi: 10.1111/csp2.280.

du Pisani, P. L. (2006). 'Direct reclamation of potable water at Windhoek's Goreangab reclamation plant', *Desalination*. Elsevier, 188(1–3), pp. 79–88. doi: 10.1016/j.desal.2005.04.104.

EPA (2010). *2009 Code of Practice for Wastewater Treatment for Single Houses*. Available at: http://www.epa.ie/pubs/advice/water/wastewater/2009 code of practice for single houses/#d.en.70640 (Accessed: 29 April 2021).

European Commission (2015). *Towards and EU Research and Innovation policy agenda for Nature-Based Solutions & Re-Naturing Cities*. Available at: https://ec.europa.eu/programmes/horizon2020/en/news/towards-eu-research-and-innovation-policy-agenda-nature-based-solutions-re-naturing-cities.

European Commission (2019). *A European Green Deal*. Available at: https://ec.europa.eu/info/strategy/priorities-2019-2024/european-green-deal_en (Accessed: 29 April 2021).

European Commission (2020a). *Eurobarometer: Protecting the environment and climate.* Available at: https://ec.europa.eu/commission/presscorner/detail/en/IP_20_331 (Accessed: 2 June 2021).

European Commission (2020b). *Innovative Ecological on-site Sanitation System for Water and Resource Savings.* Available at: https://cordis.europa.eu/project/id/689817 (Accessed: 29 April 2021).

European Commission (2020c). *Nature-Based Solutions: State of the Art in EU-funded Projects.* Available at: https://op.europa.eu/en/publication-detail/-/publication/8bb07125-4518-11eb-b59f-01aa75ed71a1/language-en (Accessed: 2 June 2021).

European Commission (2021). *Nature-based solutions.* Available at: https://ec.europa.eu/info/research-and-innovation/research-area/environment/nature-based-solutions_en (Accessed: 16 February 2021).

Finnveden, G. *et al.* (2009). 'Recent developments in Life Cycle Assessment', *Journal of Environmental Management.* Academic Press, 91(1), pp. 1–21. doi: 10.1016/j.jenvman.2009.06.018.

Franceys, R., Pickford, J. and Reed, R. (1992). *A Guide to the Development of on-Site Sanitation.* Available at: https://apps.who.int/iris/handle/10665/39313 (Accessed: 2 June 2021).

Freeman, R. E. (1984). *Strategic management: a stakeholder approach.* Boston: Pitman. Available at: https://www.worldcat.org/title/strategic-management-a-stakeholder-approach/oclc/9685996 (Accessed: 29 April 2021).

GCF (2021). *Green Climate Fund.* Available at: https://www.greenclimate.fund/ (Accessed: 29 April 2021).

Hauschild, M. *et al.* (2011). *Recommendations for Life Cycle Impact Assessment in the European context (International Reference Life Cycle Data System – ILCD handbook).* Available at: https://publications.jrc.ec.europa.eu/repository/handle/JRC61049 (Accessed: 29 April 2021).

Hrovatin, N. and Bailey, S. J. (2001). 'Implementing the European Commission's water pricing communication: Cross-country perspectives', *Utilities Policy.* Pergamon, 10(1), pp. 13–24. doi: 10.1016/S0957-1787(02)00009-7.

Istenic, D., Bodík, I. and Bulc, T. (2015). 'Status of decentralised wastewater treatment systems and barriers for implementation of nature-based systems in central and eastern Europe', *Environmental Science and Pollution Research.* Springer Verlag, 22(17), pp. 12879–12884. doi: 10.1007/s11356-014-3747-1.

JRC-IES (2012). *Characterisation factors of the ILCD Recommended Life Cycle Impact Assessment methods Database and supporting information.* Luxembourg. doi: 10.2788/60825.

Kabisch, N. *et al.* (2017). 'Nature-Based Solutions to Climate Change Adaptation in Urban Areas—Linkages Between Science, Policy and Practice', in *Theory and Practice of Urban Sustainability Transitions*. Springer, Cham, pp. 1–11. doi: 10.1007/978-3-319-56091-5_1.

Katsou, E. *et al.* (2020). 'Transformation tools enabling the implementation of nature-based solutions for creating a resourceful circular city', *Blue-Green Systems*. IWA Publishing, 2(1), pp. 188–213. doi: 10.2166/bgs.2020.929.

Kisser, J. *et al.* (2020). 'A review of nature-based solutions for resource recovery in cities', *Blue-Green Systems*. IWA Publishing, 2(1), pp. 138–172. doi: 10.2166/bgs.2020.930.

Lahnsteiner, J. and Lempert, G. (2007). 'Water management in Windhoek, Namibia', *Water Science and Technology*. IWA Publishing, 55(1–2), pp. 441–448. doi: 10.2166/wst.2007.022.

Mallett, A. (2007). 'Social acceptance of renewable energy innovations: The role of technology cooperation in urban Mexico', *Energy Policy*. Elsevier, 35(5), pp. 2790–2798. doi: 10.1016/j.enpol.2006.12.008.

Naumann, S. *et al.* (2011). *Assessment of the potential of ecosystem-based approaches to climate change adaptation and mitigation in Europe Assessment of the potential of ecosystem-based approaches to climate change adaptation and mitigation in Europe FINAL REPORT*. Available at: https://ec.europa.eu/environment/nature/climatechange/pdf/EbA_EBM_CC_FinalReport.pdf (Accessed: 29 April 2021).

NTD NGO Network (2021). *NNN |A global forum for nongovernmental organizations working together on NTDs*. Available at: https://www.ntd-ngonetwork.org/ (Accessed: 2 June 2021).

NTD NGO Network and World Health Organization (2019). *WASH and Health working together A 'HOW-TO' GUIDE FOR NEGLECTED TROPICAL DISEASE PROGRAMMES*. Available at: https://apps.who.int/wash-health-toolkit/ (Accessed: 2 June 2021).

Schwartz, S. H. (1977). 'Normative influences on altruism', *Advances in Experimental Social Psychology*. Academic Press, 10(C), pp. 221–279. doi: 10.1016/S0065-2601(08)60358-5.

Shandas, V. and Messer, W. B. (2008). 'Fostering green communities through civic engagement: Community-based enviromental stewardship in the Portland area', *Journal of the American Planning Association*. Taylor & Francis Group, 74(4), pp. 408–418. doi: 10.1080/01944360802291265.

Smith, H. M. *et al.* (2018). 'Public responses to water reuse-understanding the evidence', *Journal of Environmental Management*, 207, pp. 43–50. doi: 10.1016/j.jenvman.2017.11.021.

World Health Organization (2012). *ACCELERATING WORK TO OVERCOME THE GLOBAL IMPACT OF NEGLECTED TROPICAL DISEASES A ROADMAP FOR IMPLEMENTATION*. Available at: https://www.who.int/neglected_diseases/NTD_RoadMap_2012_Fullversion.pdf (Accessed: 2 June 2021).

DOI: 10.1561/9781680837810.ch9

Chapter 9

INNOQUA – Market Opportunities and Commercialisation

By Domenico Perfido

9.1 Introduction

This chapter explores the commercialisation journey of the INNOQUA technologies through the application of a sequence of established tools.

A top-down approach was first used to understand market potential at regional and national scales, exploring aspects such as existing sanitation provision and differences between urban and rural wastewater treatment infrastructure. The wider wastewater market landscapes in each of the INNOQUA partner countries were then explored through a PESTEL (Political, Economic, Social, Technological, Environmental, and Legal) analysis. This identified market opportunities for innovative decentralised sanitation solutions in terms of current and potential future needs.

A bottom-up approach was then used to ensure that local wastewater discharge and reuse requirements were understood, and the demonstration facilities designed and installed to test compliance against these requirements when treating wastewater from a selection of sources with different pollutant loadings.

Potential stakeholders and competitors were subsequently explored in terms of products/services offered, associated costs, market share, and strengths before the INNOQUA technologies were subjected to a SWOT (Strengths, Weaknesses, Opportunities and Threats) analysis. This informed market positioning and prioritisation which was used to develop an outline commercialisation strategy and business plan options for the INNOQUA technologies. Since the technologies still required further market testing at the end of the project, an INNOQUA Alliance was created to support continued product development and to allow the consortium to capitalise on its collective know-how in researching and commercialising other nature-based solutions.

The structured approach to market exploration, competitor analysis and product/service positioning provides a robust framework that can be readily adapted to the commercialisation needs of other innovators.

9.1.1 Market Potential

An understanding of the wider market landscape is a key aspect of any commercialisation strategy. As an international project, INNOQUA drew on trans-boundary drivers such as:

- Climate change impacts on water resource distribution and quality
- Impacts of poor/non-existent sanitation provision on health and welfare
- Impacts of wastewater on the wider environment
- Opportunities for resource efficiency, including nutrient and reuse
- The need for reduced energy and resource intensity in wastewater treatment

Several of these are considered in detail elsewhere in this book: Regulations for wastewater reuse, impacts of poor sanitation provision, water stress and resource constraints are explored in Chapter 2. More local drivers such as discharge standards and variations in regulatory approach are explored in Chapter 3. As set out in Chapter 1, there is a fundamental need to develop and rapidly implement sustainable, cost-effective sanitation solutions if the aspirations of SDG6 are to be realised.

Top-down market analysis within the INNOQUA partner countries allows opportunities to be visualised at a global level through data on sanitation provision (Figure 9.1), which immediately highlights gaps – in this case principally within India, Peru, Romania and Tanzania.

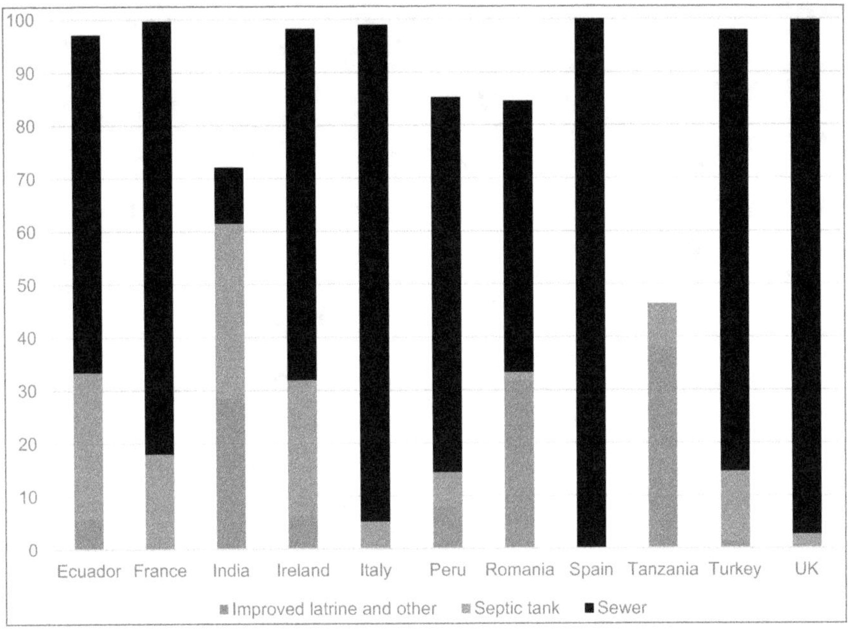

Figure 9.1. Percentage of population with access to improved sanitation in a sub-set of the INNOQUA partner countries, 2017. Improved sanitation facilities "are those designed to hygienically separate excreta from human contact, and include: flush/pour flush to piped sewer system, septic tanks or pit latrines; ventilated improved pit latrines, composting toilets or pit latrines with slabs" (UN Water, 2021).

These data do not provide insights into whether or how effectively wastewater is treated – particularly where sewer systems have been installed. For example, comparing Figure 9.2 with Figure 9.1 highlights opportunities for wastewater treatment in Ecuador, Peru and Turkey. Such treatment could take place on site (prior to discharge to sewer) or at the 'end of pipe' prior to final discharge or reuse (the classic centralised wastewater treatment approach).

In many cases there are also disparities in sanitation provision between urban and rural populations (Figure 9.3). Although decentralised wastewater treatment solutions can be installed in urban environments, land constraints and other aspects can mean that rural installations are more straightforward (see Chapter 2 for further discussion of this aspect). The headline data suggest significant opportunities for INNOQUA solutions within rural communities in Ecuador, India, Peru, Tanzania and Turkey. Rural populations are also of interest in France, Romania and the UK – where they number 13.0M, 8.9M and 10.9M inhabitants, respectively (The World Bank, 2021e).

These top-level sanitation statistics can be inputted into a PESTEL analysis, as can other readily accessible data such as anticipated sanitation expenditure (Table 9.1).

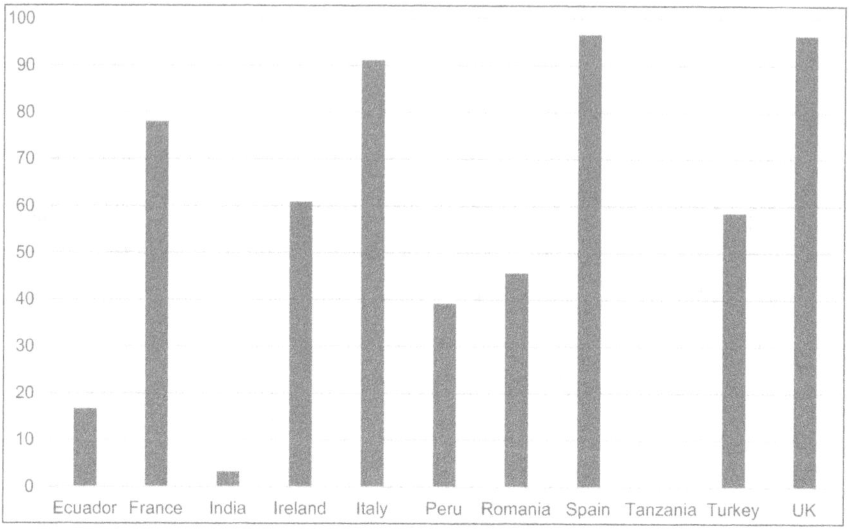

Figure 9.2. Percentage of population connected to sewer with associated wastewater treatment in a sub-set of INNOQUA partner countries, 2017. Defined as "Proportion of population using improved facilities which are not shared with other households and where excreta are transported through a sewer with wastewater and then treated off-site" (UN Water, 2021).

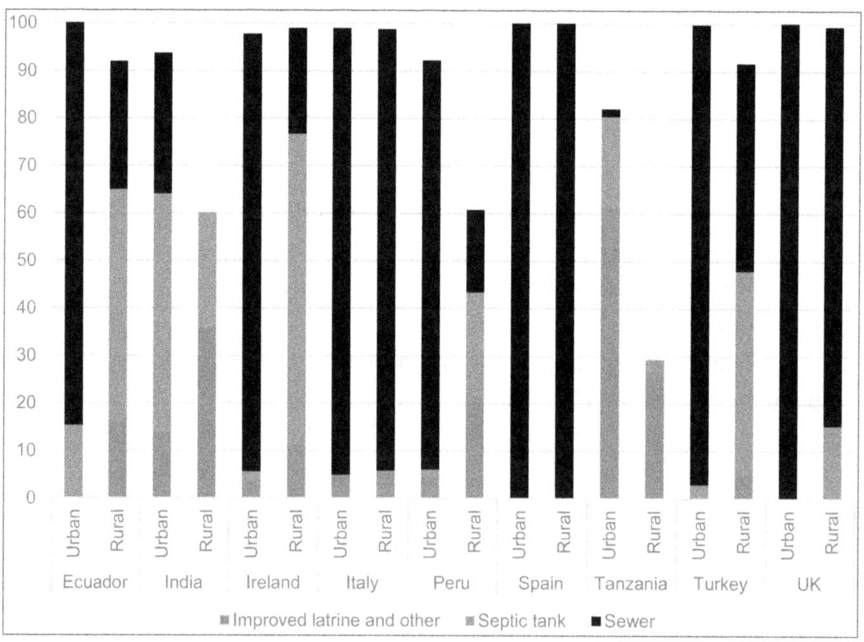

Figure 9.3. Percentage of urban and rural populations connected to different methods of improved sanitation in a sub-set of INNOQUA partner countries, 2017 (UN Water, 2021).

Table 9.1. Expected yearly investment in wastewater treatment for INNOQUA partner countries in Europe - for both new and replacement infrastructure, in 2016. Investment data from European Commission (no date), population data from The The World Bank (2021e).

Aspect Unit	Collecting System EUR per Inhabitant	Treatment Plant EUR per Inhabitant	Population Millions
Spain	172	489	46.5
France	2,750	1,550	66.7
Ireland	118	205	4.76
Italy	354	397	60.6
Romania	546	198	19.7
UK	576	904	65.6

9.2 PESTEL Analysis

9.2.1 Approach

In this section we explore the status of decentralised sanitation systems in eleven INNOQUA partner countries through a PESTEL analysis (**P**olitical, **E**conomic, **S**ocial, **T**echnological, **E**nvironmental, and **L**egal) (PESTLE Analysis, 2021). Each of the PESTEL factors was scored as set out in Table 9.2, allowing countries to be ranked in terms of likely INNOQUA market potential (Table 9.4). The PESTEL analyses were compiled and scored in collaboration with INNOQUA partners with specific expertise on wastewater markets in each country. The different aspects that they considered for each PESTEL factor are summarising in Table 9.3.

Table 9.2. PESTEL factor scoring approach.

GREEN = **score of between 7 and 10**	PESTEL factors are in line with foreseen market deployment requirements, thus the **market landscape is a driver for implementation** of the INNOQUA solutions
AMBER = **score of between 4 and 6**	PESTEL factors investigated present little or no input into foreseen market deployment requirements, thus the **market landscape is a small/non-factor**
RED = **score of between 1 and 3**	PESTEL factors present strong evidence against any foreseen market deployment requirements, thus the **market landscape is a significant barrier and strong limitation/risk**

Table 9.3. PESTEL factors and key aspects considered by INNOQUA partners.

Factor	Aspects That Were Considered
Political	• Policy/Regulatory drivers such as implementation of the EU Water Framework Directive targets for water body quality – which often require capture and particular treatment of wastewater. Aspects such as suspended solids, COD and nutrients are all commonly subject to regulatory limits • Changes in administrative structures – such as grouping smaller communities together – can help to create supportive funding environments, but may also encourage adoption of centralised wastewater treatment solutions • Water reuse may be actively encouraged or incentivised • Responsibility may be split across different branches of government and between local/central government agencies – creating barriers to co-ordinated delivery of sanitation infrastructure
Economic	• Costs of wastewater treatment and/or provision of water for drinking, irrigation and other purposes are key drivers. Where wastewater treatment solutions are already available in a given market, then the innovation needs to be cost-competitive with these solutions whether complying with discharge or reuse requirements. Where current costs are low, then this may represent a significant barrier to the adoption of INNOQUA technologies. Such costs can be highly regional within single countries • In some geographies, costs of sanitation can act as a significant barrier to their actual reuse (for example, costs for a household to connect to a new sewer network might be prohibitive for that householder and deter them from connection). This applies to some parts of both the Global North and South, and means that no matter how excellent new solutions are – they must align with the economic capabilities of target markets • Wastewater treatment can support sustainable population and economic growth, delivering high benefit-to-cost ratios. Establishing and communicating such factors may be essential when engaging with potential funders
Social	• In many countries, rural and peri-urban populations are less well served by centralised wastewater capture and treatment than urban populations – providing immediate market opportunities • Challenging terrain (for example, mountainous regions) and dispersed populations can mitigate against centralised wastewater treatment infrastructure – providing opportunities for decentralised approaches

(Continued)

Table 9.3. Continued

Factor	Aspects That Were Considered
	• Awareness of the issues arising from malfunctioning or absent sanitation can be low within communities, businesses and households – providing various opportunities for education and promotion of the INNOQUA solutions • Ensuring acceptance of wastewater reuse can be challenging • Biodiversity and sustainability may be features of national pride and discourse, indicating support for nature-based and other sanitation solutions that preserve or enhance the environment • Illness and death associated with poor sanitation represent significant drivers for change, which may be reflected at political level • Independence from centralised systems can appeal – for example, by reducing or eliminating reliance on corrupt revenue-recovery systems • Corruption at local or national level can support or hinder the deployment of new solutions
Technological	• Even where centralised wastewater treatment infrastructure is present, it requires specialist operation and maintenance by appropriately trained (and compensated) staff. Where existing works have to be enlarged or otherwise modified, the INNOQUA technologies may prove a useful low-tech alternative. The INNOQUA solutions require only periodic specialist attention. • Where the majority of a country's population already have access to a centralised sewer network then the market need for INNOQUA may appear low. However, collection of wastewater is not automatically indicative of wastewater treatment, and INNOQUA can have an important role in on-site wastewater treatment ahead of discharge to sewers that are not connected to downstream or 'end of pipe' treatment • Existing decentralised infrastructure (such as septic tanks and small package plants) requires routine intervention and may provide a favourable context for INNOQUA, as a low or zero sludge solution • Sewage and wastewater can be highly variable, depending on the local catchment and collection norms. Flexible solutions such as INNOQUA that are capable of treating concentrated and dilute (blackwater and greywater) streams provide market advantages • A lack of existing alternatives may indicate immediate market potential

(*Continued*)

Table 9.3. Continued

Factor	Aspects That Were Considered
Environmental	• Water shortages at regional or national scale can be important drivers for uptake of wastewater reuse • The need to improve the quality of water bodies by capturing and treating wastewater can be an important driver to develop and improve treatment technologies in all countries • Extremes of climate (particularly temperature) are challenging for all nature-based solutions, and adaptations have to be considered accordingly • Irrigated agriculture is a significant user of water in some areas, and can represent an important driver for wastewater reuse. The need for nutrient removal/recovery as part of the wastewater treatment should be considered in the context of potential reuse where those nutrients can have value as a resource • Climate change is leading to rapid alterations in precipitation patterns and distribution/flows of surface and ground water. This is likely to drive interest in local wastewater reuse initiatives
Legal	• In some countries new technologies must obtain Ministerial or other official approval before they can be placed on the market. Specific regulatory constraints may be applied to decentralised systems, whether discharging into aquatic or terrestrial ecosystems • International legal challenges (for example, of EU Member States by the European Court of Justice) can force the implementation of sanitation infrastructure, although may not be prescriptive with respect to centralised or decentralised approaches • Changes in regulatory approach to existing on-site sanitation can stimulate interest in alternative solutions • New legislation on aspects such as health and welfare can provide strong (indirect) drivers for the uptake of sanitation solutions – as can legislation on environmental protection in its broadest sense • Regulations can make it impossible for decentralised solutions to achieve the required discharge standards. On the other hand, regulations and/or enforcement may not exist at all

9.2.2 Country-specific Results

European countries are generally considered industrialised, an important market driver for INNOQUA deployment. However, the variety of political regimes and distinct climate zones is reflected in the range of standards and regulations across the different countries. Cities and conurbations >2000 PE within the EU are required to collect and treat their own wastewater under the Urban Waste Water Treatment Directive – while discharges from smaller communities and individual dwellings may be subject to local regulation. In Spain 350 million m^3 and in Italy 240 million m^3 of wastewater are re-used per year (Jimenez and Asano, 2008), while in France less than 1% of wastewater is reused (Plat *et al.*, 2019). Since rural, mountainous, and coastal regions in several European countries are frequently not connected to the water network and the opportunities for the re-use of natural resources increase, there are clear opportunities for the INNOQUA system.

Sanitation is a key topic for South American sustainability plans (it was even integrated into the 2016 New Urban Agenda delivered by the United Nations in Quito). On the other hand, the Latin American market is very heterogeneous and does not benefit from large common legal, cultural or organizational tools among the different countries in the continent; hence this market is much more a sum of several national markets than a large homogeneous market. However, countries like Ecuador and Colombia share many similarities, and the same is true of Peru and Bolivia (for example). Hence demonstration of the INNOQUA solutions in particular countries can be used to understand and extend market perspectives through these related countries, before extending to a broader continental scope.

As with Europe or South America, wastewater treatment in countries within the Middle East, Africa, and Asia requires unique and contextual solutions which consider metrics such as culture and climate variability, affordability of technologies, space constraints (due to growing costs of real estate) and reduced electrical loading/maintenance requirements. While these indicators are consistent with market research for INNOQUA in other regions, they become particularly important when assessing market replication potential in Turkey, Tanzania and India. During the course of the INNOQUA project a comprehensive review of each country was completed.

Based on the detailed PESTEL analyses, each studied country was ranked (Table 9.4). In some respects, this ranking may appear counterintuitive since countries such as India and Tanzania have obvious needs for additional wastewater treatment. However, the value of the PESTEL approach is that it provides a standardised, expert-driven objective approach to market review – and factors such as the availability of funding, local corruption, community interest and awareness, and absence of supportive regulatory environments can all mitigate against target markets that might otherwise appear promising. The summaries below are intended to highlight some market barriers and opportunities in each country; further detail can be viewed in the pre-market assessment (Elia *et al.*, 2017).

Table 9.4. PESTEL Scoring results.

Position	Country	Competition Metrics Score
1	Turkey	Score 44
2	Peru	Score 43
3	Italy	Score 41
4	France	Score 39
5	Ireland	Score 38
6	Spain	Score 38
7	Ecuador	Score 38
8	India	Score 38
9	UK	Score 36
10	Tanzania	Score 34
11	Romania	Score 33

9.2.2.1 Ecuador

With less than 10% of wastewater receiving treatment, Ecuador has a huge market potential for INNOQUA technologies. Over two-thirds of the population live in cities, the urbanisation rate is high, and while in urban areas the majority of houses are connected to a sewage network, most of the collected wastewater is untreated. With the expansion of cities, expansion of infrastructure is lacking. This is an opportunity for decentralised wastewater treatment solutions.

Outside urban areas, rural areas and small municipalities are also an area of interest. Local municipalities struggle to attract funding from regional or national programs and often lack the skills to implement and maintain a complex wastewater treatment system. To be successful, INNOQUA not only needs to deliver the technology, but also pay attention to relevant stakeholders such as local citizens, policy makers and local authorities, offer awareness campaigns and training sessions and provide support with respect to project and financial management. Linking to existing programs, such as green building initiatives or national programs like the Prosaneamiento project (Banco Interamericano de Desarrollo, 2014), will be required to attract the necessary funding. Partnerships with local firms will ensure buy-in from local authorities and the local community. Any system, including a nature-based system such as INNOQUA should be inexpensive and easy to operate and maintain to ensure local staff can operate the system.

Alternatively, INNOQUA can target the tourist industry – particularly in rural regions and in national parks. Resorts need to comply with strict ecological rules when they are operating in the national parks and nature-based systems such as INNOQUA can contribute positively to enabling sustainable development of tourism industries.

9.2.2.2 France

With an estimated installation of five million individual household wastewater treatment installations, the French market is of interest for decentralised nature-based solutions. The introduction of wastewater certificates in 2012 (which may not be older than three years) means that existing decentralised sanitation facilities are likely to be renovated or renewed more frequently, resulting in more opportunities for customers to consider switching to the INNOQUA system. On-site treatment units are priced in the range of € 5,000 – € 15,000 while costs for sludge removal over the lifetime of the system are around € 2,000.

9.2.2.3 India

Since its independence in 1947, India has grown to become the world's fifth largest economy, home to over 1.3 billion people. India is still considered a developing country and 400 million Indian people live in poverty (The World Bank, 2021a). Wastewater treatment is done in a highly decentralised and uncontrolled manner, while much of the wastewater remains uncollected and untreated (Schellenberg *et al.*, 2020). Despite environmental initiatives like the Clean India Mission and Clean Ganga Mission, general environmental awareness remains low. In many cases, industrial and domestic wastewater are mixed where separate treatment may be required. This can make it more challenging for nature-based solutions and indeed any technology that relies on biological treatment to deliver a standard solution for domestic environments. Lack of coordination and enforcement by local authorities has resulted in a broad range of local solutions for industry and households.

Legislation in some municipalities requires apartment buildings to install local wastewater treatment systems, although the extent of any regulatory oversight of this requirement is unclear. Where a system could meet the specific demands for this segment, such as a small footprint, low maintenance and capable of fully functioning in a basement, there are opportunities for nature-based solutions. Other areas where nature-based systems such as the INNOQUA system could add value are buildings or groups of buildings with no access to a central sewage system like schools, hospitals and informal settlements. Since the wastewater market in India is highly competitive, with many local and foreign solution providers, any system must be priced in line with competing solutions.

9.2.2.4 Ireland

Ireland has been one of the fastest growing economies in the Eurozone. Water charges, which were introduced in recent years for domestic users, have now been abandoned for political reasons. Albeit, while in place the majority of domestic users had paid the required charges.

In the past decades, wastewater treatment in urban areas has improved significantly but compliance rates for effluent quality are low compared to EU rates as a whole (European Commission, no date). In 2015, untreated wastewater from 43 areas was routinely discharged to the environment. Urban wastewater continues to be one of the principle pressures on water quality, affecting bathing water quality and contributing to river pollution. The connection rate to a centralised sewerage system is now nearly 100%, leaving little room for an INNOQUA system in an urban context in Ireland.

In rural areas, wastewater from around 500,000 dwellings is treated on-site as there is often no connection with a central sewerage system. In 2014, a National Inspection Program started. In a first round of inspections of on-site decentralised wastewater treatment systems nearly 50% failed to meet safety and health standards (epa, 2015). Public awareness about health risks related to malfunctioning septic tank systems is relatively low. Local authorities stimulate the installation of group sewage systems in rural areas. The INNOQUA system would fit this market, but would need to address also public behaviours, awareness and attitudes regarding on-site wastewater treatment systems (Hynds *et al.*, 2017).

9.2.2.5 Italy

Italy is characterised by a fragmented and complex wastewater market, with big differences in the quality of wastewater treatment plants and problems meeting EU directives. Italian legislation encourages wastewater reuse, particularly as a means of reducing primary abstractions and improving water body quality. Italian legislation for treatment and reuse of wastewater is based on the Technical Regulations on Treated Wastewater Reuse (Ministero dell'ambiente, 2006) which, in conjunction with other legislation, establish discharge and reuse quality limits. Treated wastewater can be used for the following:

- Irrigation of crops for food and feed production, as well as amenity landscaping and land used for other non-food purposes;
- Firefighting and wash-down of streets in urban centres;
- Cooling and heating;
- Toilet flushing (in buildings with dual networks: one for potable water and one for flushing only); and
- Various industrial processes (Re *et al.*, 2021).

Pressures on groundwater are particularly high in southern Italy, and regional governments add specific prescriptions to the national regulations to suit local circumstances. For example, in Sardinia the direct re-use of treated wastewater for potable purposes is allowed (Sanna, 2019). Sicily is particularly under-served with respect to wastewater treatment infrastructure, with a number of incomplete, undersized, or outdated plants that serve around 60% of the island's population. Innovations in this area are particularly needed (Re *et al.*, 2021).

9.2.2.6 Peru

According to reports from the World Bank, Peru has a healthy and fast-growing economy (The World Bank, 2021b). People have a high level of environmental awareness and there is strong political support for water management programs. Peru suffers from water shortages, especially in the coastal region, where the majority of the people live. One third of the localities do not have a system for water treatment, providing an opportunity for an INNOQUA system serving individual or groups of houses.

As the largest consumer of water and producer of wastewater, industry is an important market for decentralised wastewater treatment systems. On a per case basis the characteristics of the effluent needs to be studied in order to configure any decentralised treatment system correctly. This customisation will increase the price and also makes the offering of standardised solutions more complex.

9.2.2.7 Romania

Although the economic situation and living standards have improved since Romania became a full EU member in 2007, low income remains an important problem – particularly in rural areas where the majority of houses are not connected to drinking water or wastewater systems, and where people have little room for expenditure on sanitation systems. However, an inflow of EU funding and expected economic growth mean there is an expectation that the market for wastewater treatment technology will grow.

Local wastewater solutions for small communities and rural areas in Romania should be simple and inexpensive. Apart from the technologies themselves, INNOQUA should also offer an education program to raise awareness of responsible wastewater management and to set out how to implement a local wastewater system. One option is to offer INNOQUA as part of a regional development program, involving multiple stakeholders in a single region, such as tourist facilities, SME's and municipalities.

However, a high level of corruption means that Romania has been ranked second in the list of most corrupt countries in the developed world (Transparency International, 2017), making it more difficult for external companies to access potential markets. Licensing of INNOQUA technology to local SME's could be a way to bypass these issues.

9.2.2.8 Spain

Spain is the fourth-largest economy in the Eurozone. Spain has been emerging from recession since 2013. The unemployment rate at 18.4% is still high (as of 2016). 98% of households are connected to public sewer (OECD, 2012). The number of water-stressed regions, especially in the south of Spain, is likely to increase because of more frequent droughts due to climate change. Water conservation and wastewater reuse will become more important in the near future. In agriculture, the total irrigable area will increase. In urban areas, wastewater reuse programs have started. This is an opportunity for nature-based solutions, which can offer a local wastewater treatment system for farmers.

9.2.2.9 Tanzania

Both Tanzania's economy and population have grown steadily in the past decade. Despite the economic growth, still around 12 million people live in extreme poverty earning less than USD 0.60 per day (The World Bank, 2021c). Only 10–15% of the urban population has access to a sewerage system. Cultural norms limit the reuse of water to non-food related purposes.

Domestic wastewater is the main source of water pollution resulting in health risks. Lack of road infrastructure makes sludge removal difficult in many remote areas. This is an opportunity for the INNOQUA system to deliver a decentralised low-sludge system in rural settlements although the high poverty levels require new social business models or additional funding from donor programs or private foundations.

Another opportunity are hotels and the hospitality industry which are heavy water users and need to comply with various regulatory requirements. They risk being shut down when not meeting these standards. This is a strong incentive to invest in water management and an opportunity for INNOQUA, especially since hotels often have difficulties connecting to a centralised sewage network.

9.2.2.10 Turkey

Turkey's economy has developed strongly in the past decades, resulting in increased employment, less poverty and strong urbanisation. Lately, economic growth has slowed due to political uncertainty and terrorist attacks (The World Bank, 2021d). It is unknown what the effect will be on the further development of water and wastewater regulations since developments have largely been driven by harmonising with EU law. In the past, economic growth has prevailed above environmental performance, which is regarded as weak. Enforcement of existing laws and regulations is poor (Business Monitor International, 2015).

Opportunities exist in agriculture where water stress increases the need for farmers to reuse wastewater for irrigation, especially in the western part of Turkey.

Another opportunity is in the increasing water consumption of the growing industrial sector. The need for water reuse will increase and nature-based systems such as INNOQUA could configure their system to target eco-friendly businesses.

9.2.2.11 United Kingdom

The majority of households are connected to centralised drinking water and wastewater systems, although septic tanks are commonly used in rural areas – with estimates of 500,000 or more households using such systems for wastewater disposal (The Sunday Times, 2019). Recently introduced rules clarify the responsibilities of users of such systems and may incentivise upgrades or replacement of these decentralised units (Defra, 2015).

The UK has a highly developed wastewater management system and has established a strong regulatory framework and supporting policies, in line with the European Water Framework and Urban Waste Water Treatment Directives. The decision of the UK to leave the EU brings uncertainty with respect to the development of the UK economy and the influence on environmental policies and targets is unknown.

Where regulation with respect to nutrient and pesticide pollution is becoming stricter, the INNOQUA system could offer an inexpensive on-site solution for agriculture. For remote rural areas, an INNOQUA system providing a low sludge solution is of interest because of the difficulties such areas face with sludge removal. Overall, the UK wastewater market is a mature market with niche potential for decentralised systems in rural areas where households and small communities are reliant on septic tanks for wastewater disposal.

9.3 The Competitor Landscape

This section provides an overview of the most common decentralised/on-site sanitation systems, and some of the principle advantages and disadvantages of these are highlighted (Table 9.5) (further detail on the majority of these systems is provided in Chapter 3). Commercial examples of relevant technologies were identified for individual INNOQUA target countries and scorecards developed to summarise market knowledge for those technologies – informing future commercial positioning of INNOQUA solutions.

For each treatment type, Table 9.6 provides an overview of the presence or absence of any of the described technologies in INNOQUA target regions. This table has been developed from INNOQUA partner feedback from within each region.

Having identified existing decentralised wastewater treatment solutions for potential target countries, scorecards of key characteristics were developed for

Table 9.5. Description of existing treatment type: advantages and disadvantages (Elia *et al.*, 2017).

Treatment Type	Advantages	Disadvantages
	Aquatic Systems	
Stabilisation lagoons	• Low capital cost • Low operation and maintenance costs • Low technical manpower requirement	• Require a large area of land • May produce undesirable odours
Aerated lagoons	• Requires relatively little land area • Produces few undesirable odours	• Requires mechanical devices to aerate the basins • Produces effluents with a high suspended solids concentration
	Terrestrial Systems	
Septic tanks	• Can be used by individual households • Easy to operate and maintain • Can be built in rural areas	• Provides a low treatment efficiency • Must be pumped occasionally • Requires additional infrastructure for periodic sludge treatment/disposal
Constructed wetlands	• Removes up to 70% of solids and bacteria • Minimal capital cost • Low operation and maintenance requirements and costs	• Needs lot of land • Nutrient removal can be a challenge • Requires periodic removal of excess plant material • Best used in areas where suitable native plants are available
	Mechanical Systems	
Filtration systems	• Minimal land requirements; can be used for household-scale treatment • Relatively low cost • Easy to operate	• High maintenance • High level of preliminary treatment • Large footprint
Vertical biological reactors	• Highly efficient treatment method • Requires little land area • Applicable to small communities for local-scale treatment and to big cities for regional-scale treatment	• High cost • Complex technology • Requires technically skilled manpower for operation and maintenance • Needs spare parts (availability) • Has a high energy requirement

(*Continued*)

Table 9.5. Continued

Treatment Type	Advantages	Disadvantages
Membrane bioreactors (MBR)	• Fine screening for suspended impurities • Low space requirement • Adaptable to variable polluting loading	• Limited references in industrial application • Requires skilled operators • High capital cost • High operating cost
	Biological Systems	
Activated sludge	• Highly efficient treatment method • Requires little land area • Applicable to small communities for local-scale treatment and to big cities for regional-scale treatment	• High cost • Requires sludge disposal area (sludge is usually land-spread) • Requires technically skilled manpower for operation and maintenance
Worm Systems	• Low capital cost • Low operation cost • Low technical manpower requirement	• Need specific temperature (10°–27°C) • High maintenance cost

exemplar solutions to allow more detailed comparison with the INNOQUA solutions. The scores were weighted towards eco-sustainability aspects such as sludge production and potential for wastewater reuse after treatment. Overall scores were used to determine the detail of further investigation required for each competitor technology (Table 9.7), an example of which is presented in Table 9.8.

9.3.1 Positioning INNOQUA Within the Market (SWOT Analysis)

The four INNOQUA treatment technologies can be combined in different ways in order to achieve the most adequate sanitation system according to specific end-user/market requirements. SWOT analyses were undertaken for each of these (Table 9.9).

The SWOT analyses provide a snapshot of the status of each technology (strengths and weaknesses) as well as suggesting directions for future exploitation strategies (opportunities and threats). Taken together with the PESTEL analyses, these aspects were used as the basis for the design of value propositions and INNOQUA system business models (Figure 9.4).

Table 9.6. Mapping of decentralised wastewater treatment technologies. Shading is indicative of market knowledge: green indicates that examples of the technology are known to be used in a region; amber means that there is uncertainty as to whether that technology is in use; red means that there is a high degree of confidence that the technology is not used.

	East Africa	Eastern Europe	Western Europe	Turkey & Middle East	Latin America	South Asia
Stabilization Lagoons						
Aerated Lagoons						
Septic Tanks						
Constructed Wetlands (CW)						
Filtration Systems						
Vertical Biological Reactors (VBR)						
Activated Sludge						
Worm Systems						
Membrane Bioreactors (MBR)						

Table 9.7. Scoring metrics for competitor analysis.

Competition Metrics	Further Investigation Indicated?
GREEN CELL **score of 12 or greater**	Analysed product features or services being offered **require a detailed** cost-benefit analysis with INNOQUA solutions.
YELLOW CELL **score between 8–11**	Analysed product **requires a limited further analysis** in relation to INNOQUA solutions, further research on costs may be required
ORANGE CELL **score between 4–7**	Analysed product is not a direct INNOQUA system competitor but could occupy a market share. **Further analysis about the market potential is required**
RED CELL **score between 0–3**	Analysed product **requires no further analysis**

Table 9.8. Example competitor scorecard - in this case for the DEWATS system in India.

End-user country	India	Producer country	India			
Climate	Wide range					
Market target	Domestic					
Product Name	Company	Product Image	Product Cost	Management Cost	Level of Installation	Availability of materials
DEWATS	CDD India	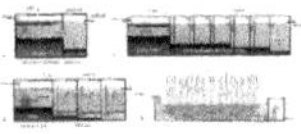	EUR800 (4 PE)	Low	Easy	Yes
Description	The decentralised wastewater treatment system is a simple design, non-dependent on energy, reliable, long lasting, tolerant towards inflow fluctuation and low in costs. It can treat organic wastewater from domestic and industrial sources. DEWATS is based on different natural water treatment techniques which are combined according to requirements such as the characteristics of wastewater, desired effluent quality and technical specifications.					
Website	http://www.cddindia.org/dewats.html					

CHARACTERISTICS[1]	THIS PRODUCT	INNOQUA SOLUTION	RESULT AND COMMENTS
Small size	1	1	
Modular tanks	1	1	
Minimum moving parts/complexity	1	1	**SCORES: 13 vs 15** The INNOQUA system aims to be more eco-sustainable due to the characteristics pertaining to low sludge.
No pumping	1	1	
Small footprint post installation	1	1	
Easy transportation	1	1	
Low weight	1	1	
Low sludge[2]	0	2	
Treated water reuse[2]	3	3	
Complete treatment process (primary, secondary and tertiary)[2]	3	3	
SCORE	**13**	**15**	
Purchase cost	EUR800		
Installation cost[3]	EUR120		
Maintenance cost[3]	EUR100		
Average cost per PE	EUR185		

NOTE (1): Characteristics agreed with INNOQUA partners and gathered from the market product research

NOTE (2): The reference scores are weighted towards eco sustainability impacts. For this reason, the score is 1 for each characteristic with exception of those more related to eco sustainability

NOTE (3): Where not available, the installation cost is estimated as 15% of the purchase cost. The maintenance cost is estimated (if not specified in the product datasheet) as 5% of the sum of purchase and installation costs

Table 9.9. SWOT analyses for each INNOQUA technology.

	Lumbrifilter
Strengths	→ Low sludge production → Domestic scale, suitable for small and medium-sized communities → Low energy usage, no external energy is required → Relatively low space requirement and buried option → Pre-produced system, no on-site construction work → Patented technology
Weaknesses	→ Only for wastewater with high organic loading (domestic wastewater) → Not suitable for climates with extreme temperatures, optimal operating temperature is between 0–30°C or need to be insulated → Further wastewater treatment needed for other uses than irrigation → It needs a pre-treatment based on fine screening
Opportunities	→ Suitable for isolated dwellings not connected to the sewer. Low cost for sludge disposal → Suitable for areas with limited or difficult space requirements, e.g. expanding urban areas
Threats	→ Industrial anaerobic wastewater systems that generate energy (bio-gas) → New industrial wastewater systems that generate only very limited amounts of sludge → Tailored biological wastewater solutions for specific application areas
	Daphniafilter
Strengths	→ Biological solution, no chemicals required → Low-cost solution → High level of *E. coli* inactivation and solids removal. Values higher than sand filter → Small footprint/size system. It can be buried → Enables domestic wastewater reuse
Weaknesses	→ Daphnia development is ensured in a water temperature range from 6°C to 26°C → Daphnia population vulnerable to increased nutrient concentrations in wastewater, especially organic matter content → Minimum HRT of 12 hours, with punctual overloads at 6 h HRT → It needs appropriate primary/secondary treatment

(Continued)

Table 9.9. Continued

Opportunities	→	Post treatment is required for safe re-use of wastewater
	→	Suitable for isolated dwellings not connected to the sewer
Threats	→	It needs appropriate primary/secondary treatment
		Bio-Solar Purification
Strengths	→	Recovers mineral nitrogen and phosphorus
	→	Enables re-use of domestic wastewater
	→	Creates fluid biomass suitable for irrigation in agriculture
	→	Patented technology
	→	Environment friendly
Weaknesses	→	Filter adapts to the contents of the wastewater therefore it is not possible to predict the performance of the filter beforehand
	→	Regular cleaning of light entrance surface required
	→	Only operational at day-time when direct sunlight is available
	→	High space required
Opportunities	→	Suitable for isolated dwellings not connected to the sewer in warm climate regions
	→	Possibility to recover and use the effluent for irrigation
Threats	→	High space required for installation
		UV purification
Strengths	→	Environment friendly
	→	Cost-effective
	→	No disinfection by-products
Weaknesses	→	During evolution, micro-organisms may adapt to UV irradiation
	→	Hard water or water that is high in iron content can absorb the UV rays before they are able to deactivate micro-organisms
Opportunities	→	UV disinfection generally accepted by the market
	→	LED prices still dropping
Threats	→	Mature market with many competitor products

9.4 INNOQUA Commercialisation Strategy

For any technology developed and demonstrated in a project such as INNOQUA to reach a stage whereby it can be commercially viable it is necessary to develop

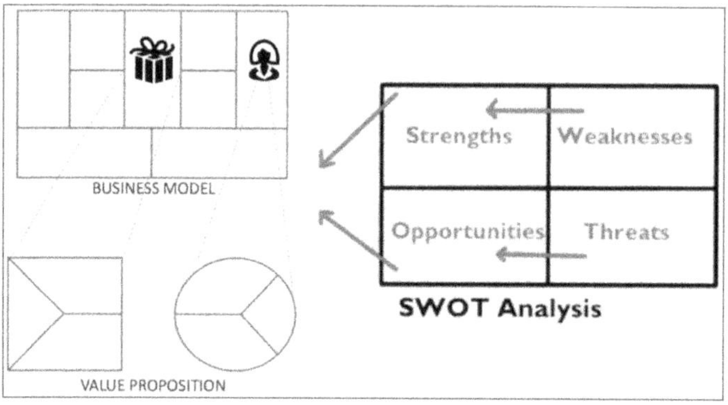

Figure 9.4. Relation between SWOT Analysis and Business Model design.

strategies that best enable the technology to be successfully deployed in any given market. One of the first steps is to decide who owns the intellectual property and how this can fit in with wider commercialisation goals. A number of valorisation options were considered within the INNOQUA project consortium, accounting for the Intellectual Property Rights (IPR) associated with each of the treatment technologies. These options are outlined below:

- **Externalisation** – The responsible INNOQUA "provider" will be one (or several) selected, external company/companies that is/are not part of the (in this case) INNOQUA consortium. The required IPR components (i.e. licenses) could either be sold to this company, or they could be transferred under special agreements and even for free, if the consortium partners owning the IPR agree.
- **Internalisation** – This strategy describes the implementation of some or all of the technologies as a product in one of the consortium partners' product portfolios. By buying or licensing parts of the IP, the responsible partner will guarantee the maintenance, marketing, augmentation with new tools and/or transfer of the platform or individual modules to industrial sectors areas not included within the original INNOQUA demonstration sites. Additionally, this strategy could be realised also by two or more partners of the consortium – each of them responsible for a different technology. If a partner requires outsourcing to deploy the system, it would be included within this strategy.
- **Spin-off/Joint Venture** – This strategy involves the consortium (or one or more partners within the consortium) creating a "spin-off" company or "Joint Venture" working with the INNOQUA technologies as main product. Therefore, the essential know-how needs to be concentrated on this new company, which would be also responsible for marketing and maintenance

activities. Some or all IP rights could be licenced to this spin-off or join venture on an exclusive or non-exclusive basis.

- **Alliance** – This strategy aims to create an association of stakeholders (researchers, industrials, software providers, service providers, etc.) working together to enable the development of products/services based on the project outcomes. Such an alliance would provide mainly specification contributions (e.g. technical or non-technical requirements), and possibly reference implementations, test suites and certification, to foster a valuable cross-industry ecosystem. Members of the alliance can make these specifications evolve through common agreement, as well as provide new reference implementations of the specifications. Technical solution manufacturers can also implement these common specifications in their own products, thus guaranteeing a certain level of compatibility and interoperability with other software products.
- **Research & Development** – This strategy describes the consortium using the current version of the technologies to build a new proposal and search for additional funding to continue research activities.

9.4.1 Developing a Business Model

The methodology utilised in order to define a common Business Model for the INNOQUA System was the Business Model Canvas (BMC) (Osterwalder and Pigneur, 2010). The canvas, its images, and its approach is intentionally visual and intended to foster creativity and interaction amongst collaborators. It uses nine building blocks to describe all aspects related to a business model. An adapted canvas from Osterwalder's approach and the key questions associated with the development of each of the building blocks is shown in Table 9.10.

To successfully bring technologies such as those developed and demonstrated in INNOQUA to the market, a combination of business actors is required. Each business actor delivers a specific part of the solution and has its own transaction mechanism. This combination of business actors and their connections is called a value network. Figure 9.5 shows an example of a value network for the INNOQUA System.

This particular value network foresees a single provider of the INNOQUA System, using components and technologies from various manufacturers and suppliers. A local contractor will manage the actual installation of the INNOQUA System at the customer's site. For most providers of technology and equipment a transaction-based model is foreseen, except for the provider of the INNOQUA software platform, which can be made available on a per customer license basis. To obtain a clearer understanding of how the INNOQUA System can be positioned into the

Table 9.10. Business Model Canvas and Questions for each aspect.

Key Partners	Key Activities	Value Propositions	Customer Relationships	Customer Segments
Who are our key partners? Who are our key suppliers? Which key resources are we acquiring from our partners? Which key activities do partners perform?	What key activities do our value propositions require? Our distribution channels? Customer relationships? Revenue streams?	What value do we deliver to the customer? Which one of our customers' problems are we helping to solve? What bundles of products and services are we offering to each segment? Which customer needs are we satisfying? What is the minimum viable product?	How do we get, keep and grow customers? Which customer relationships have we established? How are they integrated with the rest of our business model? How costly are they?	For whom are we creating value? Who are our most important customers? What are the customer archetypes?
	Key resources What key resources do our value propositions require? Our distribution channels? Customer relationships? Revenue streams?		**Channels** Through which channels do our customer segments want to be reached? How do other companies reach them now? Which ones work best? Which ones are most cost-efficient? How are we integrating them with customer routines?	
Cost structure What are the most important costs inherent to our business model? Which key resources are most expensive? Which key activities are most expensive?			**Revenue streams** For what value are our customers willing to pay? For what do they currently pay? What is the revenue model? What are the pricing tactics?	

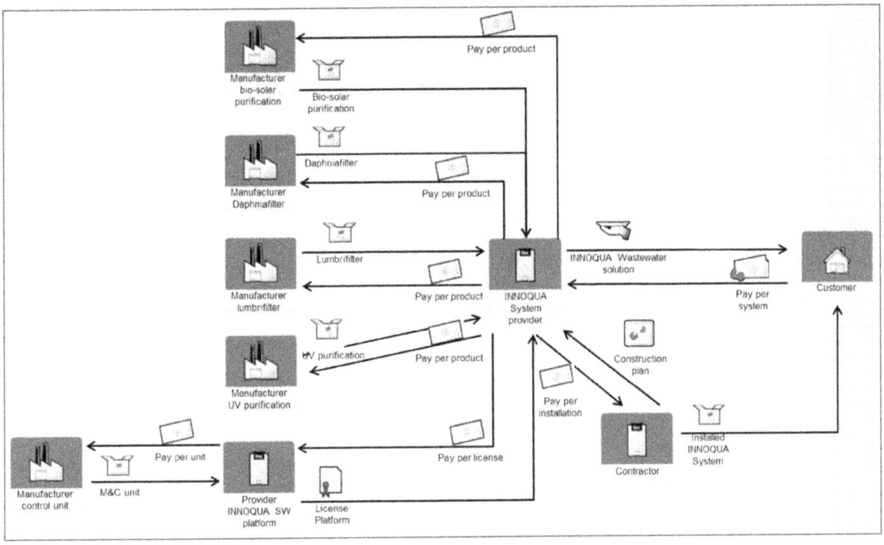

Figure 9.5. Value network of INNOQUA system.

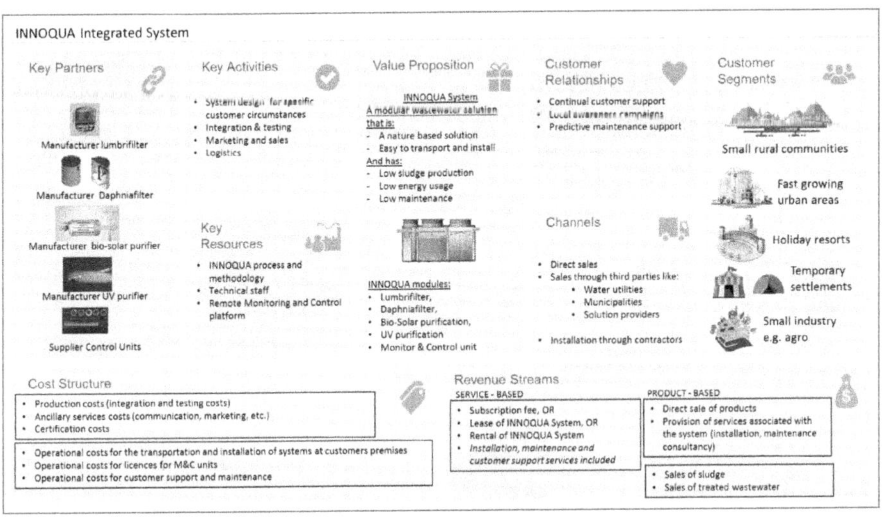

Figure 9.6. Business model canvas for an INNOQUA System Provider.

market, we zoom in to the business model of the INNOQUA System Provider. Figure 9.6 uses a business model canvas to illustrate this model.

Another key aspect of commercialisation is the readiness of the selected innovation for market entry. The European Commission, in common with many other funders and technology providers defines technology readiness on a standardised scale of Technology Readiness Levels (TRLs). These are indicators of the maturity level of particular technologies, and the measurement system is intended to provide

Table 9.11. Technology Readiness Levels as defined by the European Commission (European Commission, 2019).

Technology Readiness Level (TRL)	Description
1	Basic principles observed
2	Technology concept formulated
3	Experimental proof of concept
4	Technology validated in lab
5	Technology validated in relevant environment
6	Technology demonstrated in relevant environment
7	System prototype demonstration in operational environment
8	System complete and qualified
9	Actual system proven in operational environment

Table 9.12. TRL of the INNOQUA technologies at the end of the project.

INNOQUA Technology	TRL at completion of INNOQUA Project
Lumbrifilter	TRL 7–8
Daphniafilter	TRL 6–7
Biosolar Purification	TRL 6
UV disinfection	TRL 9
INNOQUA integrated system	TRL 7

a common understanding of technology status applicable to the entire innovation chain (European Commission, 2019) (Table 9.11).

The TRL status of the separate INNOQUA treatment modules and integrated platform at the end of the project are set out in Table 9.12. While the UV disinfection system is 'market ready' at a TRL of 9, the other modules can be described as 'pre-commercial'. Thus, a further phase of *in situ* beta testing would be required before these units are fully ready to release to the market. In this case it was not considered appropriate to develop a detailed business plan or commercialisation strategy for the integrated INNOQUA system. Instead, the emphasis has been on development of an INNOQUA Alliance – to act as an umbrella that will allow for continued development of the technologies by existing and new partners, as well as providing opportunities to address other technologies in related market sectors. The INNOQUA Alliance is introduced in Section 9.4.2.

9.4.2 The INNOQUA Alliance

A key issue at the end of large research projects is how to ensure partners can work together to continue to develop promising results. In some cases, the answer may be that the technologies are ready for market but in many cases a consortium will not be in a position to continue working on promising technologies after the project has ended. In the case of INNOQUA it was decided that the most suitable route to ensure partners could continue to collaborate and develop further the technologies and/or exploit further opportunities for technology demonstration was to develop an alliance between partners. This strategy aims at creating an association of stakeholders (researchers, industrials, software providers, service providers, etc.) working together to enable the development of products/services based on the INNOQUA platform. Such an alliance would mainly develop specifications (e.g. technical or non-technical requirements), and possibly reference implementations, test suites and certifications, to foster a valuable cross-industry ecosystem. Members of the alliance can support the evolution of these specifications through common agreement, as well as provide new reference implementations. Manufacturers can adopt the developed specifications into their own products. Marketing materials associated with the Alliance are provided in an Annex to this chapter.

The main operational principles of the INNOQUA Alliance (IA) are as follows:

Position within the value chain: INNOQUA Alliance (IA) holds the integrated system know-how as well as the technical and engineering capacity to manage wastewater treatment projects. IA is active in promoting the system in the market.

Product strategy: Product as a Service (PaaS). PaaS offers advantages to both clients and manufacturers/IA. For clients:

- PaaS transforms large capital expenses into smaller operating expenses, allowing them to amortize the cost of the product throughout its life cycle. Additionally, the client no longer assumes the risk of product failure or the responsibility for maintenance, as both are typically included with the service.
- Further, PaaS can help a client optimise its own use of a product.
- Finally, PaaS helps ensure that the client won't be stuck with obsolete equipment since the service includes upgrades (especially from an information/communication technology point of view).

On the manufacturer's/IA side:

- PaaS delivers a consistent revenue stream, which is a more sustainable business model.
- It also allows them to see how the product is being used in the field, which could offer insights into product reliability, design, and potential feature enhancements.
- Manufacturers/IA can use data analytics to find ways to enhance the value that customers receive, which ultimately provides additional revenue for both the client and the manufacturer/IA.

Business Partners:

- Water treatment operators (distributor/agent/reseller) able to take the lead of the commercial process.
- Product and various component suppliers.
- Maintenance providers.

Revenue strategy: Local operators rely on the IA for the engineering process through consultancy services (system design, sizing, commissioning carried out under IA supervision).

The ideal commercial scenario sees a Business Scouter that establishes a link with a distributor to carry out a sale:

1. The distributor complies with IA conditions for the use of the INNOQUA brand, marketing and protection of the INNOQUA system know-how (e.g. NDA frameworks signed with local operators marketing INNOQUA system).
2. Local distributor obtains the contract, directly deals with clients and lead the value chain.
3. Local distributor supplies the full system and guarantees maintenance services.
4. Local distributor relies on IA for system design and commissioning (consulting activity) by paying a fee to IA.

Regardless as to whether a single INNOQUA module or combined system are in scope, the business scouter informs the IA which will then react by identifying the partners that will be required to handle the specific sale process and defining the mechanism to share revenues. An ideal scenario of roles and revenue sharing could be:

1. IP owner/s royalties.
2. Technical partners involved in the consulting activity provided to the distributor (system design) obtain a share from the revenues.

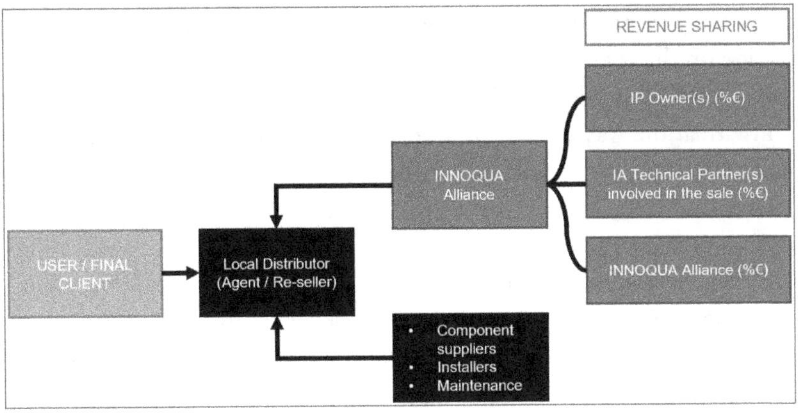

Figure 9.7. Ideal sale scenario for INNOQUA Alliance.

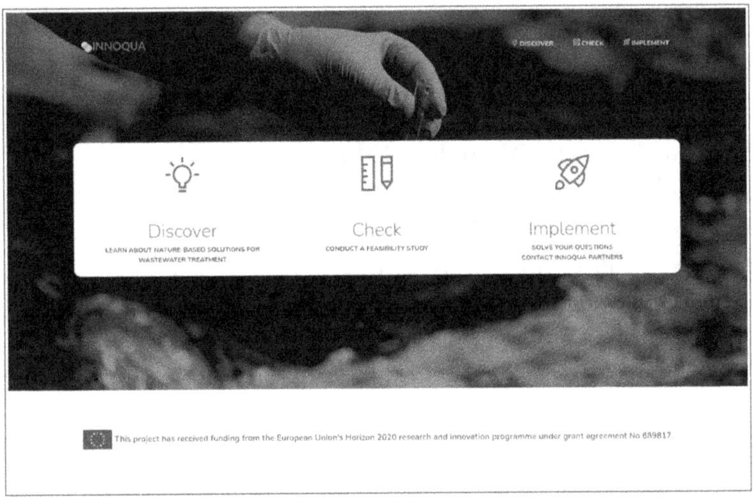

Figure 9.8. INNOQUA DMT landing page (R2M, 2021).

3. A portion of the revenues will be destined to IA to feed the collective budget aimed at covering marketing expenses.

Figure 9.7 sets out the ideal sale scenario.

Through the website, potential customers can directly contact the project coordinator and consequently activate the INNOQUA Alliance process scheme.

In order to help potential customers to decide on the configuration of the INNOQUA system relevant to their circumstances, a web-based decision-making tool (DMT) has been activated for the Italian market (R2M, 2021). The landing page is shown in Figure 9.8.

The tool aims to help potential Italian customers (designers, engineers, water utilities and public entities) to obtain a preliminary feasibility study for the installation of the INNOQUA system. Users are requested to input some data (physical information on the space available; use of the served building; number of people served; reuse water requirements etc) enabling the tool to return some important information on the applicability of the INNOQUA solution to their specific needs, which INNOQUA modules could be installed, and whether the installation will sufficiently treat the wastewater to meet the limits of Italian discharge regulations. Following initial testing in the Italian market, this tool may be adapted to serve other geographies.

The INNOQUA Project web site (INNOQUA, 2018) and DMT will be the main vehicles to handle third party interest in joining the INNOQUA Alliance and ensure that the technologies attain full commercial status and achieve the market potential identified in the PESTEL, SWOT and other tools that we have explored in this chapter.

References

Banco Interamericano de Desarrollo (2014). *EC-L1122: Programa Nacional de Inversiones en Agua, Saneamiento y Residuos Sólidos.* Available at: https://www.iadb.org/es/project/EC-L1122 (Accessed: 8 May 2021).

Business Monitor International (2015). *Turkey Water Report.*

Defra (2015). *Your sewage – Your environment.* Available at: https://consult.defra.gov.uk/water/reform-regulatory-system-small-sewage-discharges/results/your-sewage-your-environment-leaflet-july-2015.pdf (Accessed: 8 May 2021).

Elia, L. *et al.* (2017). *D1.2 – Pre-market study, including partial market surveys, social and acceptance behaviour parameters.* Available at: https://ec.europa.eu/research/participants/documents/downloadPublic?documentIds=080166e5b387c5b2&appId=PPGMS.

epa (2015). *NIP for Domestic Wastewater Treatment Systems 2013.* Available at: https://www.epa.ie/pubs/reports/water/wastewater/NIP Review Jul 2013–Dec 2014_web.pdf (Accessed: 8 May 2021).

European Commission (2019). *What are Technology Readiness Levels (TRLs) and to which Horizon 2020 call topics are they applicable?* Available at: https://ec.europa.eu/info/funding-tenders/opportunities/portal/screen/support/faq/2890 (Accessed: 8 May 2021).

European Commission (no date). *Urban Waste Water Treatment Directive: dissemination platform.* Available at: https://uwwtd.eu/stats/graphs (Accessed: 13 May 2021).

Hynds, P. *et al.* (2017). 'Efficacy of a National Hydrological Risk Communication Strategy: Domestic Wastewater Treatment Systems in the Republic of Ireland', *Journal of Hydrology*, (558), pp. 205–213. doi: 10.1016/j.jhydrol.2018.01.041.

INNOQUA (2018). *INNOQUA – innovative wastewater treatment.* Available at: https://innoqua-project.eu/ (Accessed: 1 June 2021).

Jimenez, B. and Asano, T. (2008). *Water Reuse An International Survey of current practice, issues and needs.* Available at: http://docshare04.docshare.tips/files/25777/257776745.pdf (Accessed: 30 April 2021).

Ministero dell'ambiente (2006). *Norme tecniche per il riutilizzo delle acque reflue, ai sensi dell'articolo 99, comma 1, del decreto legislativo 3 aprile 2006, n. 152., Gazzette Ufficiale.* Available at: https://www.gazzettaufficiale.it/eli/id/2006/05/11/06A04475/sg (Accessed: 14 May 2021).

OECD (2012). *Development Co-operation Report 2012: Lessons in Linking Sustainability and Development.* doi: 10.1787/dcr-2012-1-en.

Osterwalder, A. and Pigneur, Y. (2010). *Business Model Generation: A Handbook for Visionaries, Game Changers, and Challengers.* Available at: https://www.wiley.com/en-gb/Business+Model+Generation%3A+A+Handbook+for+Visionaries%2C+Game+Changers%2C+and+Challengers-p-9780470876411 (Accessed: 8 May 2021).

PESTLE Analysis (2021). *What is PESTLE Analysis?* Available at: https://pestleanalysis.com/what-is-pestle-analysis/ (Accessed: 30 April 2021).

Plat, B. *et al.* (2019). *Public water and wastewater services in France.* Available at: https://www.eureau.org/resources/publications/member-publications/5299-public-water-and-waste-water-services-in-france/file (Accessed: 1 June 2021).

R2M (2021). *Innoqua – Decision Making Tool.* Available at: https://innoqua.r2m.cloud/index.html (Accessed: 1 June 2021).

Re, I. *et al.* (2021). *Circular economy in Italy. Market analysis and business opportunities for Dutch innovative companies: Clean & Safe water and water resource recovery.* Available at: https://www.rvo.nl/sites/default/files/2021/03/CircularEconomyinItaly_2.pdf (Accessed: 14 May 2021).

Sanna, A. (2019). *RIUSO DELLE ACQUE REFLUE IN SARDEGNA.* Available at: https://www.minambiente.it/sites/default/files/archivio/allegati/CReIAMO_PA/L6WP1/sanna_riuso_delle_acque_reflue_in_sardegna_16122019.pdf (Accessed: 14 May 2021).

Schellenberg, T. *et al.* (2020). 'Wastewater Discharge Standards in the Evolving Context of Urban Sustainability – The Case of India', *Frontiers in Environmental Science.* Frontiers Media S.A., 8, p. 30. doi: 10.3389/fenvs.2020.00030.

The Sunday Times (2019). 'How to get to grips with septic tanks'. Available at: https://www.thetimes.co.uk/article/how-to-get-to-grips-with-septic-tanks-6dh8nxz9j (Accessed: 8 May 2021).

The World Bank (2021a). *India Overview.* Available at: https://www.worldbank.org/en/country/india/overview (Accessed: 8 May 2021).

The World Bank (2021b). *Peru Overview.* Available at: https://www.worldbank.org/en/country/peru/overview (Accessed: 8 May 2021).

The World Bank (2021c). *Tanzania Overview*. Available at: https://www.worldbank.org/en/country/tanzania/overview (Accessed: 8 May 2021).

The World Bank (2021d). *Turkey Overview*. Available at: https://www.worldbank.org/en/country/turkey/overview (Accessed: 8 May 2021).

The World Bank (2021e). *World Bank Open Data*. Available at: https://data.worldbank.org/ (Accessed: 13 May 2021).

Transparency International (2017). *Corruption Perceptions Index, 2016*. Available at: https://images.transparencycdn.org/images/2016_CPIReport_EN_200406_153719.pdf (Accessed: 8 May 2021).

UN Water (2021). *Indicator 6.2.1a – Sanitation*. Available at: https://www.sdg6data.org/indicator/6.2.1a (Accessed: 6 May 2021).

Annex. INNOQUA Alliance Materials

IA

INNOQUA ALLIANCE

PARTNERSHIP AGREEMENT TO BOOST COMMERCIALISATION

INNOQUA
innovation for wastewater treatment

INNOQUA BRAND

INNOQUA SYSTEM
The INNOQUA Technologies comprising a fully adaptable modular solution are indicated schematically in the figure below.

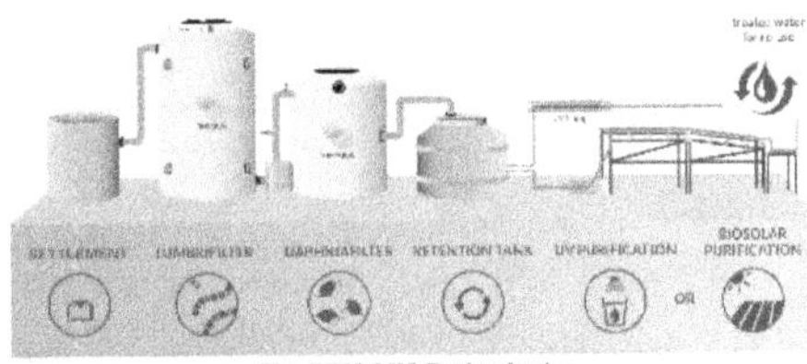

The INNOQUA Technologies

Lumbrifilter: patented system for wastewater treatment by lumbrifiltration (the use of specific earthworms)
Daphniafilter: system for wastewater treatment by means of water fleas (daphnids)
UV Purification: system designed to disinfect water with UV-exposure
BioSolar Purification: system that combines the action of microalgae and sunlight to purify wastewater.

INNOQUA TRADEMARK
INNOQUA is a registered TradeMark. The INNOQUA Alliance partners could use it for commercial and promotional activities.

INNOQUA

INNOQUA KNOW-HOW
INNOQUA System basis is funded on the corresponding 4 years EU funded project (GA 689817). The EU project gives the opportunity to test and optimize the system along the time generating a consistent know-how related to the system optimization, best practices that could

INNOQUA PROJECT HAS RECIVED FUNDING FROM THE EUROPEAN UNION'S HORIZON 2020 RESEARCH AND INNOVATION PROGRAMME UNDER GRANT AGREEMENT N. 689817

IA

INNOQUA ALLIANCE

PARTNERSHIP AGREEMENT TO BOOST COMMERCIALISATION

be exploited for commercial activities by the INNOQUA Alliance partners.

OBJECTIVE

INNOQUA Alliance aims to boost the INNOQUA system commercialization and provide consultancy services strictly linked to the INNOQUA brand (see the section INNOQUA Brand)

COST

Joining the INNOQUA Alliance is free of charge

WHY TO JOIN THE ALLIANCE?

HAVE A WORLDWIDE EXISTING NETWORK TO EXPLOIT

To exploit and build on the local markets in which the project consolidated its presence and to guarantee a preferential access to further commercial opportunities.
INNOQUA consortium partners interested in the future commercialization of the INNOQUA system will take part into the Alliance by sharing their contacts/network.

POSSIBILITY TO TAKE ADVANTAGE OF THE KNOW-HOW GENERATED BY AN EU FUNDED PROJECT

Partners who take part to the INNOQUA Alliance will have the possibility to approach better the market with their systems by taking advantage of the validation and results gathered from the field test activities carried out during the 4-year EU funded project (technical results, know-how, best practices, market studies and field test activities results).
Possibility to continue relying on the research infrastructures, technical contribution and assistance from partners that have directly developed the technologies within the INNOQUA project for further optimization and testing actions.

POSSIBILITY TO TAKE ADVANTAGE OF THE PARTNER SUPPORT FOR LOCAL COMMERCIALIZATION ACTIVITIES

Who joins the INNOQUA Alliance will have the chance to take advantage of partners and related stakeholders that are physically located in the markets where the pilot systems are installed benefiting from direct organizational, logistic and technical support with aim to consolidate O&M and training activities at local and grass-roots level and pave the way for commercialization.

POSSIBILITY TO GO TO THE MARKET ON YOUR OWN

Partners who take part to the INNOQUA Alliance will have, always, the possibility to market and commercialize the system on their own.

INNOQUA PROJECT HAS RECIVED FUNDING FROM THE EUROPEAN UNION'S HORIZON 2020 RESEARCH AND INNOVATION PROGRAMME UNDER GRANT AGREEMENT N. 689817

IA

INNOQUA ALLIANCE

PARTNERSHIP AGREEMENT TO BOOST COMMERCIALISATION

In this case, they can sell their components /System without using the INNOQUA registered trademark and, if needed, agreeing directly with the IP owners on royalties and conditions for the use and sale of the protected technologies.

WHAT WILL THE INNOQUA ALLIACE OFFER?

INNOQUA SYSTEM DESIGN
The selection of the best technological solutions to meet both users' needs and local wastewater regulations for discharge and reuse.

INNOQUA SYSTEM MAINTENACE AND OPERATIONAL FOLLOW-UP
Know how transfer, capacity building and consulting for O&M.

TARGETING WORLDWIDE MARKETS
Boosting the market penetration of the INNOQUA system by exploiting existing commercial networks in different countries in 4 continents.

HOW IT WORKS?

INNOQUA
innovation for wastewater treatment

INNOQUA ALLIANCE PARTNER:
Can use the Know-how and membership expertise generated during the EU funded project period to penetrate the market. Moreover, could take advantage of the INNOQUA brand and trademark.

INNOQUA ALLIANCE BASE CONCEPT
The ideal commercial scenario sees a Business Scouter (INNOQUA ALLIANCE partner) that establishes a link with a distributor to carry out a sale.
The distributor complies with IA conditions for the use of the INNOQUA brand, marketing and protection of the INNOQUA system know-how (e.g. NDA frameworks signed with local operators marketing INNOQUA system).
Local distributor obtains the contract, directly deals with clients and lead the value chain.
Local distributor supplies the full system and guarantees maintenance services.

INNOQUA PROJECT HAS RECIVED FUNDING FROM THE EUROPEAN UNION'S HORIZON 2020 RESEARCH AND INNOVATION PROGRAMME UNDER GRANT AGREEMENT N. 689817

IA

INNOQUA ALLIANCE

PARTNERSHIP AGREEMENT TO BOOST COMMERCIALISATION

Local distributor relies on IA for system design and commissioning (consulting activity) by paying a fee to IA.
Regardless if it is a single-module or combined system, the business scouter informs the IA that will react identifying the partners that will be required to handle the specific sale process and defining the mechanism to share revenues.
An ideal scenario of roles and revenue sharing could be:

> IP owner/s royalties
> Technical partners involved in the consulting activity provided to the distributor (system design) obtain a share from the revenues
> A portion of the revenues will be destined to IA to feed the collective budget aimed at covering marketing expenses.

In the figure here below the ideal sale scenario is depicted.

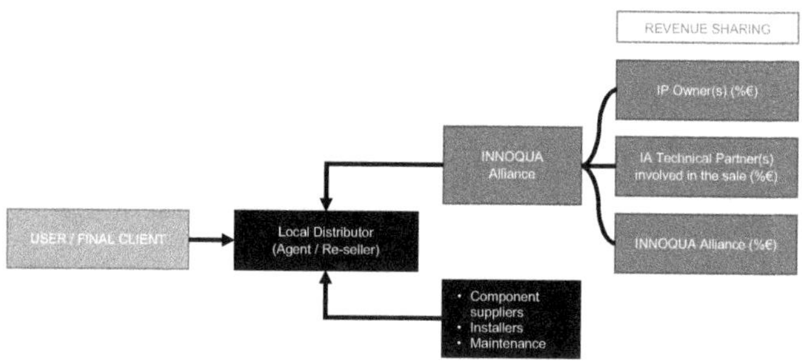

INNOQUA ALLIANCE BOARD

The INNOQUA Alliance will be managed through an internal board named IA_DMG (Innoqua Alliance - Decision Making Group). The IA-DMG will consist of one representative for each participating Company in the Alliance. The IA-DMG group will meet on a periodic (quarterly) basis or as needed when one of the participants requests it.
All DMG members will have full decision-making powers and decisions will be taken by simple majority. One of the role of IA-DMG will be to analyse case by case the commercial activities by defining: i) the partners involved in the sale; ii) the work to be carried out; iii) the percentage of fee to be shared among the partners participating in the commercial sale, IP owners and IA.

INNOQUA PROJECT HAS RECIVED FUNDING FROM THE EUROPEAN UNION'S HORIZON 2020 RESEARCH AND INNOVATION PROGRAMME UNDER GRANT AGREEMENT N. 689817

Index

About the Editors

Costel Bumbac is a senior research scientist leading the biotechnology research team within the Environmental Technology and Technology Transfer Department of ECOIND, Romania. He has 15 years of experience in research projects focused on developing and optimisation of biological wastewater/drinking water treatment technologies, biodegradability studies, soil bioremediation and waste to biofuels technologies. Within the INNOQUA project, he contributed to the development of bio-solar purification system and full-scale demonstration and served as world-wide demonstration manager.

Eoghan Clifford is a Senior Lecturer and Chartered Engineer in civil engineering. He has 18 years of experience in Environmental Engineering with a focus on wastewater and water engineering, process modelling and technology development.

He has published >80 peer-reviewed journal papers and over 200 conference publications and other contributions. He has 5 patents in the area of wastewater engineering. He is the director of the NUI Galway/EPA Water Research Facility. He was the Technical Manager of the INNOQUA project and has coordinated EU, industry and nationally funded research worth over € 13 m and a principal investigator/funded investigator on projects worth another € 18 m.

Jean-Baptiste Dussaussois was the project manager of INNOQUA. As an environment and water engineer he brought his expertise and experiences from an NGO and large company to develop nature-based wastewater treatment solutions. He has managed several international projects and is currently working with NOBATEK/INEF4, a private Research and Technology Organisation for the sustainable building sector. He has been part of several European Commission expert groups and is a member of the water parliament of the South-West of France.

Alexandre Schaal is part of the 'Tools and Systems for Energy Performance' team at NOBATEK/INEF4, where he is in charge of the development of technologies and innovative services related to water and wastewater management – such as the lumbrifilter. His multidisciplinary experiences in the field of water give him a global vision of these problems. Alexandre's involvement in environmental issues led him to create and to chair La Pagaie Sauvage, a French NGO working on microplastic pollution in freshwaters.

David Tompkins is an environmental scientist with a multi-disciplinary background encompassing horticulture, soil science, composting, anaerobic digestion, wastewater treatment, market development and policy implementation. David is a technical consultant and senior manager for Aqua Enviro, specialising in various aspects of sewage sludges, biowastes and other bioresources: hazard management, valorisation, market development, end of waste and stakeholder engagement.

www.ingramcontent.com/pod-product-compliance
Lightning Source LLC
LaVergne TN
LVHW011907200926
848484LV00002B/558
9781680837803

Contributing Authors

Costel Bumbac
National Research and Development Institute for Industrial Ecology (ECOIND), 71–73 Drumul Podul Dambovitei Street, Bucharest 060652, Romania
costel.bumbac@incdecoind.ro

Eoghan Clifford
School of Engineering and Ryan Institute, National University of Ireland Galway, H91 TK33, Ireland
eoghan.clifford@nuigalway.ie

Jordi Colomer
Department of Physics, University of Girona, M. Aurèlia Capmany, 69, 17003 Girona, Spain
jordi.colomer@udg.edu

Ziye Dai
Aqua Enviro, 8 Appleton Court, Calder Park, Wakefield, WF2 7AR, UK
ziye.dai@aquaenviro.co.uk

Jean-Baptiste Dussaussois
NOBATEK/INEF4, 67 rue de Mirambeau, 64600 Anglet, France
jbdussaussois@nobatek.com

Louise Hannon
School of Engineering and Ryan Institute, National University of Ireland Galway, H91 TK33, Ireland
louise.hannon@nuigalway.ie

Manuela Hidalgo
Department of Chemistry, University of Girona, M. Aurèlia Capmany, 69, 17003 Girona, Spain
manuela.hidalgo@udg.edu

Evelyn Herrera
Bremen Overseas Research and Development Association, BORDA e.V.;
Bauhaus Weimar University, Germany
evelynhlop@gmail.com

Elena Manea
National Research and Development Institute for Industrial Ecology (ECOIND), 71–73 Drumul Podul Dambovitei Street, Bucharest 060652, Romania
elena.manea@incdecoind.ro

Alexander Meneses-Jácome
Universidad Autónoma de Bucaramanga, Colombia
ameneses2@unab.edu.co

Domenico Perfido
R2M Solution srl, Via F.lli Cuzio 42, 27100 Pavia, Italy
domenico.perfido@r2msolution.com

Narcís Pous
Laboratory of Chemical and Environmental Engineering (LEQUiA), University of Girona, M. Aurèlia Capmany, 69, 17003 Girona, Spain
narcís.pous@udg.edu

Victoria Salvadó
Department of Chemistry, University of Girona, M. Aurèlia Capmany, 69, 17003 Girona, Spain
victoria.salvado@udg.edu

Teresa Serra
Department of Physics, University of Girona, M. Aurèlia Capmany, 69, 17003 Girona (Spain)
teresa.serra@udg.edu

Tatjana Schellenberg
Bremen Overseas Research and Development Association, BORDA e.V.;
Bauhaus Weimar University, Germany
schellenberg@borda.org;
tatjanaschellenberg@gmx.de

Olga Tiron
National Research and Development Institute for Industrial Ecology (ECOIND), 71–73 Drumul Podul Dambovitei Street, Bucharest 060652, Romania
olga.tiron@incdecoind.ro

David Tompkins
Aqua Enviro, 8 Appleton Court, Wakefield WF2 7AR, UK
David.Tompkins@aquaenviro.co.uk